FOUNDATIONS
of ANALYSIS

The Appleton-Century Mathematics Series

Raymond W. Brink and John M. H. Olmsted, Editors

FOUNDATIONS
of ANALYSIS
with an introduction to
Logic and Set Theory

DOUGLAS A. CLARKE
BARRON BRAINERD
RODERICK A. ROSS
all of the
University of Toronto

H. EDWIN TOTTON
York County Board of Education

MORRIS J. LIEBOVITZ
Southwood Secondary School

GEORGE A. SCROGGIE
Ontario Department of Education

APPLETON-CENTURY-CROFTS

EDUCATIONAL DIVISION
MEREDITH CORPORATION New York

Copyright © 1971 by

MEREDITH CORPORATION

731–1

Library of Congress Card Number: 73–136217

PRINTED IN THE UNITED STATES OF AMERICA

390–19756–4

723337

PREFACE

The basic ingredients of real analysis are

1. the real numbers R (or subsets thereof),

2. functions from R to R, and

3. certain operations on these functions.

By the phrase "foundations of analysis" is usually meant a detailed, rigorous, logical reduction of these concepts to the primitive intuitively conceived notions of "natural number" and "set".

This book carries out such a program in an easy-going way, emphasizing the underlying logical and set–theoretical principles involved. Topics are discussed using the so-called spiral technique. They are first introduced in an elementary way and then re-examined a number of times later in gradually more sophisticated settings. This is all done with extensive use of examples and counter examples but without undue emphasis on technical arguments.

The book has been written for the benefit of the reader who desires an insight into the basic ideas of analysis but who does not wish to submit to the rigorous discipline needed for the more advanced university courses designed more for aspiring research mathematicians.

In a sense this book is the spiritual descendant of Felix Klein's book *Elementary Mathematics from an Advanced Standpoint*. Virtually everything treated in the book is necessary background for the secondary school teacher of mathematics. Klein's book, written over 60 years ago, attempted to broaden the outlook of contemporary teachers. At that time, with the static school curriculum, the need was less obvious. Today, because of the extensive changes taking place at all levels of our educational system the need is more urgent. We have confined our attention to those aspects of analysis most relevant to present secondary school courses. So, for example, we spend a great deal of time on limits but hardly mention topological ideas such as compactness or connectedness. The latter are downplayed, not because they are intrinsically unimportant, but because they are less important to the teacher of secondary school mathematics. For the same reason we do nothing with functions of more than one variable. Since this is not a calculus book in the traditional sense, we omit most of the physical or geometrical applications found in such books. The bibliography lists a number of these if the reader wishes to investigate such topics. It might be helpful if the reader had a previous acquaintance with elementary calculus. Nevertheless the text is self-contained and written for self-study as much as for the classroom.

In addition to its attempt to fill the needs of teachers (or teachers in training) the book is also suitable for general mathematics courses for students of the Humanities or Social Sciences in Universities and Colleges. For this purpose it could be used in a number of ways to suit a variety of one- or two-semester courses.

Note to the Reader

For self-study it is recommended that this book be read in the order in which it is written, with no omissions. However, since Chapter 10 is in some respects peripheral to the central theme of the book it could be omitted (but see the opening remarks on p. 164). On the other hand the reader whose main interest is in logic could read Chapter 10 any time after he reads Chapter 3.

Note to the Instructor

The book can be divided very roughly into five parts, which overlap to some extent, but the instructor who can supply the necessary "bridging" material can fit these together in various ways. The five parts of the book are the following.

(1) A technically undemanding (but not entirely elementary) introduction to sets, logic, and the nature of mathematical proof (Chapters 1, 3, 4 and—to a lesser extent—Chapter 2; cf. also Chapter 10).

(2) An elementary formal development of the number systems of real analysis; the natural numbers, integers, rationals, and reals (Chapters 2, 5, 6, 7, and 8). The treatment in these chapters stresses the algebraic structure of the number systems. Thus both the rationals and the reals are developed as instances of ordered fields. The important concepts of sequence and limit of a sequence are first developed completely in terms of the rationals. This makes the discussion of the same concepts

in the context of the reals easier for the reader to absorb. A further advantage in our approach to sequences is that we consider order convergence before convergence in the usual sense. The former is intuitively easier to grasp than the latter. Eventually they are proved to be equivalent.

(3) A treatment of functions and relations. The basic notions of elementary logic in Chapter 3, and the introductory ideas on functions and relations in Chapter 7, are used, together with the insights gained from the study of number systems, to delve more deeply into functions and relations in Chapters 9 and 10.

(4) A discussion of sequences and series and their limits. This account brings out the dualism between discrete and continuous systems. It also gives a short introduction to the calculus of finite differences and applications to the summation of finite series (Chapters 11 and 12).

(5) A formal development of the differential and integral calculus of a single variable. These concepts are introduced as the continuous analogs of finite difference and sum of a series. As mentioned above we are concerned more with the conceptual rather than the computational, so we have omitted most of the usual geometrical or physical applications (Chapters 13 and 14).

CONTENTS

7. RELATIONS AND FUNCTIONS

8. REAL NUMBERS

9. THE ALGEBRA OF REAL FUNCTIONS

10. LOGIC

11. SEQUENCES AND SERIES

12. LIMITS AND CONTINUITY OF REAL FUNCTIONS

13. DIFFERENTIATION

14. INTEGRATION AND ANTIDIFFERENTIATION

FOUNDATIONS
of ANALYSIS

1

Sets

1.1 THE NOTION OF SET

Perhaps the most basic notion in mathematics is that of a **set**. Just as a single geometric point is a basic concept which is undefined, so the notion of a set is undefined. However, we all have an intuitive understanding of what is meant by a "point"; intuitively, we understand a "set" to be a collection of objects. These may be any objects whatever, but we shall be primarily interested in collections of mathematical objects, for example, sets of points or sets of numbers.

A word or two is in order to clarify some of the notions introduced above.

When we refer to a "mathematical object", we are talking about anything that is properly studied under the broad meaning of mathematics. There is no intention that "object" should signify something concrete; indeed, all mathematical objects are abstract. Without delving into this matter further, we list a few mathematical objects: 1, number, real number, π, point, the point whose coordinates are $(1, \pi)$, line, equation, graph, the line which is the graph of the equation $y = \pi x + 1$, plane, region, set, the set of rational numbers, the set of rational numbers less than π, function, tetrahedron, vector, matrix.

It is a basic principle of mathematics that any mathematical object introduced into discussion be clearly, precisely, and exactly defined in terms of already known objects. This process must, however, stop somewhere at some fundamental level, and a few objects must be left undefined, for example, point or set. The collection of all such objects (for example, the collection of all sets) is an undefined concept. Undefined concepts are taken to be intuitively understood, and the objects which belong to the concepts have

1

certain properties which we simply accept. At a more sophisticated level, undefined concepts are *characterized* by a set of axioms, but this approach is beyond the level of this discussion.

1.2 MEMBERSHIP

We adopt the point of view that a particular set A is determined if and only if we can, at least in principle, decide whether or not a specific object x **belongs to** set A. If the object x does belong to A, we call x a **member,** or **element,** of A and write this symbolically as $x \in A$. If x is not a member of A, we write $x \notin A$. For example, if N stands for the set of natural numbers, then $12 \in N$, $13^{11} \in N$, $5/13 \notin N$, and $x/2 \in N$, if and only if x is an even natural number.

What we mean by "at least in principle" is illustrated by the following example.

Let A be the set of all real numbers which have a consecutive sequence of at least ten 6's somewhere to the right of the decimal point. Then the rational number 2.6 (that is, 2.666...) belongs to A, and the rational number 2.6 does not. Now we wish to investigate whether π does or does not belong to A. Certainly one or the other is the case. If, in fact, $\pi \in A$, then we can in principle *in a strong sense* verify this by computing decimal places of π until we recognize ten consecutive 6's. If, in fact, $\pi \notin A$, then we can never verify this in the *strong sense above* by simply computing its decimal expansion, because the computation of all decimal places is a non-terminating procedure and π, being irrational, has a non-repeating decimal expansion. However, the non-appearance of ten consecutive 6's after 100^{100} decimal places does not imply that they will not occur eventually. Accordingly, we place a weaker interpretation on "at least in principle" and assume that there is some way of answering the question, even if we do not at present know how. In this sense, the set A is determined or *well defined*.

(There is a small group of mathematicians, called Intuitionists, who would not agree with the foregoing argument and would claim that the set A is not well defined.)

1.3 EQUALITY OF SETS

Two sets A and B are said to be **equal,** written $A = B$, if and only if every element of A is an element of B, and conversely, every element of B is an element of A. In other words, two sets are equal if and only if they are identically the same set in the sense that they contain the same elements. This definition of equality of sets is a natural consequence of the point of view adopted in 1.2.

The need for the definition above arises from the fact that the same set can be described in different ways. Consider the following three descriptions:

(i) The set of all rational numbers x such that $x^2 - 3x + 2 = 0$,

(ii) The set of all natural numbers x such that $1 \leqslant x < 3$,

(iii) The set containing exactly the elements 1, 2.

These three sets are equal by the definition above.

1.4   SET NOTATION

There are a number of different conventional ways of naming or describing sets.

1. Any set which is referred to frequently is assigned some specific letter as a name, and this letter is reserved exclusively for this set (throughout some specific context). We shall adopt this practice with the following four sets of numbers. It is assumed that the reader has some knowledge of the numbers that belong to these sets. (We shall, however, discuss them all later on.)

$N = \{\text{the natural numbers}\} = \{\text{the non-negative integers}\}$.

$I = \{\text{the integers}\}$.

$Q = \{\text{the rational numbers}\}$.

$R = \{\text{the real numbers}\}$.

We shall introduce additional standard names from time to time.

It should be noted that we include zero among the natural numbers; some authors do not. This is simply a matter of taste. As a further point of interest, whether the set of natural numbers and the set of non-negative integers are equal or "isomorphic" (cf. 7.9) depends on the degree of sophistication and the level of abstraction intended.

2. Another way to describe or name a set is to write out some description which will be understood and place it between "{" and "}". We have already done this above.

3. A third method is to list the elements of a set between "{" and "}"; for example, the set of (iii) in 1.3 is $\{1, 2\}$. This is possible, even in principle, for finite sets only, and frequently with finite sets it is not practical due to limitations of space. These difficulties can sometimes be avoided by devices which we illustrate below.

(i) The set of all natural numbers not greater than 10^{10} could be written $\{0, 1, 2, ..., 10^{10}\}$.

(ii) $N = \{0, 1, 2, ...\}$.

(iii) $I = \{..., -2, -1, 0, 1, 2, ...\}$
$\qquad = \{0, \pm 1, \pm 2, ...\}.$

(iv) $E = \{\text{the even integers}\}$
$\qquad = \{..., -4, -2, 0, 2, 4, ...\}$
$\qquad = \{0, \pm 2, \pm 4, ...\}.$

4. The next method is the most condensed and powerful. Let us illustrate it.

(i) Set (i) of 1.3 is written as $\{x \in Q \mid x^2 - 3x + 2 = 0\}$.

(ii) Set (ii) of 1.3 is written as $\{x \in N \mid 1 \leqslant x < 3\}$.

(iii) $E = \{x \in I \mid x \text{ is divisible by } 2\}$ (Note: 0 is divisible by 2.)
$\qquad = \{x \in I \mid x = 2y \text{ for some } y \in I\}$
$\qquad = \{x \in I \mid \text{there exists a } y \in I \text{ such that } x = 2y\}.$

The general form of this notation is $\{x \in A \mid \mathcal{P}(x)\}$ where A is some previously specified set and $\mathcal{P}(x)$ is a statement (called an *open sentence* or *predicate*) containing the *variable* x. "$\{x \in A \mid \mathcal{P}(x)\}$" is read "the set of all x in A such that $\mathcal{P}$ of x", and more precisely means "the set of all those objects x which are simultaneously elements of the set A and make the sentence $\mathcal{P}(x)$ true". Occasionally the set A is omitted thus $\{x \mid \mathcal{P}(x)\}$, when it is clearly understood from the context which set A is implicitly intended.

There are certain conventions connected with this notation which must be carefully understood. They are frequently abused, even in textbooks.

(a) $\{x \in A \mid \mathcal{P}(x), \mathcal{Q}(x)\}$ means $\{x \in A \mid \mathcal{P}(x) \text{ and } \mathcal{Q}(x)\}$.

If, in contrast, we wish to describe the (quite different) set $\{x \in A \mid \mathcal{P}(x) \text{ or } \mathcal{Q}(x)\}$ then we must write this out in full. Remember, the comma between $\mathcal{P}(x)$, $\mathcal{Q}(x)$ in the first line means "and"!

(b) $\{x \in A \mid \mathcal{P}(x, y)\}$ means $\{x \in A \mid \mathcal{P}(x, y) \text{ holds for all } y\}$; that is, we are describing the set of all those elements x of A which make $\mathcal{P}(x, y)$ true no matter what substitution is made for y, with the provision that the variable y itself will be restricted to some set which will either be clear from context, or explicitly stated, for example, $\{x \in A \mid \mathcal{P}(x, y) \text{ holds for all } y \in B\}$.

If, in contrast, we wish to describe the (quite different) set $\{x \in A \mid \text{there exists a } y \in B \text{ such that } \mathcal{P}(x, y)\}$, then we must write this out in full. A synonymous description is $\{x \in A \mid \mathcal{P}(x, y) \text{ for some } y \in B\}$. An example of this appears in paragraph 4 (iii).

Occasionally one sees: $E = \{x \in I \mid x = 2y, y \in I\}$. The reader automatically supplies the desired interpretation once he understands what is meant, but strictly speaking this is a gross abuse of the notation. In fact, $\{x \in I \mid x = 2y, y \in I\}$ means $\{x \in I \mid \text{for all } y \in I, x = 2y\}$ and is the empty set (see 1.6 below); no integer is twice every integer!

As an example of the correct use of the notation in (b), the set of prime numbers is described as follows:

$$\{x \in N \mid (\text{If } y \in N, \ y > 1, \ y \neq x, \text{ then } y \text{ does not divide } x), \ x > 1\}.$$

One further point should be emphasized. The "x" in $\{x \in A \mid \mathscr{P}(x)\}$ is what might be called a "dummy" variable (usually called a *bound* variable). Rather than try to define this concept we shall instead point out its two basic consequences and urge the reader to think about them carefully.

Substitution of specific objects for such a dummy or bound variable x is not allowed; indeed it leads to nonsense. $\{x \in R \mid x^2 - 3x + 2 = 0\}$ means something; but $\{3 \in R \mid 3^2 - 3 \cdot 3 + 2 = 0\}$ does not.

Replacement of x by some other variable not already in use leaves everything unchanged. $\{y \in R \mid y^2 - 3y + 2 = 0\}$ makes perfectly good sense and describes the same set as $\{x \in R \mid x^2 - 3x + 2 = 0\}$. This point will be clarified if one considers the fact that $y \in \{x \in R \mid x^2 - 3x + 2 = 0\}$ if and only if $y \in R$ and $y^2 - 3y + 2 = 0$.

By contrast, the "x" in "$x^2 - 3x + 2 = 0$" is a *free* variable. Substitution of (suitable) objects for x makes perfectly good sense; "$3^2 - 3 \cdot 3 + 2 = 0$" is meaningful (although false). On the other hand, we cannot say that "$x^2 - 3x + 2 = 0$" is true exactly when "$y^2 - 3y + 2 = 0$" is true; substituting 1 for x and 3 for y makes the first true and the second false.

1.5 SUBSETS

A set A is said to be a **subset** of a set B if and only if every element of A is also an element of B; this is written $A \subseteq B$. Thus,

$$N \subseteq I, \quad I \subseteq Q, \quad Q \subseteq R, \quad \text{and} \quad E \subseteq I.$$

Also,

$$\{x \in R \mid x \leqslant 0\} \subseteq \{x \in R \mid x < 3\}.$$

From this definition it follows that every set A is a subset of itself; $A \subseteq A$.

Here is an alternative formulation of the definition of subset.

Definition 1. $A \subseteq B$ *if and only if* $x \in A$ *implies* $x \in B$ *for all* x.

This formulation emphasizes the fact that *in order to prove* $A \subseteq B$ *we first assume* $x \in A$ *(for arbitrary* x*) and then deduce from this that* $x \in B$.

The definition of equality of sets in 1.3 can now be restated as follows:

Definition 2. $A = B$ *if and only if* $A \subseteq B$ *and* $B \subseteq A$.

Hence, in order to prove that two sets A, B are equal, we must prove two things: $A \subseteq B$, $B \subseteq A$.

To illustrate these points, consider the following sets.

$$A = \{x \in I \,|\, x = 4m \text{ for some } m \in I\},$$
$$B = \{y \in I \,|\, y = 2n \text{ for some } n \in I\},$$
$$C = \{z \in I \,|\, z \text{ is divisible by 2}\}. \quad \text{(Note, 0 is divisible by 2.)}$$

Show that:

(i) $A \subseteq B$,   (ii) $A \neq B$,   (iii) $B = C$.

Solution:

(i) Suppose $x \in A$. Then $x = 4m$ for some $m \in I$. So $x = 2(2m)$, and thence (since $2m \in I$), $x \in B$. Therefore $A \subseteq B$.

(ii) To show $A \neq B$, it is sufficient to find some $x \in A$ such that $x \notin B$ or some $y \in B$ such that $y \notin A$. But we know that $A \subseteq B$, so we must find a $y \in B$ such that $y \notin A$. Now $6 = 2 \cdot 3$ and $6 \in B$, but $6 \notin A$ (since $6 = 4m$ for *no* $m \in I$). Hence, $A \neq B$.

(iii) We must show that (a) $B \subseteq C$ and (b) $C \subseteq B$.

(a) Let $x \in B$. Then $x = 2n$ for some $n \in I$. But this means $2 \,|\, x$ (a symbolic way of saying 2 divides x) and hence $x \in C$. Therefore, $B \subseteq C$.

(b) Conversely, if $x \in C$, then $2 \,|\, x$; then $x = 2n$ for some $n \in I$. Hence $x \in B$. Therefore $C \subseteq B$.

We conclude that $B = C$.

The set A above is said to be a **proper subset** of B. It follows that A is also a proper subset of C. We write this as $A \subset B$. Every set is a subset of itself, but not a proper one. We define a proper subset as follows:

Definition 3. $A \subset B$ *if and only if* $A \subseteq B$ *and* $A \neq B$.

Of course, $N \subset I$, $I \subset Q$, and $Q \subset R$.

1.6   THE EMPTY SET

There is one special set with which we need to deal from time to time. This is the **empty set,** the set which contains no elements at all! (The empty set is sometimes called the null set or the void set.) The empty set is almost always denoted by the symbol $\emptyset$.

Beginners are sometimes surprised to learn that the empty set is considered to be a set. However, the empty set satisfies the criterion of 1.2 quite trivially. For any x we can always answer the question "Does $x \in \emptyset$?" The answer is always "No". There is only one empty set (by convention); the empty set of integers is the same set as the empty set of oranges.

A still greater surprise is the fact that *the empty set is a subset of every set*. This is not a convention; we shall state it as a theorem and prove it in 4.5.

The empty set can be described in a multitude of ways:

$$\emptyset = \{x \in I \mid x \neq x\}$$
$$= \{y \in R \mid y < y\}$$
$$= \{x \in N \mid x \text{ is an odd prime divisible by 2}\}$$
$$= \{\text{all people each of whom can sit in his own lap}\}.$$

1.7 EXERCISES

1. Describe each of the following sets in the form $\{x \in A \mid \mathcal{P}(x)\}$.
   (i) The odd integers.
   (ii) The prime numbers.
   (iii) The rational numbers each of which is the positive square root of some rational number.
   (iv) The irrational numbers each of which is a cube root of some negative integer.
   (v) The real numbers each of which is a square root of some negative integer.
   (vi) The set containing exactly π and $\sqrt{\pi}$.
   (vii) The real numbers greater than π and less than $\sqrt{\pi}$.
   (viii) The real numbers greater than π or less than $\sqrt{\pi}$.
   (ix) The real numbers less than π and greater than $\sqrt{\pi}$.
   (x) The real numbers less than π or greater than $\sqrt{\pi}$.
2. Consider the following sets: $N, I, Q, R,$
   $E = \{\text{even integers}\}$, $O = \{\text{odd integers}\}$,
   $P = \{\text{positive integers}\}$, $Pr = \{\text{prime numbers}\}$,
   $L = \{x \in Q \mid -1 < x < 5/2\}$, $M = \{x \in R \mid -\pi < x \leqslant 3\pi\}$,
   $\quad\quad S = \{x \in R \mid x = a + b\sqrt{2} \text{ for some } a, b \in Q\}$,
   $\quad\quad T = \{x \in Q \mid x = 1/2^k \text{ for some } k \in N\}$.
   For each pair of sets A, B from this list, prove or disprove the following:
   $$A \subset B, \quad A \subseteq B, \quad B \subset A,$$
   $$B \subseteq A, \quad A = B, \quad A \neq B.$$
3. List all the subsets of the set $\{1, 2, 3, 4\}$. (Remember that $\emptyset$ is a subset of every set.) How many subsets does a set of n elements have?
4. Sets can be elements of sets. For example, let $F = \{\{1\}, \{1, 2\}, N\}$. Then $\{1\} \in F$, $\{1, 2\} \in F$, and $N \in F$ are true, but $\{1\} \subset F$ is false. List all the subsets of F. How many of these subsets contain 1 as an element? Be careful! How many contain $\{1\}$ as an element?

2

The Natural Numbers

2.1   A BASIC SET

One of the most interesting and fundamental sets with which, we assume,
the reader is already familiar is the **set of natural numbers**, $N = \{0, 1, 2, 3, ...\}$.
Now, N is interesting both because it is a set, and because it is, indeed, the
set with which we count things. In addition, the set N possesses much useful
and interesting mathematical structure which we shall now examine in detail.

2.2   OPERATIONS, ADDITION AND MULTIPLICATION

There are two familiar **operations** defined on the set N; addition and multiplication. When we say, for example, that "$+$" is an operation on N we mean
the following: if $a,b \in N$, then $a + b \in N$. In other words, the result of applying the operation on two elements of N yields an element of N. This is known
as the **closure property** of an operation.

It is possible to define many operations on the set N. Of these, some can
be expressed in terms of the basic operations of addition and multiplication. For example, if we define (temporarily) $*$ by the equation $a * b =
(3a) + b$, then $*$ is an operation on N, and $5 * 6 = (3 \times 5) + 6 = 21$.

There exist operations on N which are not defined (nor definable) in
terms of addition and multiplication alone. For example, if we define $\vee$ by

$$a \vee b = \begin{cases} \text{the larger of } a, b & \text{if } a \neq b, \\ a & \text{if } a = b, \end{cases}$$

8

then $\vee$ is an operation on N. In contrast, division is not an operation on N because $11/119 \notin N$, and the closure property fails.

Now let us return to addition and multiplication. These operations enjoy a number of important properties in addition to that of closure. We shall list the basic ones without proof. Although it is possible to start with fundamental definitions of N, $+$, $\times$, and prove these properties as theorems, we shall treat them here as axioms.

Axioms.   *For all $m,n,p \in N$, the following hold*:

A1. $m + n = n + m$.   *Commutative Property of Addition*

A2. $(m + n) + p = m + (n + p)$.   *Associative Property of Addition*

A3.   There exists a (unique)   *Property of Additive Identity*
      element $0 \in N$ such that
      $0 + m = m$.

M1. $mn = nm$.   *Commutative Property of Multiplication*

M2. $(mn)p = m(np)$.   *Associative Property of Multiplication*

M3.   There exists a (unique)   *Property of Multiplicative Identity*
      element $1 \in N$ such that
      $1m = m$.

D. $m(n + p) = mn + mp$ *Distributive Property*

C1.   If $m + n = m + p$,   *Cancellation Property of Addition*
      then $n = p$.

C2.   If $mn = mp$ and $m \neq 0$,   *Cancellation Property of Multiplication*
      then $n = p$.

Axioms A1 and M1 assert that the order of the elements in addition and in multiplication, respectively, is immaterial; either order yields the same number. These axioms may seem vacuous to the student who has limited experience with mathematical systems. It is, however, the case that there exist operations which are not commutative. To give a simple (but not very useful) example, let the operation r be defined on N by the equation

$$m \, r \, n = n, \text{ for all } m,n \in N;$$

that is, we take the right-hand member of the pair m, n. Then r is certainly an operation because it has the closure property, but r is not commutative; $1 \, r \, 2 = 2$, but $2 \, r \, 1 = 1$, so $1 \, r \, 2 \neq 2 \, r \, 1$.

The operation m^n is also non-commutative. Observe that this is truly an operation on N (if we temporarily adopt the convention that $0^0 = 0$); for any $m,n \in N$, $m^n \in N$. Now $3^2 \neq 2^3$, so the commutative property fails.

Axioms A2 and M2 assert that the order in which successive additions and successive multiplications are respectively performed is immaterial. In particular, this means that we can omit the brackets and write $m + n + p$ and $m\,n\,p$, without ambiguity.

Operations which are non-associative exist. For example, $(m^n)^p$ is not in general equal to $m^{(n^p)}$; $(2^3)^2 = 64$, but $2^{(3^2)} = 512$. By convention, $m^{n^p} = m^{(n^p)}$.

Axiom A3 asserts the existence of a unique element, in this case *zero*, which is called **the identity element of addition.** Similarly, M3 asserts the existence of an **identity element of multiplication,** in this case *one*. (We put the word "unique" in parentheses because it can in fact be omitted. See Exercise 5 in 2.7.)

Axiom D is self-explanatory and no doubt familiar to the reader. It is of interest to note that the other distributive property fails; that is, we do not in general have $m + (np) = (m + n)(m + p)$. This remark is not so foolish as it may appear at first sight; there do exist pairs of operations on N with respect to which *both* distributive properties hold. For an example, see Exercise 3 in 2.7.

The Cancellation Properties, Axioms C1, C2 are, so to speak, the best we can do in N in the direction of subtraction and division. If we could subtract in N, C1 would be provable by subtracting m from both sides of the equation $m + n = m + p$. However, subtraction is not an operation on N; for example, $2 - 3 \notin N$. Similarly, if we could divide (by non-zero numbers) C2 would be provable. Still, C1 and C2 are properties of the natural numbers despite the fact that we cannot always subtract or divide.

There is one further axiom which is usually left unstated. It is implicit in all of mathematics, and is probably best thought of as a principle of logic.

If $m = p$, *then in any mathematical statement containing m, any occurrence of m may be replaced by p.* This is called the **replacement property of equality.**

The converses of Axioms C1 and C2 need not be postulated; they follow directly from the replacement property of equality. We shall provide a proof of this for C1 and leave the case for C2 as an exercise (Exercise 6 of 2.7). The reader may feel that this theorem and its proof are either obvious or trivial. We include them primarily as an illustration of techniques of careful logical reasoning.

Theorem 1.   *For all $m,n,p \in N$, if $n = p$, then*

$$\text{(a)}\quad m + n = m + p, \quad and \quad \text{(b)}\quad mn = mp.$$

Proof of (a).

(1) Assume $n = p$.

(2) Trivially, $m + n = m + n$.

(3) From (1), (2), and the replacement property, $m + n = m + p$. (Here we have substituted p for the second occurrence only of n in $m + n = m + n$.)

(4) From (1), (3); if $n = p$, then $m + n = m + p$. We assumed the *hypothesis* $n = p$, and derived the *conclusion* $m + n = m + p$. (cf. 3.6.)

There are many properties of addition and multiplication which derive from the axioms. The following is an example.

Theorem 2.   *For all* $m \in N$, $m0 = 0$.

Proof.   (1) $0 + 0 = 0$ A3

(2) $m0 = m0$

(3) $m(0 + 0) = m0$ (1), (2), Replacement

(4) $m0 + m0 = m0$ (3), D, Replacement

(5) $m0 = 0 + m0$ A3

(6) $m0 + m0 = 0 + m0$ (4), (5), Replacement

(7) $m0 = 0$ (6), C1

From now on we shall usually take the liberty of omitting "replacement".

Corollary.   *For all* $m \in N$, $0m = 0$.

Proof.   By Theorem 2 and M1.

2.3   EXPONENTIATION

For the sake of completeness, let us recall briefly what the reader already knows about exponentiation. Let $m, n, p \in N$, $m > 0$.

Definition 1.
$$m^n = \begin{cases} m \cdot m \cdot \ldots \cdot m, \text{ with } n \text{ factors}, & \text{if } n > 0, \\ 1, & \text{if } n = 0. \end{cases}$$

The following laws of exponents are well known, and easily proved. (Exercise 8 of 2.7).

E1. $m^n \cdot m^p = m^{n+p}$.

E2. $(m^n)^p = m^{np}$.

E3. $m^p n^p = (mn)^p$.

In m^n, m is called the **base** and n is called the **exponent**.

2.4 RELATIONS, ORDER

By a **relation** on a set A we shall mean an open sentence containing exactly two variables which take their values in A. An open sentence is an expression containing variables which becomes a sentence when the variables are replaced by specific elements of A: "$m = n$" is an open sentence; $1 = 2$ is a sentence. Open sentences are also called predicates. (The distinction between "$=$" and "$m = n$" is a moot one which need not concern us here.) Relations and predicates are discussed in detail in Chapters 7 and 10.

The reader is already familiar with the "less than" relation on N: $m < n$. Geometrically, $m < n$ means m is to the left of n on the number line for N. (See Figure 2–1.)

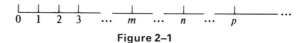

Figure 2–1

There is also an arithmetic definition. Let $m, n \in N$.

Definition 1. $m < n$ *if and only if there is a* $k \in N$, $k \neq 0$, *such that* $m + k = n$.

Of course, $k = n - m$. But, as we observed before, subtraction is not an operation on N; indeed $(n - m) \in N$ when and only when $m < n$ or $m = n$.

The *relation* $<$ is defined in terms of the operation $+$ and the relation $=$ as indicated in Definition 1. At this point we raise the question "What is the difference between an operation and a relation?" The difference between a relation and an operation is this (using $<$ and $+$ as examples): $m < n$ is either true or false, whereas $m + n$ is an element of N. Let us describe this situation in more detail. On the one hand, both $<$ and $+$ have the property that the variables m, n in the expressions $m < n$ and $m + n$ take their values in the set N. On the other hand, the *value* of the expression "$m < n$" for particular m and n is either *true* or *false*, whereas the *value* of "$m + n$" for specific m and n is *an element of* N; "$2 < 1$" is false, but "$2 + 1$" is another name for 3.

We write "$m \not< n$" for "$m < n$ is false," and "$m > n$" for "$n < m$".

The *relation* $<$ *has the following properties* which are geometrically obvious from Figure 2–1. Let $m, n, p \in N$.

O1. $m \not< m$.	*Irreflexive Property*
O2. If $m < n$, then $n \not< m$.	*Asymmetric Property*
O3. If $m < n$ and $n < p$, then $m < p$.	*Transitive Property*
O4. Exactly one of $m < n$, $m = n$, $n < m$ holds.	*Total (linear or trichotomy) Property*

More generally, let $\mathscr{R}$ be any relation on a set A. Recall that this means $\mathscr{R}(x, y)$, or $x \mathscr{R} y$ as we shall write it here, is a predicate (open sentence) with two variables ranging over the set A. If $x \mathscr{R} y$ is false, we write $x \not{\mathscr{R}} y$.

Suppose $\mathscr{R}$ satisfies O1; that is, $x \not{\mathscr{R}} x$ for all $x \in A$. Then we say that $\mathscr{R}$ is irreflexive. For example, if A is the set of real numbers, and $\mathscr{R}$ is defined by $x \mathscr{R} y$ if and only if $x = \sqrt{y}$, then $\mathscr{R}$ is irreflexive; $5 \neq \sqrt{5}$.

Similarly, if $x \mathscr{R} y$ implies $y \not{\mathscr{R}} x$ for all $x, y \in A$, we say $\mathscr{R}$ satisfies O2 and is asymmetric. Using the relation $\mathscr{R}$ from the preceding paragraph, $7 = \sqrt{49}$ but $49 \neq \sqrt{7}$. And so on for O3 and O4.

Any relation $\mathscr{R}$ which satisfies O1 to O4, that is, is irreflexive, asymmetric, transitive, and total, is called a **total** (or **linear**) **ordering**. In particular then, the relation $<$ is a total ordering on N.

It is interesting to note that there are ways of totally ordering N other than the natural order $<$.

Consider the order relation $\mathscr{R}_1$ whereby every even number is "less than" every odd number, but otherwise the numbers are ordered according to the natural order; more precisely, $m \mathscr{R}_1 n$ if and only if (m is even and n is odd) or (m and n are even and $m < n$) or (m and n are odd and $m < n$). This ordering is illustrated in Figure 2–2. We leave the verification of O1 to O4 to the reader as an exercise (Exercise 9 of 2.7).

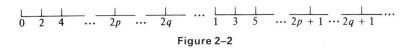

Figure 2–2

Note that this ordering is a transfinite ordering; there are, for example, infinitely many p such that $p \mathscr{R}_1 1$.

The relation $>$ may be thought of as a total ordering differing from $<$; that is, if we temporarily think of "greater than" as a new kind of "less than", we get a total ordering of N which looks like Figure 2–3.

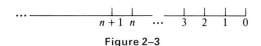

Figure 2–3

As another example of total ordering, let $x \mathscr{R}_2 y$ if and only if ($x = 0$ and $y \neq 0$) or ($x \neq 0$, $y \neq 0$, and $x > y$); this is illustrated in Figure 2–4.

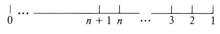

Figure 2–4

Any relation which satisfies O1 to O3 is called a **partial ordering**. Of course, every total ordering is a partial ordering, but there are partial orderings which are not total. An example on N is the following: $m \mathcal{R}_3 n$ if and only if (m and n are even and $m < n$) or (m and n are odd and $m < n$). This relation, unlike $\mathcal{R}_1$, is not total, because we have neither $1 \mathcal{R}_3 2$ nor $2 \mathcal{R}_3 1$. $\mathcal{R}_3$ is illustrated in Figure 2–5.

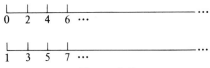

Figure 2–5

Partial orderings are very important in mathematics, but most of the partial orderings we shall meet here will be total orderings.

The reader should entertain himself by constructing more total (or partial) orderings of N (and of I, Q, R, as well). He should also think about relations which are *not* order relations. We shall have more to say about other relations later.

If we let "$m \leqslant n$" stand for "$m < n$ or $m = n$", then $\leqslant$ is a relation on N which has the following properties. Let $m,n,p \in N$.

O'1. $m \leqslant m$.   *Reflexive Property*

O'2. If $m \leqslant n$ and $n \leqslant m$, then $m = n$.   *Antisymmetric Property*

O'3. If $m \leqslant n$ and $n \leqslant p$, then $m \leqslant p$.   *Transitive Property*

O'4. Either $m \leqslant n$ or $n \leqslant m$.   *Total Property*

Properties O'1 to O'4 for $\leqslant$ follow directly from the properties O1 to O4 of $<$ and the definition of $\leqslant$. Conversely, if we start with $\leqslant$ and the properties O'1 to O'4, and then define "$m < n$" to mean "$m \leqslant n$ and $m \neq n$", properties O1 to O4 can be shown to hold for $<$. (See Exercise 10 of 3.8.)

Note that property O'4 uses the inclusive "or"; that is, it means either "$m \leqslant n$" is true, or "$n \leqslant m$" is true, or both. (In contrast, property O4 meant "exactly one".) It follows that both $m \leqslant n$ and $n \leqslant m$ can hold, but of course by property O'2, then $m = n$.

Sometimes the properties O'1 to O'4 are taken as the defining properties of a linear order (and O'1 to O'3 for a partial order). By this we mean, as before, if $\mathcal{R}$ is such that for all x, y, z (i) $x \mathcal{R} x$, (ii) $x \mathcal{R} y$ and $y \mathcal{R} x$ implies $x = y$, (iii) $x \mathcal{R} y$ and $y \mathcal{R} z$ implies $x \mathcal{R} z$, and (iv) either $x \mathcal{R} y$ or $y \mathcal{R} x$, then $\mathcal{R}$ is called a total or linear order. In this case $\mathcal{R}$ must be interpreted as defining an ordering of the "equal-to-or-less-than" variety, so to speak. Then the corresponding "strictly-less-than" variety can be defined as we did two paragraphs above, and such a relation will always obey O1 to O4 (or O1 to O3 if it is partial).

A very basic example of a partial order on N satisfying O'1 to O'3, but not O'4, is $m|n$, "m divides n" or "n is a multiple of m" (Exercise 10 of 2.7). As another example, consider the relation $\mathcal{R}_4$ on N defined by $x\,\mathcal{R}_4\,y$ if and only if ($x|y$ and $x \neq y$). This satisfies O1 to O3, but not O4. In other words, $x\,\mathcal{R}_4\,y$ is the "strictly less than" version of $x|y$. To see that $\mathcal{R}_4$ satisfies O1, note that $x \neq x$ is false, so ($x|x$ and $x \neq x$) is false; $x\,\not\mathcal{R}_4\,x$ and O1 holds. As for O4, note that $5\,\not\mathcal{R}_4\,7$ and $7\,\not\mathcal{R}_4\,5$; hence O4 fails. We leave the verification of O2 and O3 to the reader.

Theorem. *The order relation $<$ on N is compatible with $+$, $\times$, in the following sense. Let $m,n,p \in N$.*

> OA. $m + n < p + n$ *if and only if* $m < p$.
>
> OM. *Provided* $n > 0$, $mn < pn$ *if and only if* $m < p$.

Proof of OA. We must prove two implications:

(i) $m + n < p + n$ implies $m < p$.　(ii) $m < p$ implies $m + n < p + n$.

Proof of (i) Assume $m + n < p + n$. Prove $m < p$.

Suppose $m = p$.

Then $m + n = p + n$.	Theorem 1 of 2.2

This and our assumption contradict O4.
Therefore $m \neq p$.

Suppose $m > p$.

Then $m = p + k$ for some $k \in N$, $k \neq 0$.	Definition 1
Then $m + n = (p + k) + n$,	Theorem 1 of 2.2
$\quad\quad\quad = (p + n) + k$.	A1, A2
Therefore $m + n > p + n$.	Definition 1

This and our assumption contradict O4.
Therefore $m \not> p$.

It follows that $m < p$.	O4

Proof of (ii) Assume $m < p$. Prove $m + n < p + n$.

Suppose $m + n = p + n$.

Then $m = p$.	C1

This and our assumption contradict O4.
Therefore $m + n \neq p + n$.

Suppose $m + n > p + n$.

Then $m + n = (p + n) + k$,	Definition 1
$\quad\quad\quad = (p + k) + n$.	A2, A1
Then $\quad\quad m = p + k$.	C1
Then $\quad\quad m > p$.	Definition 1

This and our assumption contradict O4.
Therefore $m + n \not> p + n$.

It follows that $m + n < p + n$.	O4

We leave the proof of OM as Exercise 11 of 2.7; the steps are similar to those in the proof of OA.

We shall now discuss an additional property possessed by $<$.

Let $M \subseteq N$. The **least element** of M (with respect to $<$) is that natural number in M, if it exists, which is less than every other number in M. For example, 0 is the least element of N itself.

Definition 2. *If A is a set with a total ordering $<$, and if $B \subseteq A$, then x is the **least element** of B provided*: (i) $x \in B$ *and* (ii) *for any* $y \in B$, $x \leqslant y$.

The least element of the set of prime numbers is 2. The least element of the set of odd natural numbers is 1. In fact,

OW. *For any $M \subseteq N$ such that $M \neq \emptyset$, M has a least element (with respect to the natural order).*

Property OW is an additional property possessed by $<$ and is called the **well ordering property**. Any total ordering which additionally satisfies OW is called a **well ordering**.

As we have just stated (as an axiom), $<$ is a well ordering of N. The ordering of Figure 2–2 is also a well ordering, but those of Figures 2–3 and 2–4 are not.

In this discussion, we have assumed that properties O1 to O4 were geometrically obvious, and we have treated them as axioms, just as we did the basic properties of $+$, $\times$ in 2.2. However, the reader should note that we could have started with more basic definitions of N, $+$, $\times$ and proved O1 to O4. Property OW, on the other hand, does not fall into this category; it is (a version of) one of the fundamental defining properties of N, and, for this reason, it is of considerable importance. The importance of property OW is perhaps best revealed by the fact that it is equivalent to the *Principle of Mathematical Induction*, which we shall discuss in the next section.

2.5 MATHEMATICAL INDUCTION

Mathematical induction (sometimes called finite induction) is a principle or method of proof. It is applicable to a predicate $\mathscr{P}(n)$ containing a variable n which takes its values in N. We can often use mathematical induction when we wish to prove "$\mathscr{P}(n)$ is true for all $n \in N$".

Mathematical induction depends on the unique nature of the natural order of N. There is a least element 0, after each element n there is a next, $n + 1$, and there are no other natural numbers than those thereby obtained. In other words, the natural order of N is a well ordering, but not a transfinite

well ordering such as that of Fig. 2–2 of 2.4. Think of a ladder extending up-
wards indefinitely. Let 0 represent the point of ground on which the ladder
rests; let 1 represent the first rung, 2 the second, and so on. If you can stand
at the bottom of the ladder (at 0), and if from any position (that is, 0 or any
rung) you are able to step up to the next, then you can reach any rung of the
ladder you like, at least in principle. This is not to say that you can reach the
top of the ladder, for our mythical ladder has no top. But, given sufficient
time, you can reach any rung of the ladder you care to choose. This, in the
form of a crude analogy, is the:

Principle of Mathematical Induction. Let $\mathscr{P}(n)$ be any predicate. In order
to prove that $\mathscr{P}(n)$ is true for all $n \in N$, it is sufficient to prove the following
two statements.

(A) $\mathscr{P}(0)$ is true.

(B) For any $m \in N$, the truth of $\mathscr{P}(m)$ implies the truth of $\mathscr{P}(m+1)$.

In other words, if we can (A) stand at the bottom of our ladder, and if we
can (B) always proceed from one rung to the next, then we can reach any rung
we wish.

Statement (A) is called the **basis** of the induction, and Statement (B) is
called the **induction step.**

Let us illustrate this principle by using it to prove a formula with which
the reader is undoubtedly familiar and has developed by some other method.
Prove that

(1) $$0 + 1 + 2 + \ldots + n = \frac{n(n+1)}{2}.$$

Here $\mathscr{P}(n)$ is Formula (1).

(A) **Basis.** $\mathscr{P}(0)$ is

$$0 = \frac{0 \cdot 1}{2},$$

which is obviously true; so the Basis is proved.

(B) **Induction step.**   Assume $\mathscr{P}(m)$ for any m; that is,

(1a) $$0 + 1 + 2 + \ldots + m = \frac{m(m+1)}{2}.$$

We must show that $\mathscr{P}(m+1)$ is true,

(1b) $$0 + 1 + 2 + \ldots + m + (m+1) = \frac{(m+1)(m+2)}{2}.$$

Formula (1a) is called the **induction hypothesis**. Beginners often have the uncomfortable feeling that when we say "assume $\mathscr{P}(m)$, the induction hypothesis" we are assuming that which we are trying to prove. This *not* the case. Essentially we are trying to prove the *hypothetical* statement (or implication): if, for any m, $\mathscr{P}(m)$ is true, then so is $\mathscr{P}(m+1)$. The way to prove such a hypothetical statement is to assume $\mathscr{P}(m)$ *temporarily* and then determine whether we can deduce $\mathscr{P}(m+1)$ from it. (Cf. 3.6.)

Now let us apply this to the case in hand, and try to derive (1b) from the Induction Hypothesis (1a).

$$0 + 1 + 2 + \ldots + m + (m+1)$$
$$= (0 + 1 + 2 + \ldots + m) + m + 1$$
$$= \frac{m(m+1)}{2} + (m+1) \text{ by (1a), Induction Hypothesis}$$
$$= \frac{m(m+1) + 2(m+1)}{2}$$
$$= \frac{(m+1)(m+2)}{2}.$$

Thus, by assuming (1a), we have deduced (1b) completing the Induction Step.

Hence, by the Principle of Mathematical Induction, Formula (1) is true for all $n \in N$.

Now let us write out a proof of the formula

(2) $$1 + 3 + 5 + \ldots + (2n+1) = (n+1)^2$$

leaving out all the intervening discussion.

Basis. For $n = 0$, Formula (2) reduces to $1 = 1$; this is trivially true.

Induction step. Assume as the induction hypothesis $\mathscr{P}(m)$,

(2a) $$1 + 3 + 5 + \ldots + (2m+1) = (m+1)^2.$$

We must prove

(2b) $$1 + 3 + 5 + \ldots + (2m+1) + (2m+3) = [(m+1)+1]^2.$$

Left side $= 1 + 3 + 5 + \ldots + (2m+1) + (2m+3)$
$$= (m+1)^2 + (2m+3) \text{ by (2a), Induction Hypothesis}$$
$$= (m^2 + 2m + 1) + (2m+3)$$
$$= m^2 + 4m + 4 = (m+2)^2 = [(m+1)+1]^2.$$

By assuming (2a), we have deduced (2b). In other words, by assuming $\mathscr{P}(m)$, we have proved $\mathscr{P}(m + 1)$. This completes the induction.

One fundamental property of N implicit in some of the foregoing discussion is that every $n \in N$ other than zero is the unique next element or **successor** of some other $m \in N$. We assume this is obvious, and state it as an axiom.

S.   For any $n \in N$, $n \neq 0$, there exists a unique $m \in N$ such that $n = m + 1$. (And hence, of course, $m < n$ by Definition 1 of 2.4.)

Let us now derive the Principle of Mathematical Induction from Axioms S and OW. We shall prove this by the indirect method, or **reductio ad absurdum.** To do this, we shall assume that the Principle is not a valid principle of proof and derive from this assumption a contradiction.

Suppose that the Principle is not a valid principle of proof. This means that there must exist at least one predicate $\mathscr{Q}(n)$ with the properties:

(a) $\mathscr{Q}(0)$ is true,

(b) For all $m \in N$, if $\mathscr{Q}(m)$ is true, then $\mathscr{Q}(m + 1)$ is true, and

(c) $\mathscr{Q}(n)$ is *not* true for all $n \in N$.

Now consider the set $P = \{x \in N \mid \mathscr{Q}(x) \text{ is false.}\}$.

(d) $0 \notin P$ by (a)

(e) $P \neq \emptyset$ by (c)
    Then P contains a least element.   (e), OW
    Call this element l, where $l \neq 0$.   (d)

(f) Then $\mathscr{Q}(l)$ is false. $l \in P$

(g) Now choose $m < l$. Then $\mathscr{Q}(m)$ is true.

(Recall l was chosen as the least element such that $\mathscr{Q}(x)$ is false.)
    By axiom S (recall that $l \neq 0$) there is an $m \in N$ such that
    $l = m + 1$, (and $m < l$).
    Since $\mathscr{Q}(m)$ is true, $\mathscr{Q}(m + 1)$ is true.   (b)
    Then $\mathscr{Q}(l)$ is true, which contradicts (f).

Having obtained our contradiction, we conclude that the Principle of Mathematical Induction is a valid principle of proof.

2.6   LEAST AND GREATEST ELEMENTS, UPPER AND LOWER BOUNDS

We shall now introduce some notions, concerning totally ordered sets, which will be useful later on. These notions are particularly simple in terms of N, $<$. We introduce them now so that the reader may become familiar with them in a simple context. On the other hand, because it is so simple, this study is not especially interesting; bear with us.

The notion of a *least element* was defined in Definition 2 of 2.4. Similarly,

Definition 1. *If A is a set with a total ordering $<$, and if $B \subseteq A$, then x is the **greatest element** of B provided*: (i) $x \in B$ *and* (ii) *for any $y \in B$, $y \leqslant x$.*

Nothing is said in the definition of least (or greatest) element to suggest that it need exist for every subset $B \subseteq A$. However, when A is N, Axiom OW is a fundamental defining property of N which asserts that a least element exists. On the other hand, not every subset of N has a greatest element. Indeed N itself has no greatest element; nor does the set of even numbers, nor the set of prime numbers. All the finite subsets, and only these, have greatest elements.

Least (or greatest) elements of a set, when they exist, are always unique; that is, there is never more than one. If we suppose m, n are both least (or greatest) elements of some set, then by the definition $m \leqslant n$ and $n \leqslant m$. Hence, by O'2 of 2.4, $m = n$.

If N is ordered as in Figure 2–3 of 2.4, then every subset has a greatest element, but only the finite subsets have a least element. (Looking ahead a bit, there are subsets of the rationals Q which have neither a least nor a greatest element; some which have one and not the other; some which have both. All this is with respect to the natural order, of course.)

Definition 2. *Let A be any set totally ordered by a relation $<$. Let $B \subseteq A$. Then*

(a) *x is **an upper bound** of B provided that every $y \in B$ satisfies $y \leqslant x$,*

(b) *x is **a lower bound** of B provided that every $y \in B$ satisfies $y \geqslant x$.*

Definition 2 does not assert that an upper (or lower) bound need exist. If it does exist, it need not belong to B, nor need it be unique.

Let us restrict our attention to N, $<$. Consider the set $B_1 = \{2, 4, 6\}$. The number 6 is an upper bound of B_1 and $6 \in B_1$. On the other hand, 7 is an upper bound of B_1 and $7 \notin B_1$. In fact, the set $U_1 = \{x \mid x \geqslant 6\}$ contains exactly all the upper bounds of B_1. Similarly, $L_1 = \{x \mid x \leqslant 2\} = \{0, 1, 2\}$ contains exactly all the lower bounds of B_1, and $2 \in B_1$, but $0,1 \notin B_1$.

Consider the set $B_2 = \{0, 2, 4, 6, \ldots\}$. B_2 has no upper bounds in N. The set $L_2 = \{0\}$ contains all of its lower bounds; that is, 0 is the only lower bound of B_2 (and $0 \in B_2$).

In general, for any totally ordered set A, a subset $B \subseteq A$ need not have any upper (or lower) bounds at all, but if it does it may have many. Let U be the set of all upper bounds of B, and let L be the set of all lower bounds. If $U \neq \emptyset$, it may have a *least* element (which must necessarily be unique);

this element is called the **least upper bound** of B; similarly, if $L \neq \emptyset$ and has a *greatest* element, this element is called the **greatest lower bound** of B. (These are often abbreviated "l.u.b." and "g.l.b." respectively.)[1] A greatest element is always the least upper bound. (Prove this. Exercise 14 of 2.7.) Similarly, a least element is always the greatest lower bound.

There exist subsets of totally ordered sets which do not have a greatest (or least) element but do have a l.u.b. (or g.l.b.). Consider, for example, the set $\{x \in Q \mid x = 1/n$ for some $n \in N, n \neq 0\}$, ordered by the natural order; this set has no least element, but 0 is its g.l.b. On the other hand, 1 is both the greatest element and the l.u.b. This situation does not occur with respect to N and $<$ which is a particularly simple ordering.

We can summarize the relevant facts about N and $<$ as follows.

(1a) Not every subset of N has an upper bound.

(1b) Not every subset of N has a greatest element.

(2a) Every subset $M \subseteq N$ which does have an upper bound, has a l.u.b., l, and l is also the greatest element of M.

(2b) Conversely, every subset $M \subseteq N$ which has a greatest element g also has an upper bound (for example, g itself) and g is the l.u.b. of M.

(3a) Every subset $M \subseteq N$ has a lower bound. This is so because

(3b) every subset $M \subseteq N$ has a least element l (by OW), and l is necessarily a lower bound; in fact it is the g.l.b. of M.

2.7   EXERCISES

1. Which of the following are operations on N?
   - (i) $a\,\mathbf{1}\,b = (a + b)^2$.
   - (ii) $a\,\mathbf{2}\,b = \sqrt{(a + b)}$.
   - (iii) $a\,\mathbf{3}\,b = a + (a \vee b)$, where $\vee$ is as defined in 2.2.
   - (iv) $a\,\mathbf{4}\,b = 3 + a$.
   - (v) $a\,\mathbf{5}\,b = 0$.
   - (vi) $a\,\mathbf{6}\,b = \dfrac{(a + b)}{2}$.
   - (vii) $a\,\mathbf{7}\,b =$ the smallest $x \in N$ such that $x > \dfrac{(a + b)}{2}$.

2. (a) Find three examples (other than those in the text) of operations on N which are not commutative.
   (b) Find three examples (other than those in the text) of operations on N which are not associative.

---

[1] It has become quite common in mathematics to call a least upper bound a *supremum* (abbreviated *sup*) and a greatest lower bound an *infimum* (abbreviated *inf*). We shall use l.u.b. and g.l.b. because they are more suggestive.

3. Let $\vee$ be as defined in 2.2. Let $\wedge$ be similarly defined by

$$a \wedge b = \begin{cases} \text{the smaller of } a, b & \text{if } a \neq b, \\ a & \text{if } a = b. \end{cases}$$

Show that both distributive laws hold with respect to $\vee$ and $\wedge$; that is, for any $a, b, c \in N$,

$$a \wedge (b \vee c) = (a \wedge b) \vee (a \wedge c),$$
$$a \vee (b \wedge c) = (a \vee b) \wedge (a \vee c).$$

4. Find two operations on N with respect to which neither distributive law holds.

5. In Axioms A3 and M3, the word "unique" can be omitted, and the uniqueness of 0 and 1 can be proved on the basis of the axioms (so modified). Prove this uniqueness. (For example, suppose $0 + m = m$ and $0' + m = m$, then show that $0 = 0'$.)

6. Prove: if $n = p$, then $mn = mp$ (the converse of C_2 except that we disregard $m \neq 0$).

7. In Axiom O4, the word "exactly" can be omitted and then established on the basis of the axioms (so modified). Let O4 be stated in this manner: that *at least* one of $m < n$, $m = n$, $m > n$ holds. Then show that *at most* one holds.

8. Prove E1, E2, E3 of 2.3 by mathematical induction.

9. Verify that the orderings of Figures 2–2, 2–3, 2–4 are total orderings of N. Verify that the ordering of Figure 2–5 is a partial ordering of N.

10. Prove that $m|n$ is a partial, but not a total, ordering of N.

11. Prove OM of the Theorem of 2.4.

12. Prove each of the following by mathematical induction.

   (i) $1^n = 1$ for all $n \in N$.

   (ii) $1 + 8 + 27 + \ldots + n^3 = \left[\dfrac{n(n+1)}{2} \right]^2$ for all $n \in N$.

   (iii) For real numbers $x_1, x_2, \ldots, x_n$,
   $$x_1^2 + x_2^2 + \ldots + x_n^2 > 0 \text{ unless } x_1 = x_2 = \ldots = x_n = 0.$$

13. Which of the following relations on N are total orderings? For those that are not, explain why they are not.

   (i) $m \mathscr{P}_1 n$ if and only if $m < 2n$.

   (ii) $m = n$.

   (iii) $m \mathscr{P}_3 n$ if and only if $(n = 27 \text{ and } m \neq 27)$ or $(n \neq 27 \text{ and } m < n)$.

   (iv) $m \mathscr{P}_4 n$ if and only if $(n = 27)$ or $(m \leqslant n \text{ and } n \neq 27)$.

14. Let A be any totally ordered set. Prove that the greatest element of A, if it exists, is its least upper bound.

15. (a) Let N be ordered by the ordering of Figure 2–2 of 2.4. Prove that every subset of N has a least element. Find subsets A, B of N which have no greatest element, but such that A has an upper bound and B does not. Does A have a least upper bound? What is it?

   (b) Let N be ordered by the ordering of Figure 2–4 of 2.4. What is the least element of N? Find a subset of N with no least element. Prove that every subset has a greatest element. Find a subset with no least element, but with a g.l.b.

3

Logical Connectives

3.1 SENTENCES

By a **sentence** (or **statement**) we shall mean a written or verbal assertion which is either **true** or **false**, but not both. This description is adequate for our purposes, but of course would be unsatisfactory to a teacher of English who would insist that imperative sentences and questions are also sentences even though they cannot be characterized as being true or false. Thus our sentences are actually declarative sentences.

For the most part we deal with sentences about mathematical objects; for example:

(1) $2 + 2 = 4$;

(2) $1 + 1 = 2$;

(3) $n^2 > n$ for all $n \in N$;

(4)   for all $n, m, p \in N$, $n(p + m) = np + nm$;

(5) $1 + 5 = 7$.

Each of these is a sentence about the natural numbers. All are true except (3) and (5). Sentence (5) is obviously false, and (3) is false because there is at least one natural number n such that n^2 is not greater than n; for example, $n = 1$.

For a given sentence $\mathscr{S}$, the **truth value** of $\mathscr{S}$ is defined to be $\mathscr{T}$ if $\mathscr{S}$ is true, and $\mathscr{F}$ otherwise (that is, in case $\mathscr{S}$ is false).

We shall assume, in our discussion, that certain sentences are **simple**. By this we mean only that we do not analyze their structure in this context. Letters such as $\mathscr{P}, \mathscr{Q}, \mathscr{R}, \mathscr{S}$ will be used to designate simple sentences. From simple sentences we shall construct **compound** sentences such as "$\mathscr{P}$ and $\mathscr{Q}$"

or "If $\mathscr{P}$, then $\mathscr{Q}$", with the help of connectives. Whether a specific sentence is simple or compound will depend on our point of view. For example, "$n^2 > n$ for all $n \in N$" will be treated as a simple (false) sentence in the present context, because we are not now analyzing assertions of the kind "for all ...". The latter will be treated in Chapter 10.

3.2 CONNECTIVES

In mathematical discussions we often make compound sentences involving one or more of the **connectives**: "It is false that ...", "... and ...", "... or ...", and "If ..., then ...".

(1)   *It is false that $n^2 > n$ for all $n \in N$.*

(2) $1 + 1 = 2$ *and* $2 + 2 = 4$.

(3) $1 + 1 = 2$ *or* $1 + 1 \neq 2$.

(4)   *If $n^2 > n$ for all $n \in N$, then $1 > 1$.*

Other connectives (for sentences) are often used, but careful analysis of the arguments in which they appear will show that their meaning can be reduced to some combination of the above connectives.

Of course our problem now is how to determine the truth value of a sentence composed of two or more sentences with known truth values. In succeeding sections we consider the problem for each of the four connectives given above.

3.3 IT IS FALSE THAT

If $\mathscr{S}$ is a sentence, we often verbally abbreviate "It is false that $\mathscr{S}$" as "not $\mathscr{S}$" and write $\neg \mathscr{S}$. If $\mathscr{S}$ is a sentence, then $\neg \mathscr{S}$ is called the **negation** of $\mathscr{S}$.

We adopt the reasonable convention of defining the truth value of $\neg \mathscr{S}$ to be (i) $\mathscr{T}$ if $\mathscr{S}$ is false, and (ii) $\mathscr{F}$ is $\mathscr{S}$ is true. Thus we make the following definition.

Definition. $\neg \mathscr{S}$ *is true if and only if $\mathscr{S}$ is false.*

3.4 AND

For sentences $\mathscr{S}_1$ and $\mathscr{S}_2$, the compound sentence "$\mathscr{S}_1$ and $\mathscr{S}_2$" is written symbolically $\mathscr{S}_1 \wedge \mathscr{S}_2$. It is intuitively obvious that we should make the following definition.

Definition. *$\mathscr{S}_1 \wedge \mathscr{S}_2$ is true if and only if both $\mathscr{S}_1$ is true and $\mathscr{S}_2$ is true.*

Indeed, if someone asserts that "Our class president is a boy and our class treasurer is a girl," this assertion is correct (true) only when both clauses are themselves correct.

At this point a word of caution should be introduced. Consider the sentence

(1)   "It is false that $4 > 5$ and it is false that $8 < 6$."

The sentence is essentially ambiguous in its written form. It could mean (i) that the statements "$4 > 5$" and "$8 < 6$" are both false, or (ii) that the statement "$4 > 5$ and it is false that $8 < 6$" is false. By the use of parentheses this ambiguity can be disposed of. Indeed, if we write $\mathscr{P}$ for "$4 > 5$" and $\mathscr{Q}$ for "$8 < 6$," then (i) clearly has the meaning

(2) $(\neg \mathscr{P}) \wedge (\neg \mathscr{Q})$

and (ii) clearly has the meaning

(3) $\neg [\mathscr{P} \wedge (\neg \mathscr{Q})]$.

We can write (2) and (3) *more simply* as

(2') $\neg \mathscr{P} \wedge \neg \mathscr{Q}$

and

(3') $\neg (\mathscr{P} \wedge \neg \mathscr{Q})$

and adhere to the following convention:

Unless otherwise indicated by parentheses, $\neg$ governs only the simple sentence which immediately follows it.

Finally, when referring to statements of the form $\mathscr{S}_1 \wedge \mathscr{S}_2$ we use the term **conjunction**. That is to say $\mathscr{S}_1 \wedge \mathscr{S}_2$ is the conjunction of $\mathscr{S}_1$ and $\mathscr{S}_2$. $\mathscr{S}_1$ and $\mathscr{S}_2$ are called the **arguments** of the conjunction $\mathscr{S}_1 \wedge \mathscr{S}_2$. Analogously, $\mathscr{S}$ is called the argument of the negation $\neg \mathscr{S}$.

3.5   OR

The meaning of "or" is not so clear cut as are the meanings of "not" and "and". There are two possible meanings of "or": the inclusive "or" and the exclusive "or". As an example of the inclusive "or", consider Axiom O′4 in section 2.4,

"For any pair m, n of natural numbers either $m \leqslant n$ or $n \leqslant m$."

If we analyze this statement, we might write it "$\mathscr{S}_1$ or $\mathscr{S}_2$" where $\mathscr{S}_1$ is "$m \leqslant n$" and $\mathscr{S}_2$ is "$n \leqslant m$". We would then say "$\mathscr{S}_1$ or $\mathscr{S}_2$" is true if either $\mathscr{S}_1$ is true or $\mathscr{S}_2$ is true. However, if $m = n$, then both $\mathscr{S}_1$ and $\mathscr{S}_2$ are true. In this case most people would wish to accept "$\mathscr{S}_1$ or $\mathscr{S}_2$" as true. We use the word "inclusive" to indicate that "$\mathscr{S}_1$ or $\mathscr{S}_2$" is true if either $\mathscr{S}_1$ is true, or $\mathscr{S}_2$ is true, *or both*. This "or" is abbreviated "$\vee$", and a sentence of the form $\mathscr{S}_1 \vee \mathscr{S}_2$ is called a **disjunction** with arguments $\mathscr{S}_1$ and $\mathscr{S}_2$.

Definition. *$\mathscr{S}_1 \vee \mathscr{S}_2$ is true if and only if at least one of the sentences $\mathscr{S}_1$, $\mathscr{S}_2$ is true.*

Now consider the exclusive "or", for which there is no generally accepted notation. An example of the use of the exclusive "or" is the first sentence of this chapter.

Thus, "$\mathscr{S}_1$ *(exclusive) or* $\mathscr{S}_2$" is true if exactly one of $\mathscr{S}_1$ and $\mathscr{S}_2$ is true, and it is false otherwise.

In mathematical discussions "or" and "$\vee$" are used to signify only the inclusive "or". When the exclusive "or" is intended, the words "but not both", or some equivalent, must be added. (Cf. Exercise 3 of 3.8.)

3.6   IF . . . , THEN . . .

In the course of a mathematical demonstration, the following sequence of sentences might occur:

A well-known property of natural numbers asserts that for $m, n, p \in N$,

(1) if $n > p$ and $m \neq 0$, then $nm > pm$.

We have shown above that

(2) $n > p$ and $m \neq 0$,

and so we can conclude that

(3) $nm > pm$.

Arguments of this type are frequently encountered in mathematical texts. They have the general form:

We know that

(1') if $\mathscr{S}_1$, then $\mathscr{S}_2$,

and that

(2') $\mathscr{S}_1$ is true.

Therefore we can assert that

(3') $\mathscr{S}_2$ is true.

Thus from the truth of (1') and the truth of (2'), the truth of (3') follows. Schematically, we can write

(4)
$$
\begin{cases}
\text{If } \mathscr{S}_1, \text{ then } \mathscr{S}_2 & \text{(true)} \\
\mathscr{S}_1 & \text{(true)} \\
\hline
\text{Thence } \mathscr{S}_2 & \text{(true).}
\end{cases}
$$

This technique for proving $\mathscr{S}_2$ is called the **rule of detachment** or **modus ponens** and is an indispensable tool of mathematical reasoning.

It is usual to employ the notation $\mathscr{S}_1 \Rightarrow \mathscr{S}_2$ for the sentence "If $\mathscr{S}_1$, then $\mathscr{S}_2$". Thus (4) becomes

(4')
$$
\begin{cases}
\mathscr{S}_1 \Rightarrow \mathscr{S}_2 & \text{(true),} \\
\mathscr{S}_1 & \text{(true),} \\
\hline
\mathscr{S}_2 & \text{(true).}
\end{cases}
$$

A sentence $\mathscr{S}_1 \Rightarrow \mathscr{S}_2$ is called an **implication** or **conditional sentence**. $\mathscr{S}_1$ is called the **antecedent** and $\mathscr{S}_2$ the **consequent** of the implication.

Now suppose in the implication $\mathscr{S}_1 \Rightarrow \mathscr{S}_2$ we know that $\mathscr{S}_2$ is false. It is clear from (4') that if we are to preserve modus ponens, then either $\mathscr{S}_1$ is not true or $\mathscr{S}_1 \Rightarrow \mathscr{S}_2$ is not true. (Indeed if both were true, then of necessity, by modus ponens, $\mathscr{S}_2$ would be true.) If the implication $\mathscr{S}_1 \Rightarrow \mathscr{S}_2$ were true, then we could *not have*

$\mathscr{S}_1$ true and $\mathscr{S}_2$ false, that is, "$\mathscr{S}_1$ and $\neg \mathscr{S}_2$" true.

We *would have*

(5)　　　$\neg(\mathscr{S}_1 \wedge \neg \mathscr{S}_2)$　true.

Thus if $\mathscr{S}_1 \Rightarrow \mathscr{S}_2$ is true, sentence (5) is true.

Now consider the following schematic argument which is similar in form to (4) and (4').

(4'')
$$
\begin{cases}
\neg(\mathscr{S}_1 \wedge \neg \mathscr{S}_2) & \text{(true),} \\
\mathscr{S}_1 & \text{(true),} \\
\hline
\mathscr{S}_2 & \text{(true).}
\end{cases}
$$

That this is a valid argument follows easily from the definitions in 3.3 and 3.4. If we have $\neg(\mathscr{S}_1 \wedge \neg \mathscr{S}_2)$ true, then we have $\mathscr{S}_1 \wedge \neg \mathscr{S}_2$ false. Hence, at least one of $\mathscr{S}_1$, $\neg\mathscr{S}_2$ is false. But $\mathscr{S}_1$ is true, so $\neg\mathscr{S}_2$ is false; that is, $\mathscr{S}_2$ is true.

In mathematics we adopt the convention that $\mathscr{S}_1 \Rightarrow \mathscr{S}_2$ and $\neg(\mathscr{S}_1 \wedge \neg\mathscr{S}_2)$ are **logically equivalent**, that is, the first sentence ($\mathscr{S}_1 \Rightarrow \mathscr{S}_2$) is true if and only if the second ($\neg(\mathscr{S}_1 \wedge \neg\mathscr{S}_2)$) is true. This means that for any pair of sentences, $\mathscr{S}_1$ and $\mathscr{S}_2$, the truth value of $\mathscr{S}_1 \Rightarrow \mathscr{S}_2$ is exactly the truth value of $\neg(\mathscr{S}_1 \wedge \neg\mathscr{S}_2)$. Thus under our convention, $\mathscr{S}_1 \Rightarrow \mathscr{S}_2$ is interpreted as $\neg(\mathscr{S}_1 \wedge \neg\mathscr{S}_2)$.

To determine the truth values of $\mathscr{S}_1 \Rightarrow \mathscr{S}_2$ in terms of the truth values of $\mathscr{S}_1$ and $\mathscr{S}_2$, we need only look at the truth values of (5). From the Definition in 3.4, we deduce that $\mathscr{S}_1 \wedge \neg\mathscr{S}_2$ is true only when both $\mathscr{S}_1$ and $\neg\mathscr{S}_2$ are true. From the Definition in 3.3, $\neg\mathscr{S}_2$ is true only when $\mathscr{S}_2$ is false. Thus $\mathscr{S}_1 \wedge \neg\mathscr{S}_2$ is true only when $\mathscr{S}_1$ is true and $\mathscr{S}_2$ is false. Hence, by the Definition in 3.3 again, we conclude that:

(6) $\neg(\mathscr{S}_1 \wedge \neg\mathscr{S}_2)$ is false exactly when $\mathscr{S}_1$ is true and $\mathscr{S}_2$ is false.

Now since we interpret $\mathscr{S}_1 \Rightarrow \mathscr{S}_2$ as $\neg(\mathscr{S}_1 \wedge \neg\mathscr{S}_2)$ we must have the following.

Definition 1. $\mathscr{S}_1 \Rightarrow \mathscr{S}_2$ *is false exactly when* $\mathscr{S}_1$ *is true and* $\mathscr{S}_2$ *is false.*

It is of interest to note that if $\mathscr{S}_1$ is replaced by $\neg\mathscr{P}_2$ and $\mathscr{S}_2$ by $\neg\mathscr{P}_1$, then, by Definition 1,

(7) $\neg\mathscr{P}_2 \Rightarrow \neg\mathscr{P}_1$ is false if and only if $\neg\mathscr{P}_2$ is true and $\neg\mathscr{P}_1$ is false.

Thus, from the Definition in 3.3, we deduce that

(8) $\neg\mathscr{P}_2 \Rightarrow \neg\mathscr{P}_1$ is false if and only if $\mathscr{P}_2$ is false and $\mathscr{P}_1$ is true.

However, by Definition 1, $\mathscr{P}_1 \Rightarrow \mathscr{P}_2$ is false exactly when $\mathscr{P}_2$ is false and $\mathscr{P}_1$ is true. Thus $\neg\mathscr{P}_2 \Rightarrow \neg\mathscr{P}_1$ is false exactly when $\mathscr{P}_1 \Rightarrow \mathscr{P}_2$ is false. This means of course that $\mathscr{P}_1 \Rightarrow \mathscr{P}_2$ and $\neg\mathscr{P}_2 \Rightarrow \neg\mathscr{P}_1$ are logically equivalent. $\neg\mathscr{P}_2 \Rightarrow \neg\mathscr{P}_1$ is called the **contrapositive** of $\mathscr{P}_1 \Rightarrow \mathscr{P}_2$.

It is often easier (and by the discussion above, just as satisfactory) to prove the contrapositive of an implication than it is to prove the implication itself. Consider the following sentence concerning natural numbers.

(9)   If n^2 is even, then n is even.

It seems a formidable task to prove this directly. Now consider its contrapositive.

(10)   If n is odd, then n^2 is odd.

This is easily proved as follows. Assume n odd; then there is a natural number k such that $n = 2k + 1$. Then

$$n^2 = (2k + 1)^2 = 4k^2 + 4k + 1,$$

and

$$n^2 = 2(2k^2 + 2k) + 1.$$

Thus n^2 is odd. (It has the form $2j + 1$ where $j = 2k^2 + 2k$.) Since (10) has been proved, and (10) is the contrapositive of (9), then (9) is true.

Note the method of proof in (10). We are required to prove an implication. In the technique employed we assume the antecedent of the implication (temporarily) and deduce from it the consequent. This amounts to showing that when $\mathscr{S}_1$ is true, $\mathscr{S}_2$ is not false and so, by Definition 1, $\mathscr{S}_1 \Rightarrow \mathscr{S}_2$ is not false.

The sentence $\mathscr{P}_2 \Rightarrow \mathscr{P}_1$ (which is logically equivalent to $\neg \mathscr{P}_1 \Rightarrow \neg \mathscr{P}_2$) is called the **converse** of the implication $\mathscr{P}_1 \Rightarrow \mathscr{P}_2$ and should not be confused with the contrapositive of $\mathscr{P}_1 \Rightarrow \mathscr{P}_2$. As the following example shows, an implication is *not* logically equivalent to its converse.

(11)   If n is divisible by 4, then n is divisible by 2.

This is clearly true because 4 is divisible by 2. Now study the converse.

(12)   If n is divisible by 2, then n is divisible by 4.

This is not true. As a counterexample, let $n = 6$. We see that $2|6$ while $\neg 4|6$.

There is one more connective used quite often in mathematics which can be defined in terms of the connectives described above.

Definition 2. $\mathscr{S}_1 \Leftrightarrow \mathscr{S}_2$ *means* $(\mathscr{S}_1 \Rightarrow \mathscr{S}_2) \wedge (\mathscr{S}_2 \Rightarrow \mathscr{S}_1)$. *Thus* $\mathscr{S}_1 \Leftrightarrow \mathscr{S}_2$ *is true if and only if both* $\mathscr{S}_1 \Rightarrow \mathscr{S}_2$ *and* $\mathscr{S}_2 \Rightarrow \mathscr{S}_1$ *are true.*

A careful check of Definition 1 and the Definition in 3.4 will yield the result that $\mathscr{S}_1 \Leftrightarrow \mathscr{S}_2$ is true if and only if $\mathscr{S}_1$ and $\mathscr{S}_2$ have the same truth value; that is if and only if either both $\mathscr{S}_1$ and $\mathscr{S}_2$ are true or both are false. Thus $\mathscr{S}_1 \Leftrightarrow \mathscr{S}_2$ has the intuitive meaning "$\mathscr{S}_1$ if and only if $\mathscr{S}_2$". Note that $\mathscr{S}_1 \Leftrightarrow \mathscr{S}_2$ is defined as the conjunction of $\mathscr{S}_1 \Rightarrow \mathscr{S}_2$ and its converse. The sentence $\mathscr{S}_1 \Leftrightarrow \mathscr{S}_2$ is called a **biconditional sentence.**

Since mathematical discourse is composed, to a large extent, of sentences of the form $\mathscr{S}_1 \Rightarrow \mathscr{S}_2$ and $\mathscr{S}_1 \Leftrightarrow \mathscr{S}_2$, mathematicians have developed many ways of expressing these two sentences in order to avoid being dull and repetitious. Table 3.1 lists some of the possibilities.

Table 3.1

Symbolic sentence	Sentence expressed in words
$\mathscr{S}_1 \Rightarrow \mathscr{S}_2$	If $\mathscr{S}_1$, then $\mathscr{S}_2$. $\mathscr{S}_2$ if $\mathscr{S}_1$. $\mathscr{S}_1$ only if $\mathscr{S}_2$. $\mathscr{S}_1$ is a sufficient condition for $\mathscr{S}_2$. $\mathscr{S}_2$ is a necessary condition for $\mathscr{S}_1$.
$\mathscr{S}_1 \Leftrightarrow \mathscr{S}_2$	$\mathscr{S}_1$ if and only if $\mathscr{S}_2$. $\mathscr{S}_2$ if and only if $\mathscr{S}_1$. $\mathscr{S}_1$ is a necessary and sufficient condition for $\mathscr{S}_2$. $\mathscr{S}_2$ is a necessary and sufficient condition for $\mathscr{S}_1$.

3.7   TRUTH TABLES

The definitions of the truth values of $\neg \mathscr{S}_1$, $\mathscr{S}_1 \wedge \mathscr{S}_2$, $\mathscr{S}_1 \vee \mathscr{S}_2$, $\mathscr{S}_1 \Rightarrow \mathscr{S}_2$, and $\mathscr{S}_1 \Leftrightarrow \mathscr{S}_2$ in terms of the truth values of the arguments $\mathscr{S}_1$ and $\mathscr{S}_2$ can be exhibited in the form of **truth tables**. The truth table for "$\neg$" appears in Figure 3–1. All possible truth values of $\mathscr{S}$ are listed under $\mathscr{S}$ and the corresponding truth values for $\neg \mathscr{S}$ are listed beside them.

$\mathscr{S}$	$\neg \mathscr{S}$
$\mathscr{T}$	$\mathscr{F}$
$\mathscr{F}$	$\mathscr{T}$

Figure 3–1

In the cases $\mathscr{S}_1 \wedge \mathscr{S}_2$, $\mathscr{S}_1 \vee \mathscr{S}_2$, all possible combinations of the truth values of the arguments $\mathscr{S}_1$ and $\mathscr{S}_2$ are listed. (There are four.) The corresponding values for $\mathscr{S}_1 \wedge \mathscr{S}_2$ and $\mathscr{S}_1 \vee \mathscr{S}_2$ are then listed in the appropriate places. According to the Definitions in 3.4 and 3.5 the truth tables for $\mathscr{S}_1 \wedge \mathscr{S}_2$ and $\mathscr{S}_1 \vee \mathscr{S}_2$ must be as indicated in Figure 3–2.

$\mathscr{S}_1$	$\mathscr{S}_2$	$\mathscr{S}_1 \wedge \mathscr{S}_2$	$\mathscr{S}_1 \vee \mathscr{S}_2$
$\mathscr{T}$	$\mathscr{T}$	$\mathscr{T}$	$\mathscr{T}$
$\mathscr{T}$	$\mathscr{F}$	$\mathscr{F}$	$\mathscr{T}$
$\mathscr{F}$	$\mathscr{T}$	$\mathscr{F}$	$\mathscr{T}$
$\mathscr{F}$	$\mathscr{F}$	$\mathscr{F}$	$\mathscr{F}$

Figure 3–2

Since $\mathscr{S}_1 \Rightarrow \mathscr{S}_2$ is logically equivalent to $\neg(\mathscr{S}_1 \wedge \neg \mathscr{S}_2)$, $\mathscr{S}_1 \Rightarrow \mathscr{S}_2$ will have the same truth table as $\neg(\mathscr{S}_1 \wedge \neg \mathscr{S}_2)$. We can construct the truth table for $\neg(\mathscr{S}_1 \wedge \neg \mathscr{S}_2)$ from the truth tables of "$\neg$" and "$\wedge$" already constructed, as indicated in Figure 3–3. In Figure 3–3, column three is deter-

$\mathscr{S}_1$	$\mathscr{S}_2$	$\neg \mathscr{S}_2$	$\mathscr{S}_1 \wedge \neg \mathscr{S}_2$	$\neg(\mathscr{S}_1 \wedge \neg \mathscr{S}_2)$
$\mathscr{T}$	$\mathscr{T}$	$\mathscr{F}$	$\mathscr{F}$	$\mathscr{T}$
$\mathscr{T}$	$\mathscr{F}$	$\mathscr{T}$	$\mathscr{T}$	$\mathscr{F}$
$\mathscr{F}$	$\mathscr{T}$	$\mathscr{F}$	$\mathscr{F}$	$\mathscr{T}$
$\mathscr{F}$	$\mathscr{F}$	$\mathscr{T}$	$\mathscr{F}$	$\mathscr{T}$

Figure 3–3

mined from column two by the truth table for "$\neg$"; column four is determined from columns one and three by the truth table for "$\wedge$", and so on. Thus the truth table for $\mathscr{S}_1 \Rightarrow \mathscr{S}_2$ is given by Figure 3–4. It should be noted that the data in Figure 3–4 coincide exactly with Definition 1 of 3.6; the sentence $\mathscr{S}_1 \Rightarrow \mathscr{S}_2$ is false exactly in the case where $\mathscr{S}_1$ is true and $\mathscr{S}_2$ is false. The construction of the truth table for $\mathscr{S}_1 \Leftrightarrow \mathscr{S}_2$ is left to the reader.

$\mathscr{S}_1$	$\mathscr{S}_2$	$\mathscr{S}_1 \Rightarrow \mathscr{S}_2$
$\mathscr{T}$	$\mathscr{T}$	$\mathscr{T}$
$\mathscr{T}$	$\mathscr{F}$	$\mathscr{F}$
$\mathscr{F}$	$\mathscr{T}$	$\mathscr{T}$
$\mathscr{F}$	$\mathscr{F}$	$\mathscr{T}$

Figure 3–4

Truth tables are useful for verifying that certain sentences are **tautologies**, that is, are universally true no matter what the truth values of their arguments may be. Consider, for example, the sentence $\mathscr{S} \vee \neg\mathscr{S}$. This is true independently of the truth value of the argument $\mathscr{S}$ as can be seen by exhibiting the truth table (Figure 3–5). Similarly $(\mathscr{S} \wedge \mathscr{P}) \Rightarrow \mathscr{S}$ is also a tautology (Figure 3–6). Other tautologies are discussed in the exercises.

$\mathscr{S}$	$\neg\mathscr{S}$	$\mathscr{S} \vee \neg\mathscr{S}$
$\mathscr{T}$	$\mathscr{F}$	$\mathscr{T}$
$\mathscr{F}$	$\mathscr{T}$	$\mathscr{T}$

Figure 3–5

$\mathscr{S}$	$\mathscr{P}$	$\mathscr{S} \wedge \mathscr{P}$	$(\mathscr{S} \wedge \mathscr{P}) \Rightarrow \mathscr{S}$
$\mathscr{T}$	$\mathscr{T}$	$\mathscr{T}$	$\mathscr{T}$
$\mathscr{T}$	$\mathscr{F}$	$\mathscr{F}$	$\mathscr{T}$
$\mathscr{F}$	$\mathscr{T}$	$\mathscr{F}$	$\mathscr{T}$
$\mathscr{F}$	$\mathscr{F}$	$\mathscr{F}$	$\mathscr{T}$

Figure 3–6

An understanding of the nature and meaning of the foregoing connectives is an important first step, but only a first step, toward an understanding of the logical reasoning which underlies all mathematical thought. The use of truth tables to determine the truth values of compound sentences (and, in particular, to determine whether or not a sentence is a tautology) is of limited value. In all this, what is more important is for the reader to learn to recognize valid forms of reasoning whereby we *prove* sentences of the forms illustrated above or *use* such sentences in the proofs of others. Furthermore, sentence structure must be analyzed in more detail before we can do this job adequately. Roughly speaking, we must analyze sentence structure into the "subject" and "predicate" of grammar. This leads to open sentences (or predicates) such as "$x < 10$", and thence to sentences of the form "for all $x, x < 10$" or "there exists an x such that $x < 10$". We shall discuss these matters in Chapter 10, to which the reader may turn now if he wishes.

3.8   EXERCISES

1. For each of the following sentences define $\mathscr{S}$, $\mathscr{P}$, $\mathscr{Q}$ and express the sentence in appropriate symbolic form.

   (a) The numbers 7, 5, and 3 are all greater than 2.

   (b) If 7 is greater than 5 and 3 is greater than 2, then 12 is greater than 8.

   (c) It is false that $6 + 5 = 11$ or if $1 + 0 = 0$, then $5 = 25$.

   (d) If $6 + 1 = 8$ or $7 + 5 = 11$, then $1 = 0$.

2. Using the definitions of $\neg$, $\wedge$, $\vee$, $\Rightarrow$ find the truth values of each of the sentences in Exercise 1 above.

3. Express the sentence "$\mathscr{S}_1$ exclusive or $\mathscr{S}_2$" using the connectives $\neg$, $\wedge$, $\vee$.

4. Show that $\mathscr{S}_1 \vee \mathscr{S}_2$ is logically equivalent to $\neg(\neg \mathscr{S}_1 \wedge \neg \mathscr{S}_2)$. Then show that $\mathscr{S}_1 \wedge \mathscr{S}_2$ is logically equivalent to $\neg(\neg \mathscr{S}_1 \vee \neg \mathscr{S}_2)$.

5. (a) Suppose you are given that $\mathscr{S}_1 \vee \mathscr{S}_2$ is true and $\mathscr{S}_1$ is true. What can be said about the truth value of $\mathscr{S}_2$?

   (b) Suppose $\mathscr{S}_1 \wedge \mathscr{S}_2$ is true. What can be said about the truth values of $\mathscr{S}_1$ and $\mathscr{S}_2$?

   (c) Suppose $\mathscr{S}_1 \wedge (\mathscr{S}_2 \vee \mathscr{S}_3)$ is true and $\mathscr{S}_3$ is false; what can be said about $\mathscr{S}_1$ and $\mathscr{S}_2$?

6. Define $\mathscr{S}$ and $\mathscr{Q}$ appropriately for each of the following sentences and express them in the form of a conditional or biconditional sentence.

   (a) If m and n are natural numbers, then $m + n$ is a natural number.

   (b) A necessary and sufficient condition for two triangles to be congruent is that two sides and an included angle of one be equal to two sides and an included angle of the other.

   (c) Dogs eat cats only if cats are available.

   (d) Dogs eat cats if cats are available.

   (e) Dogs eat cats if and only if cats are available.

7. Make a truth table for

   (a) the biconditional, and

   (b) the exclusive or.

8. Show that two compound sentences $\mathscr{P}$ and $\mathscr{Q}$ involving the symbols $\mathscr{S}_1$, $\mathscr{S}_2$, ..., $\mathscr{S}_n$ are logically equivalent if and only if $\mathscr{P} \Leftrightarrow \mathscr{Q}$ is a tautology.

9. Use the results of Exercise 8 to verify formulas (1) to (8) of 10.8.

10. (a) Given the axioms O1 to O4 for "$<$" from Chapter 2 (that is, given that they are each true), show that the axioms O'1 to O'4 for "$\leq$" from Chapter 2 follow.

   (b) Show also the converse of part (a), that is, show that if O'1 to O'4 are true, then O1 to O4 are true.

11. Determine which of the following are tautologies. (Try using verbal arguments based on the definitions instead of, or in addition to, the evaluation of a truth table.) For each one that is not a tautology, find a *counterexample*; that is, values of $\mathscr{P}$, $\mathscr{Q}$, $\mathscr{R}$ that make the given sentence false.

   (a) $((\mathscr{P} \vee \mathscr{Q}) \wedge \mathscr{Q}) \Leftrightarrow \mathscr{Q}$.

   (b) $((\mathscr{P} \vee \mathscr{R}) \wedge \mathscr{Q}) \Leftrightarrow \mathscr{Q}$.

   (c) $((\mathscr{P} \wedge \mathscr{Q}) \vee \mathscr{Q}) \Leftrightarrow \mathscr{Q}$.

   (d) $((\mathscr{P} \wedge \mathscr{Q}) \vee \mathscr{Q}) \Leftrightarrow \mathscr{P}$.

4

The Algebra of Sets

4.1 OPERATIONS ON SETS

In the previous section, operations called the logical connectives enabled us to build compound sentences from given sentences. In much the same way we shall now define operations on sets, enabling us to form new sets from given sets. The fundamental operations we shall discuss are called *union*, *intersection*, and *complement*, and they are defined in 4.2, 4.3, and 4.4.

4.2 UNIONS

The **union** of two sets A and B, written $A \cup B$, and read "A union B", is defined as follows.

Definition. *$A \cup B$ is the set of all elements that are members of A or B.*

It should be noted that the inclusive "or" is used in this definition. An alternative way of expressing this definition is to say that $x \in A \cup B$ is true if and only if $x \in A$ or $x \in B$, and this form can be related to the definition of 3.5 in the following way. Let $\mathscr{S}_1$ be the sentence $x \in A$, and let $\mathscr{S}_2$ be the sentence $x \in B$. Then the sentence $x \in A \cup B$ can be written as $\mathscr{S}_1 \vee \mathscr{S}_2$, for from the definition of 3.5, $\mathscr{S}_1 \vee \mathscr{S}_2$ is true if and only if at least one of the sentences $\mathscr{S}_1$, $\mathscr{S}_2$ is true, and this is precisely the definition above. To illustrate this definition consider the following examples.

34

Example 1. If $A = \{1, 5, 7, 8\}$, $B = \{1, 2, 7, 11\}$, then
$$A \cup B = \{1, 2, 5, 7, 8, 11\}.$$

Example 2. If $A = \{a, b, c, d\}$, $B = \{a\}$, then $A \cup B = \{a, b, c, d\} = A$.

Example 3. If $A = \{0, 2, 4, 6, 8, \ldots\}$, $B = \{1, 3, 5, 7, \ldots\}$, then
$$A \cup B = \{0, 1, 2, 3, \ldots\} = N.$$

Example 4. Let E represent the Euclidean plane with a Cartesian coordinate system imposed on it. Let each point in E be designated by its coordinates (x, y). If

$$A = \{(x, y) \in E \mid x^2 + y^2 \leqslant 1\},$$
$$B = \{(x, y) \in E \mid (0 \leqslant x \leqslant 2) \wedge (-3 \leqslant y \leqslant 3)\},$$

then

$$A \cup B = \{(x, y) \in E \mid (-1 \leqslant x \leqslant 0 \wedge -\sqrt{1-x^2} \leqslant y \leqslant \sqrt{1-x^2})$$
$$\vee (0 \leqslant x \leqslant 2 \wedge -3 \leqslant y \leqslant 3)\}.$$

This somewhat complicated example can best be illustrated with a diagram. The set A consists of all points on or inside the unit circle with center the origin. The set B consists of all points on or inside a rectangle with sides $x = 0$, $x = 2$, $y = -3$, $y = +3$. Then the set $A \cup B$ consists of all points in both these regions—in other words the shaded region in Figure 4–1, including the boundaries.

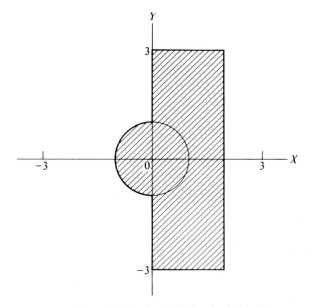

Figure 4–1 $A \cup B$ is the shaded region

4.3 INTERSECTIONS

The **intersection** of two sets A and B, written $A \cap B$, and read "A intersection B", is defined as follows.

Definition.   *$A \cap B$ is the set of all elements that are members of both A and B.*

This definition can be written in a somewhat longer fashion as $x \in A \cap B$ if and only if $x \in A$ and $x \in B$. If we let $\mathcal{S}_1$ be the sentence $x \in A$ and $\mathcal{S}_2$ be the sentence $x \in B$, then according to the definition of 3.4 the sentence $x \in A \cap B$ can be written as $\mathcal{S}_1 \wedge \mathcal{S}_2$.

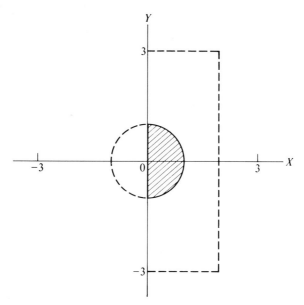

Figure 4–2 $A \cap B$ is the shaded region (including its boundary)

As illustrations, consider the following examples.

Example 1.   If $A = \{1, 5, 7, 8\}$, $B = \{1, 2, 7, 11\}$, then $A \cap B = \{1, 7\}$.

Example 2.   If $A = N$, $B = \{x \in N \mid 3 \text{ divides } x\}$, then

$$A \cap B = \{0, 3, 6, 9, 12, \ldots\} = B.$$

Example 3. If A is the set of all even integers and B is the set of all odd integers, then $A \cap B = \varnothing$.

In this example there are no elements in the intersection; in other words, the intersection is the empty set. This concept, which was introduced in 1.6, will be discussed at greater length in 4.5.

Example 4. Let A and B be the two sets of Example 4 of 4.2. Then $A \cap B$ is illustrated in Figure 4–2.

4.4 RELATIVE COMPLEMENTS

The **relative complement** of a set B with respect to a set A, written $A - B$, and read "A minus B", is defined as follows.

Definition. $A - B$ *is the set of all elements of A that are not elements of B.*

This definition can be written as

$$A - B = \{x \in A \mid x \notin B\}.$$

If we let $\mathscr{S}_1$ be the sentence $x \in A$ and $\mathscr{S}_2$ be the sentence $x \in B$, then according to the definitions of 3.3 and 3.4 the sentence $x \in A - B$ can be written as $\mathscr{S}_1 \wedge \neg \mathscr{S}_2$.

Consider the following examples.

Example 1. If $A = \{1, 5, 7, 8\}$, $B = \{1, 2, 7, 11\}$, then

$$A - B = \{5, 8\}, \qquad B - A = \{2, 11\}.$$

Example 2. If $A = \{(x, y) \in E \mid x^2 + y^2 \leqslant 1\}$,
$$B = \{(x, y) \in E \mid 0 \leqslant x \leqslant 1 \wedge 0 \leqslant y \leqslant 1\},$$

then

$$A - B = \{(x, y) \in E \mid -1 \leqslant x < 0 \wedge -\sqrt{1 - x^2} \leqslant y \leqslant \sqrt{1 - x^2}\}$$
$$\cup \{(x, y) \in E \mid 0 \leqslant x \leqslant 1 \wedge 0 > y \geqslant -\sqrt{1 - x^2}\}.$$

This set is illustrated by the shaded region in Figure 4–3. Note, for example, that $(0, 0)$, $(0, \frac{1}{2})$ and $(1, 0)$ are not in $A - B$. Also,

$$B - A = \{(x, y) \in E \mid 0 \leqslant x \leqslant 1 \wedge \sqrt{1 - x^2} < y \leqslant 1\}.$$

This set is illustrated by the shaded region in Figure 4–4a. Are $(0, 1)$, $(1, 0)$ in $B - A$?

Now let

$$C = \{(x, y) \in E \,|\, x^2 + y^2 < 1\} \qquad \text{(the "interior" of } A).$$

Then $B - C = \{(x, y) \in E \,|\, 0 \leqslant x \leqslant 1 \,\wedge\, \sqrt{1 - x^2} \leqslant y \leqslant 1\}$ as illustrated in Figure 4–4b. (The part of the boundary on the circle is now included.)

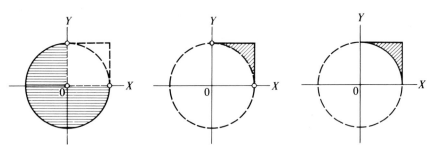

Figure 4–3 The shaded region is $A - B$ Figure 4–4a The shaded region is $B - A$ Figure 4–4b The shaded region is $B - C$

4.5 THE UNIVERSAL SET AND THE EMPTY SET

If all the sets that are considered in a certain context are subsets of a set U, then U is called the **universe** or **universal set** in that context.

Thus in a discussion concerning exclusively the natural numbers, the universe is the set of all natural numbers. If the discussion involves the set of points in a circle, the universe is the set of all points in the circle. The **absolute complement** of a set A, written $\bar{A}$, is defined as follows.

Definition 1. *$\bar{A}$ is the set of all elements in the universe which are not in A.*

An alternative way of defining $\bar{A}$ is to write

$$\bar{A} = \{x \in U \,|\, x \notin A\},$$
$$\text{or } \bar{A} = U - A \qquad \text{by the definition of 4.4.}$$

In 1.6 the empty set $\varnothing$ was defined as the set containing no elements, and it was stated that the empty set is a subset of every set. We shall now prove this result.

Theorem. *The empty set is a subset of every set.*

Proof. We shall prove that $\varnothing \subseteq A$ for all A by Definition 1 of 1.5. To show that $x \in \varnothing \Rightarrow x \in A$ is true (for any x), it is sufficient, by Figure 3–4, to show that $x \in \emptyset$ is false (for any x). But by the definition of $\emptyset$, $x \in \emptyset$ *is* false for any x.

The introduction of the empty set enables us to define **disjoint** sets.

Definition 2.   *Two sets A and B are **disjoint** if and only if $A \cap B = \emptyset$.*

If two sets are not disjoint then they are said to **intersect**. That is, two sets A and B intersect if and only if $A \cap B \neq \emptyset$.

Definitions 1 and 2 are illustrated by the following examples.

Example 1.   If $A = N$, $B = \{1/2, 3/2, 5/2, 7/2, ...\}$ then $A \cap B = \emptyset$. That is A and B are disjoint.

Example 2.   If $A = \{(x,y) \in E \mid (x + 2)^2 + y^2 \leqslant 1\}$,
$$B = \{(x,y) \in E \mid (x-1)^2 + y^2 \leqslant 1\},$$

then $A \cap B = \emptyset$ (see Figure 4–5).

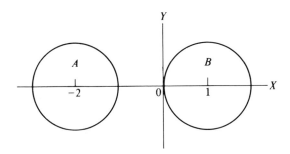

Figure 4–5 $A \cap B = \emptyset$

Example 3.   If $A = \{(x,y) \in E \mid (x + 1)^2 + y^2 \leqslant 1\}$,
$$B = \{(x,y) \in E \mid (x-1)^2 + y^2 \leqslant 1\},$$

then $A \cap B = \{(0,0)\}$.

In other words, in this example the intersection of the two sets is the single point, the origin (see Figure 4–6).

Example 4.   If $A = N$, $B = I$, then $A \cap B = \{0, 1, 2, ...\} = N$. In this example A and B intersect.

Example 5.   If $A = Q$, $U = R$, then $\bar{Q} = R - Q$. The set $\bar{Q}$ is the set of irrational numbers.

It will be recalled that in 1.4 the notation $\{x \in A \mid \mathscr{P}(x)\}$ was introduced, and was defined to be the set of all x in A such that the sentence $\mathscr{P}(x)$ is true. We can use this notation, together with the idea of the universal set and the logical connectives of conjunction and disjunction, to give another definition of union and intersection. We write

$$A \cup B = \{x \in U \mid x \in A \lor x \in B\},$$

$$A \cap B = \{x \in U \mid x \in A \land x \in B\}.$$

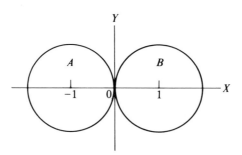

Figure 4–6 $A \cap B = \{(0, 0)\}$

4.6 VENN DIAGRAMS

To help in visualizing the operations defined in the previous sections, there is a useful device called the **Venn diagram**. In a Venn diagram the elements of a set are represented by the points inside a closed curve (usually a circle). Thus, if A and B are two sets represented by the circles in Figure 4–7, then $A \cup B$ is represented by the shaded region. Using the same representation of A and B, $A \cap B$ is represented by the shaded area in Figure 4–8. The universal set is usually represented by a rectangle; thus $\overline{A}$ is the shaded region in Figure 4–11.

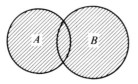

Figure 4–7 $A \cup B$ is shaded

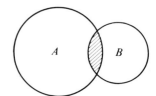

Figure 4–8 $A \cap B$ is shaded

Disjoint sets can be shown by non-overlapping circles, as in Figure 4–12, where A and B represent non-intersecting sets. If a set A properly contains a set B, then the circle representing B in the Venn diagram will be completely inside the circle representing A, as in Figure 4–13. A more complicated example is shown in Figure 4–14 where the shaded region represents $\bar{A} \cap B$. (See also Figure 4–10.)

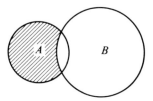

Figure 4–9 $A - B$ is shaded
This is the relative complement
of A with respect to B

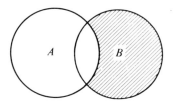

Figure 4–10 $B - A$ is shaded

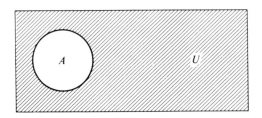

Figure 4–11 $\bar{A} = U - A$ is shaded

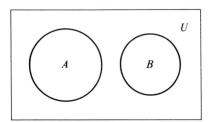

Figure 4–12 $A \cap B = \emptyset$

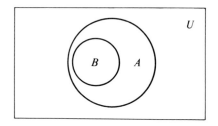

Figure 4–13 $B \subset A$

The main value of a Venn diagram is to provide a geometrical picture of the relationship between sets. However, with such diagrams we can often analyze complicated expressions involving a number of sets and operations.

As an example, we shall show that $\overline{A \cap \bar{B}} \cup B = \bar{A} \cup B$, using a Venn diagram. $A \cap \bar{B}$ is the shaded region in Figure 4–15, and $\overline{A \cap \bar{B}}$ is the unshaded region inside the rectangle. Now $\overline{A \cap \bar{B}} \cup B$ is obviously the same unshaded region. This completes the illustration, if we observe that $\bar{A} \cup B$ is also this same unshaded region.

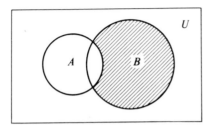

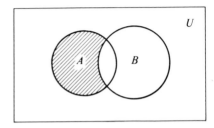

Figure 4–14 $\bar{A} \cap B$ is shaded   **Figure 4–15** $\overline{A \cap \bar{B}}$ is the unshaded region

We must be careful to avoid thinking that we have actually *proved* that $\overline{A \cap \bar{B}} \cup B = \bar{A} \cup B$ using a Venn diagram. We have merely made the result plausible and at the same time provided a picture of the various sets. This procedure does not constitute a proof, because not all sets can be represented by Venn diagrams; for example, the set N of all natural numbers, or the set of all permutations of 10 different objects. The Venn diagram is simply a heuristic device. Once the result seems plausible using this approach, we prove the result in the usual way. To prove that $\overline{A \cap \bar{B}} \cup B = \bar{A} \cup B$ we must proceed as follows. We first establish a preliminary result.

Lemma.   (i) $x \in A \Leftrightarrow x \notin \bar{A}$,   (ii) $x \notin A \Leftrightarrow x \in \bar{A}$.

Although these results seem obvious, we can prove them from our Definitions using a contrapositive argument.

Proof. We shall prove $x \in A \Rightarrow x \notin \bar{A}$ by proving that $\neg\, x \notin \bar{A} \Rightarrow \neg\, x \in A$. Now $\neg\, x \notin \bar{A}$ is the same as $x \in \bar{A}$, and, by Definition 1 of 4.5, $x \in \bar{A} \Rightarrow (x \in U \wedge x \notin A)$. But $x \notin A$ is the same as $\neg\, x \in A$ and the result $x \in A \Rightarrow x \notin \bar{A}$ follows.

The converse is an immediate deduction from Definition 1 of 4.5. The proof of (ii) will be left as an exercise for the reader.

We now return to the proof of our example: If A and B are any two sets, $\overline{A \cap \bar{B}} \cup B = \bar{A} \cup B$.

Proof. By Definition 2 of 1.5 we must establish

(a) $\overline{A \cap \overline{B}} \cup B \subseteq \overline{A} \cup B$ and

(b) $\overline{A} \cup B \subseteq \overline{A \cap \overline{B}} \cup B$.

(a) Let $x \in \overline{A \cap \overline{B}} \cup B$. Then $x \in \overline{A \cap \overline{B}}$ or $x \in B$.   (Definition of 4.2)

    If $x \in B$ then $x \in \overline{A} \cup B$.   (Definition of 4.2)

    If $x \in \overline{A \cap \overline{B}}$ then $x \in U$ and $x \notin A \cap \overline{B}$.   (Definition 1 of 4.5)

    Thus $x \notin A$ or $x \notin \overline{B}$.   (Definition of 4.3)

    That is $x \in \overline{A}$ or $x \in B$.   (Lemma)

    Hence $x \in \overline{A} \cup B$.   (Definition of 4.2)

    This completes the proof of (a).

(b) Let $x \in \overline{A} \cup B$. Then $x \in \overline{A}$ or $x \in B$.   (Definition of 4.2)

    If $x \in B$ then $x \in \overline{A \cap \overline{B}} \cup B$.   (Definition of 4.2)

    If $x \in \overline{A}$ then $x \notin A$ and $x \notin A \cap \overline{B}$.   (Lemma and Definition of 4.3)

    Thus $x \in \overline{A \cap \overline{B}}$.   (Lemma)

    Finally $x \in \overline{A \cap \overline{B}} \cup B$, and this completes the proof of (b).   (Definition of 4.2)

    Thus $\overline{A \cap \overline{B}} \cup B = \overline{A} \cup B$.

4.7   ALGEBRA OF SETS

In working with sets, it is useful to prove a number of basic properties. Then we use these properties of sets as tools to handle more complex problems. The reader will be familiar with this procedure since it is the method used, for example, in synthetic geometry. The basic properties which we shall establish are the following.

Theorem B. *Let A, B, C be any subsets of a universal set U, and let $\emptyset$ be the empty set. Then the following results are true.*

B1(a)	$A \cup (B \cup C) = (A \cup B) \cup C$	*Associative Properties*
B1(b)	$A \cap (B \cap C) = (A \cap B) \cap C$	
B2(a)	$A \cup (B \cap C) = (A \cup B) \cap (A \cup C)$	*Distributive Properties*
B2(b)	$A \cap (B \cup C) = (A \cap B) \cup (A \cap C)$	
B3(a)	$A \cup B = B \cup A$	*Commutative Properties*
B3(b)	$A \cap B = B \cap A$	
B4(a)	$A \cup \emptyset = A$	
B4(b)	$A \cap U = A$	
B5(a)	$A \cup \overline{A} = U$	
B5(b)	$A \cap \overline{A} = \emptyset$	

Proof. The proof is straightforward and follows the method used in 4.6. As examples we shall prove B4(a) and B2(a).

B4(a) We first establish that $A \cup \varnothing \subseteq A$.

If $x \in A \cup \varnothing$, then $x \in A$ or $x \in \varnothing$.   (Def. of 4.2)

But $\varnothing$ has no members, hence $x \in A$.

We now prove that $A \subseteq A \cup \varnothing$.

Let $x \in A$, then $x \in A \cup \varnothing$.   (Def. of 4.2)

Thus $A \cup \varnothing = A$.

B2(a) We first prove that

$A \cup (B \cap C) \subseteq (A \cup B) \cap (A \cup C)$.

If $x \in A \cup (B \cap C)$, then $x \in A$ or $x \in B \cap C$.   (Def. of 4.2)

If $x \in A$ then $x \in A \cup B$ and $x \in A \cup C$,   (Def. of 4.2)

hence $x \in (A \cup B) \cap (A \cup C)$.   (Def. of 4.3)

If $x \in B \cap C$ then $x \in B$ and $x \in C$.   (Def. of 4.3)

Hence $x \in A \cup B$ and $x \in A \cup C$.   (Def. of 4.2)

Thus $x \in (A \cup B) \cap (A \cup C)$.   (Def. of 4.3)

We now prove that

$(A \cup B) \cap (A \cup C) \subseteq A \cup (B \cap C)$.

If $x \in (A \cup B) \cap (A \cup C)$, then $x \in A \cup B$ and $x \in A \cup C$.   (Def. of 4.3)

Now if $x \notin A$, then $x \in B$ and $x \in C$;   (Def. of 4.2)

hence, $x \in B \cap C$, and it follows that   (Def. of 4.3)

$x \in A \cup (B \cap C)$.   (Def. of 4.2)

But if $x \in A$, $x \in A \cup (B \cap C)$.   (Def. of 4.2)

Thus, $A \cup (B \cap C) = (A \cup B) \cap (A \cup C)$.

The proofs of the remaining parts of Theorem B will be left as exercises for the reader.

We shall now adopt a different point of view. We proved Theorem B by making use of the idea of set membership. Suppose we now regard the theorem simply as a set of axioms to which the objects A, B, C, ... (which we shall continue to call sets) are subject. Then Theorem B will be transformed into what we shall call Axiom B, as follows.

Axiom B. *If A, B, C are any sets, the following properties will be assumed.*

B1(a) $A \cup (B \cup C) = (A \cup B) \cup C$

B1(b) $A \cap (B \cap C) = (A \cap B) \cap C$ (*Associative Axioms*)

B2(a) $A \cup (B \cap C) = (A \cup B) \cap (A \cup C)$

B2(b) $A \cap (B \cup C) = (A \cap B) \cup (A \cap C)$ (*Distributive Axioms*)

If A, B are any sets the following properties will be assumed.

B3(a) $A \cup B = B \cup A$

B3(b) $A \cap B = B \cap A$ (*Commutative Axioms*)

There exist sets Ø, U such that for any set A

B4(a) $A \cup \varnothing = A$,

B4(b) $A \cap U = A$.

If A is any set, there exists a set $\bar{A}$ such that

B5(a) $A \cup \bar{A} = U$,

B5(b) $A \cap \bar{A} = \varnothing$.

We can now prove theorems by using *only Axiom B and no other assumptions.* Axiom B comprises the basic assumptions of what is known as a Boolean Algebra (named after George Boole 1815–1864). For the rest of this chapter any references such as B1(a) will be to Axiom B not Theorem B. (Hence, the theorems apply to any Boolean Algebra.)

As examples of this approach we shall prove the following theorems.

Theorem 1. *The sets Ø and U are unique.*

Proof. (a) Suppose $\varnothing_1$ and $\varnothing_2$ both satisfy B4(a) for all A, that is, there are two empty sets.

Then $\varnothing_1 \cup \varnothing_2 = \varnothing_1$ and $\varnothing_2 \cup \varnothing_1 = \varnothing_2$.   B4(a)

But $\varnothing_1 \cup \varnothing_2 = \varnothing_2 \cup \varnothing_1$.   B3(a)

Hence $\varnothing_1 = \varnothing_2$ and Ø is unique.

(b) The proof that U is unique is similar to the proof of (a) and will be left to the reader.

Theorem 2. *For any A, $\bar{A}$ is unique.*

Proof. Assume $\bar{A}_1$, $\bar{A}_2$ are both complements of A.

Then $\bar{A}_1 = \bar{A}_1 \cup \varnothing$ B4(a)

$= \bar{A}_1 \cup (A \cap \bar{A}_2)$ B5(b)

$= (\bar{A}_1 \cup A) \cap (\bar{A}_1 \cup \bar{A}_2)$ B2(a)

$= (A \cup \bar{A}_1) \cap (\bar{A}_1 \cup \bar{A}_2)$ B3(a)

$= U \cap (\bar{A}_1 \cup \bar{A}_2)$ B5(a)

$= (\bar{A}_1 \cup \bar{A}_2) \cap U$ B3(b)

$= (\bar{A}_1 \cup \bar{A}_2)$.   B4(b)

Thus $\bar{A}_1 = \bar{A}_1 \cup \bar{A}_2$.

Similarly, we may show

$\bar{A}_2 = \bar{A}_2 \cup \bar{A}_1$.

Hence

$\bar{A}_2 = \bar{A}_1$,   B3(a)

and $\bar{A}$ is unique.

Corollary 1. *If, for any* C, $A \cup C = U$ *and* $A \cap C = \emptyset$, *then* $C = \bar{A}$.

Proof. This follows immediately from Theorem 2, B5(a), and B5(b).

Corollary 2. $\bar{\bar{A}} = A$.

Proof. We consider a set $\bar{A}$. Now the complement of $\bar{A}$, that is, $\bar{\bar{A}}$, is unique, and, by Corollary 1, if $\bar{A} \cup C = U$ and $\bar{A} \cap C = \emptyset$, then $C = \bar{\bar{A}}$. But by B5(a) and B5(b) (with B3(a) and B3(b)), $C = A$; hence $\bar{\bar{A}} = A$.

Theorem 3. *For any* A, B (i) $\overline{A \cup B} = \bar{A} \cap \bar{B}$,

(ii) $\overline{A \cap B} = \bar{A} \cup \bar{B}$.   (De Morgan's Laws)

Proof. We shall prove (i) and leave the proof of (ii) as an exercise for the reader.

If we prove that

(a) $(A \cup B) \cup (\bar{A} \cap \bar{B}) = U$ and

(b) $(A \cup B) \cap (\bar{A} \cap \bar{B}) = \emptyset$,

then the result follows by Corollary 1 of Theorem 2. Now for (a) we have

$$
\begin{aligned}
(A \cup B) \cup (\bar{A} \cap \bar{B}) &= [(A \cup B) \cup \bar{A}] \cap [(A \cup B)] \cup \bar{B}] && \text{B2(a)}\\
&= [(A \cup \bar{A}) \cup B] \cap [A \cup (B \cup \bar{B})] && \text{B1(a)}\\
&= (U \cup B) \cap (A \cup U) && \text{B5(a)}\\
&= U \cap U = U && \text{B4(b) and}\\
& && \text{Exercises}\\
& && \text{3(a), 3(d)}
\end{aligned}
$$

Similarly (b) can be proved and the proof of (i) is complete.

It will be noticed by the reader that in the proofs of the preceding three theorems, only the ten axioms B1(a)–B5(b) have been used. Actually the ten axioms are more than are necessary. It can be shown,[1] for example, that the associative laws are redundant, in the sense that they can be derived from the other eight axioms. The remaining eight axioms are independent of one another.

We are now in a position to give a very brief proof of the example of 4.6: For any A, B, $(\overline{A \cap \bar{B}}) \cup B = \bar{A} \cup B$.

Proof.
$$
\begin{aligned}
\overline{A \cap \bar{B}} &= \bar{A} \cup \bar{\bar{B}} && \text{Theorem 3, (ii)}\\
&= \bar{A} \cup B. && \text{Theorem 2, Corollary 2}
\end{aligned}
$$

Hence
$$
\begin{aligned}
(\overline{A \cap \bar{B}}) \cup B &= (\bar{A} \cup B) \cup B\\
&= \bar{A} \cup (B \cup B) && \text{B1(a)}\\
&= \bar{A} \cup B. && \text{Exercise 3(c)}
\end{aligned}
$$

[1] *Boolean Algebra* by R. L. Goodstein, The Macmillan Co., New York.

4.8   PARTIAL ORDERING OF A BOOLEAN ALGEBRA

The idea of set inclusion has been defined in 1.5 in terms of the concept of set membership. We wish to give an alternative definition directly connected to our axioms for a Boolean Algebra.

Definition. $A \subseteq B$ *if and only if* $A = A \cap B$.

Lemma.   *For all* A, $\emptyset \subseteq A$ *and* $A \subseteq U$.

> *Proof.*   Since $\emptyset = \emptyset \cap A$ Exercise 3(b)
> $\emptyset \subseteq A$ for all A.   Def. of 4.8
>   Also $A = U \cap A$ B4(b)
>   Thus $A \subseteq U$ for all A.

Theorem.   *A Boolean Algebra is partially ordered by inclusion.*

     Proof.   The partial ordering referred to here is of the equal-to-or-less-than variety, defined in O'1, O'2, O'3 of 2.4. We must prove

(a) $A \subseteq A$. Since $A \cap A = A$ by Exercise 3(d), the result follows from the definition.

(b) If $A \subseteq B$ and $B \subseteq A$, then $A = B$. We have $A = A \cap B$ and $B = B \cap A = A \cap B$. Thus $B = A$.

(c) If $A \subseteq B$ and $B \subseteq C$ then $A \subseteq C$. We have $A = A \cap B$, $B = B \cap C$.
Then $A \cap C = (A \cap B) \cap C = A \cap (B \cap C)$.
Thus $A \cap C = A \cap B = A$.   B1(b)

     Example.   Consider all the subsets of $U = \{1, 2, 3\}$. These are $\emptyset$, $\{1\}$, $\{2\}$, $\{3\}$, $\{1, 2\}$, $\{1, 3\}$, $\{2, 3\}$, $\{1, 2, 3\}$. Now $\emptyset$ is a subset of all the sets (Theorem of 4.5). Also the set $\{1\}$ is a subset of itself, $\{1, 2\}$, and $\{1, 2, 3\}$, and so on. The relation of the various sets to each other can be made clearer by a diagram. (Figure 4–16).

     In the diagram the various sets are indicated by the labeled points and any set contained in another can be joined to it by moving upward along the line segments. The set $\{2\}$ is contained in $\{1, 2\}$ and $\{2, 3\}$, but not in $\{1, 3\}$. It will be left as an exercise for the reader to show that these sets form a Boolean Algebra.

Definition.   *A totally ordered subset of a partially ordered set is called a* **chain.**

Thus, in the example, $\varnothing \subset \{1\} \subset \{1, 3\}$; $\varnothing \subset \{2\} \subset \{2, 3\} \subset \{1, 2, 3\}$; $\{1\} \subset \{1, 3\} \subset \{1, 2, 3\}$ are three chains. From Definition 2 of 2.4 the least elements of these three chains are respectively $\varnothing$, $\varnothing$, $\{1\}$. By considering this example and the ones given in the exercises, it should be clear to the reader that Boolean algebras (although partially ordered), in contrast to the natural numbers, are not in general linearly ordered.

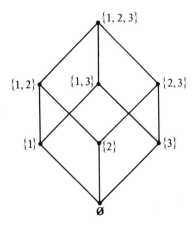

Figure 4–16   Partial ordering of the subsets of $\{1, 2, 3\}$

4.9 EXERCISES

1. In the following examples, (i) check the validity with a Venn diagram, (ii) prove using the set membership approach, (iii) prove using axioms B.

   (a) $A - (A - B) = A \cap B$.

   (b) $(A - B) \cap C = (A \cap C) - B$.

   (c) $A \cup B = (A \cap B) \cup (A \cap \bar{B}) \cup (\bar{A} \cap B)$.

   (d) $(A \cup B) \cap (\bar{A} \cup C) = (\bar{A} \cap B) \cup (A \cap C)$.

   (e) $\overline{A \cap B} = \bar{A} \cup \bar{B}$.

2. Prove Theorem B of 4.7 using the set membership approach.

3. Using Axioms B prove that, for all A,

   (a) $A \cup U = U$,

   (b) $A \cap \varnothing = \varnothing$,

   (c) $A \cup A = A$, } *Idempotent Laws*

   (d) $A \cap A = A$,

   (e) $A \cap (A \cup B) = A$, } *Absorption Laws*

   (f) $A \cup (A \cap B) = A$.

4. Prove that U is unique.

5. (i) Draw a partial ordering diagram for
       (a) the Boolean algebra of all subsets of a set of two elements.
       (b) the Boolean algebra of all subsets of a set of four elements.
    (ii) In each of these cases identify the chains and the least elements.

6. Prove that $\bar{U} = \emptyset$, $\bar{\emptyset} = U$.

7. (a) Prove that a Boolean Algebra is a partially ordered set with a greatest element U and a least element $\emptyset$.
    (b) Prove that every pair of elements A, B has a l.u.b. $A \cup B$ and a g.l.b. $A \cap B$.

8. Show that the logic of sentences of Chapter 3 is a Boolean algebra with the following correspondences: $\mathscr{F}$ for $\emptyset$; $\mathscr{T}$ for U; $\vee$ for $\cup$; $\wedge$ for $\cap$; $\neg$ for $-$; $\Rightarrow$ for $\subseteq$.

9. Three subsets A, B, C (possibly empty) of a universal set U can intersect each other in a number of ways; for example, $\emptyset \neq A \subset B \neq U$ and $B \cap C = \emptyset$.
    (a) Draw Venn diagrams illustrating 15 different ways.
    (b) Prove that there are 241 more! (Hint: $241 + 15 = 2^8$.)

    This illustrates one of the shortcomings of Venn diagrams. So does the next exercise. And so does Example 2 of 4.4. Why?

10. (a) Draw a Venn diagram showing four non-empty sets intersecting each other. (Hint: see the conclusion of part (d) below).
    (b) Prove that $n^2 - n + 2 < 2^n$ for all $n \in N$, $n > 3$.
    (c) There are n circles in a plane. Each circle intersects every other circle and no circle passes through a common point of two other circles. Show that the plane is thereby divided into $n^2 - n + 2$ regions.
    (d) Prove that the most general Venn diagram for more than three sets cannot be drawn with circles alone.

5

Integers

5.1 THE SET *I*

Most readers have encountered the integers previously and know that

$$I = \{0, \pm 1, \pm 2, \pm 3, ...\}.$$

This set satisfies axioms A1 to A3, M1 to M3, D, C2, and O1 to O4, together with A4.

A4. For each $a \in I$, there is a unique element $-a \in I$ such that $a + (-a) = 0$.

We sum up this axiom by saying that every element of I possesses an **additive inverse**. To verify axiom A4 for the intuitive integers, we need only note that

(i) for a positive integer $a \in I$, there is always $-a \in I$ such that $a + (-a) = a - a = 0$;

(ii) for a negative integer $a \in I$, $a = -b$ for some positive integer b, and $b = -(-b)$. Thus, by using A1,

$$0 = b + (-b) = (-b) + b = (-b) + [-(-b)] = a + (-a);$$

(iii) for $a = 0$, $0 + 0 = 0$.

Therefore, A4 is *valid*.

By employing axioms A1 and A4, it follows that $(-a) + a = 0$.

Further, it can be shown that the cancellation law C1 for the set of integers is a consequence of A4. Indeed, if

$$a + b = a + c,$$

then $(-a) + (a + b) = (-a) + (a + c)$.

Thus $[(-a) + a] + b = [(-a) + a] + c$.   A2

Thus $0 + b = 0 + c$ A4

and $b = c$.   A3

It is important to note that the natural numbers $N = \{0, 1, 2, ...\}$ appear as a subset of I, and that the operations of addition and multiplication defined in Chapter 2 for the natural numbers are the same as the addition and multiplication induced on N in its role as a subset of I.

The extension of N to the larger set I, as you may remember, is carried out in order to provide a solution x for equations of the form $a + x = b$. In the set of natural numbers, if $a > b$, then no natural number x exists such that $a + x = b$. Thus N is enlarged to I to provide solutions for all possible choices of a and b in I.

In the following section we give an explicit construction of I from N by providing special elements which are solutions of the equation $a + x = b$. On first reading the reader may omit this section (5.2) without impairing his understanding of following sections.

5.2   DEFINITION OF THE INTEGERS FROM THE NATURAL NUMBERS

The following construction of I from N is not the one usually encountered in texts. However, it is basically simpler than the usual one. (The latter requires the notion of partitioning a set into equivalence classes; see 7.5.) Furthermore, the approach adopted here closely resembles that used in constructing the complex numbers from the real numbers.

Given the set N with operations "+" and "·" we wish to provide the missing solutions of $a + x = b$. We might begin by inventing a solution to the simplest such equation, viz., $1 + x = 0$. Let us call it "minus one" or "$^-1$". Since our new system, to be useful, must be closed under "+" and "·" we must then provide for the powers of minus one, $(^-1)^n$, and natural number multiples of these powers. Thus a general element in our augmented set, call it N^*, will be expressible in the form

$$m_0 + m_1(^-1) + m_2(^-1)^2 + ... + m_n(^-1)^n$$
$$\text{where } n \in N, \text{ and } m_i \in N \text{ for } i = 0, 1, ..., n.$$

Since we wish N^* to have the properties of N plus the new property discussed above, we shall impose the condition that N^* satisfy A1 to A3, M1 to M3, C1, C2, and D, and proceed to examine some of the consequences. We start with

(1)		$1 + (^-1) = 0$	Assumption
		Then $n[1 + (^-1)] = n0 = 0$	Theorem 2 of 2.2
(2)	and	$n(1) + n(^-1) = 0.$	D
	If $n = 0,$	$0(1) + 0(^-1) = 0$	
	and	$0 + 0(^-1) = 0$	Theorem 2 of 2.2
(3)	or	$0(^-1) = (^-1)0 = 0.$	A3, M1
(4)	Also	$1(^-1) = (^-1)1 = {}^-1.$	M1, M3
(5)	Then	$(^-1)[1 + (^-1)] = (^-1)0 = 0.$	(1), (3)
(6)	But	$(^-1)[1 + (^-1)] = (^-1)1 + (^-1)(^-1).$	D
(7)	Thus	$(^-1) + (^-1)^2 = 0.$	(4), (5), (6)
	Then	$1 + (^-1) = (^-1)^2 + (^-1)$	(1), (7), A1
(8)	and	$1 = (^-1)^2.$	C1

By repeated applications of (4) and (8),

(8a) $(^-1)^{2n} = 1$ for all $n \in N,$

(8b) $(^-1)^{2n+1} = {}^-1$ for all $n \in N.$

This result reveals to us that a general element of N^* need not involve explicitly powers of $^-1$. Indeed, N^* can be described as

$$N^* = \{m + n(^-1) \mid m,n \in N\}$$

where addition is carried out as follows,

$$[m_1 + n_1(^-1)] + [m_2 + n_2(^-1)] = (m_1 + m_2) + (n_1 + n_2)(^-1) \quad \text{A1, A2, D}$$

and multiplication, as follows,

$$[m_1 + n_1(^-1)] [m_2 + n_2(^-1)]$$
$$= m_1 m_2 + m_1 n_2(^-1) + n_1(^-1)m_2 + [n_1(^-1) n_2(^-1)]$$
$$= (m_1 m_2 + n_1 n_2) + (m_1 n_2 + n_1 m_2)(^-1). \quad \text{D, M1, (8)}$$

Now the question arises: Is N^* large enough to supply a solution to $a + x = b$ for every pair $a,b \in N^*$? Suppose[1] $a = m + (^-1)n$ and $b = p + (^-1)q$. Then $a + x = b$ becomes

(9) $\qquad\qquad m + (^-1)n + x = p + (^-1)q.$

---

[1] It is sometimes technically advantageous to express the elements of N^* in the form $m + (^-1)n$. In Exercise 11 the reader is asked to show that $m + n(^-1) = m + (^-1)n$.

If we add $n + (^-1)m$ to both sides, we find that the left-hand side of (9) becomes

$$[n + (^-1)m] + [m + (^-1)n] + x$$
$$= [n + (^-1)n] + [m + (^-1)m] + x \qquad \text{A1, A2}$$
$$= 0 + 0 + x \qquad\qquad\qquad\qquad\qquad \text{A4}$$
$$= x,$$

while the right-hand side becomes

$$[n + (^-1)m] + [p + (^-1)q]$$
$$= (n + p) + (m + q)(^-1). \qquad \text{A1, A2, D}$$

Hence
$$x = (n + p) + (m + q)(^-1)$$

is a solution of (9). We see that N^* is indeed a set such that any equation of the form $a + x = b$ with $a,b \in N^*$ has a solution in N^*.

Now N^* can be recognized as the set I of integers if we make the following observations:

(i) if $m > n$, then $m = n + k$ for some $k \in N$ and so
$$m + (^-1)n = k + n + (^-1)n = k + 0 = k,$$

(ii) if $m = n$, then $m + (^-1)n = 0$,

(iii) if $m < n$, then $n = m + j$ for some $j \in N$, and
$$m + (^-1)n = m + (^-1)m + (^-1)j = 0 + (^-1)j = (^-1)j.$$

We now identify $^-1$ with -1 in which case $(^-1)n$ becomes $-n$, and N^* becomes the set of integers.

5.3 ADDITIVE GROUP OF INTEGERS

In mathematics we give special names to systems which arise frequently, and one such system is the group. A **group** is a set G of objects provided with an operation which we shall indicate here by "o" such that if $a,b,c \in G$:

G1 $a \circ (b \circ c) = (a \circ b) \circ c$;

G2   there is a special element $e \in G$ such that $e \circ a = a \circ e = a$;

G3   for each $a \in G$, there is a unique $a^{-1} \in G$, such that

$$a \circ a^{-1} = a^{-1} \circ a = e.$$

One of the first examples of a group that a student encounters is **the additive group of integers**. If we interpret "o" as "$+$", e as 0, and a^{-1} as $-a$, then G1 becomes A2, G2 becomes A3, and G3 becomes A4. To repeat, if we identify "$+$" with "o", "0" with "e", and $-a$ with a^{-1}, then the Axioms A2, A3, and A4, all of which are satisfied by integers, assert that the integers under addition form a group. This group is called the **additive group of integers.**

It is significant to note that if a general mathematical system is given in terms of axioms it follows that *any* system satisfying the axioms also satisfies all theorems that can be proved from the axioms.

Consider, for example, the following theorem.

Theorem.   *Let* a, b, c *belong to a group* G. *Then* $a \circ b = a \circ c$ *if and only if* $b = c$; *and* $b \circ a = c \circ a$ *if and only if* $b = c$.

Proof.   (i) If $b = c$, then from the properties of equality $a \circ b = a \circ c$.

(ii) On the other hand, if $a \circ b = a \circ c$,

then $a^{-1} \circ (a \circ b) = a^{-1} \circ (a \circ c)$,

$\qquad (a^{-1} \circ a) \circ b = (a^{-1} \circ a) \circ c$,   G1

$\qquad e \circ b = e \circ c$, and   G3

$\qquad b = c$.   G2

The proof of the second part is identical with the proof of the first except that a is on the right side of b and c. (Note that we cannot use commutativity, for we have postulated no such axiom.)

Thus in any group the cancellation property is valid.

In the sequel we shall observe that certain systems are groups. It is then an immediate consequence of the Theorem that for such systems cancellation is valid.

5.4   ORDER IN THE INTEGERS

Most readers have studied the relation of order in the integers and know that the integers are totally ordered under the relation:

(1) $m < n$ if and only if there is a $k \in \{1, 2, 3, ...\}$ such that $m + k = n$, or equivalently,

(2) $m < n$ if and only if $n - m \in \{1, 2, 3, ...\}$.

Note that (1) is in form very similar to Definition 1 in 2.4. It is therefore easy to verify that Axioms O1 to O4 and OA in that section are valid for integers.

In addition note that the order relation defined in $N \subseteq I$ by virtue of Definition (1) coincides with the order defined on N in Chapter 2.

Statement OM can be seen to be valid for I also.

Suppose $m < p$ and $n > 0$ where $m,n,p \in I$.

Then $p - m > 0$ and $n > 0$ where $p - m,n$ can be considered as elements of N.

Thus $pn - mn > 0$,

and $mn < pn$.

The proof of the converse is left to the reader.

In addition to OM, I also satisfies OM'.

OM'.   If $n < 0$, then $mn < pn$ if and only if $p < m$.

       If $n < 0$, $-n > 0$.

Suppose $p < m$.

Then $-np < -nm$

that is $-nm + np > 0$

that is $np - nm > 0$

or $nm < np$.

Conversely, if $p < m$, the reader can show that $mn < pn$, provided that $n < 0$.

     The axiom OW, however, is not satisfied by the set of integers, because I itself is an example of a subset of I which has no least element. Thus I is not well ordered. We can, however, prove the following theorem.

Theorem.   *If $S \subseteq I$ has a lower bound, then S has a least element.*

     Proof.   Let q be a lower bound of S.

         Thus, for each $s \in S$, $q \leqslant s$.

         If $0 \leqslant q$,

         then $S \subseteq N$ and has a least element because N is well ordered.

         If, on the other hand, $q < 0$,

         then $0 < -q$ and $q + (-q) = 0$.

         Now since $q \leqslant s$ for all $s \in S$, we can apply OA to obtain

         $q + (-q) = 0 \leqslant s - q$ for all $s \in S$.

         Thus the set $S^* = \{s - q \mid s \in S\}$ has a least element $j \in S^*$, because $S^* \subseteq N$.

         Now by virtue of its membership in S^*, $j = s_0 - q$ for some $s_0 \in S$, and by virtue of its role as the least element

         $s_0 - q \leqslant s - q$ for all $s \in S$.

         We now apply OA to obtain $s_0 \leqslant s$ for $s \in S$.

         Hence s_0 is the least element of S.

     Two corollaries follow immediately from this result.

Corollary 1.   *If $S \subseteq I$ has a lower bound, S has a greatest lower bound.*

Proof. From a remark in 2.6 the least element of a set (if there is one) is also a greatest lower bound.

Corollary 2.   *If $S \subseteq I$ has an upper bound, then S has a greatest element.*

    Proof.   First observe

        (i) an element $s_0 \in S$ is the greatest element of S if and only if $-s_0$ is the least element of $S' = \{-s \,|\, s \in S\}$, and

        (ii) S has an upper bound if and only if S' has a lower bound. (The reader is advised to take a particular set S and check (i) and (ii) for that S.)

Since S has an upper bound by hypothesis, we can apply modus ponens to *the hypothesis and (ii)*, thereby arriving at the result that S' has a lower bound.

Theorem 1 ensures that S' has a least element s_1.

By referring back to observation (i), we see that the element $-s_1 \in S$ is the greatest element of S because $-(-s_1) = s$, is the least element of S'.

Corollary 3.   *If $S \subseteq I$ has an upper bound, then S has a least upper bound.*

5.5   MULTIPLICATION IN THE INTEGERS

The main result derived in this section is the so called **fundamental theorem of arithmetic** which states that every integer is equal to plus or minus a product of prime numbers and that this product is unique except for the order of its factors. A **prime number** is a positive integer p, different from 1, which has no other factors except p and 1. Thus 2, 3, 5, 7, 11, 13, 17, ... are prime numbers while $4 = 2 \cdot 2$, $6 = 2 \cdot 3$, $21 = 3 \cdot 7$ are not.

    In order to prove this fundamental theorem we need a number of preliminary theorems. Some of these will be well known to the reader although their proofs may not be.

Theorem 1.   *If b is an integer and a is a positive integer, then there exist integers k and γ with $-a < \gamma < a$ such that*

$$(1) \qquad\qquad\qquad b = ka + \gamma.$$

    Proof.   Suppose we prove this theorem for any number $b \geqslant 0$.

        Then the proof holds for any $b < 0$.

        Indeed, if $b < 0$,

        then $-b > 0$.

Thus there exist k' and γ' such that $-b = k'a + \gamma'$ and $-a < \gamma' < a$.

We multiply by (-1) to obtain
$b = (-k')a + (-\gamma')$.

If we let $k = -k'$ and $\gamma = -\gamma'$, we find that
$b = ka + \gamma$ and $-a < \gamma < a$.

Therefore we need prove our theorem for $b \geqslant 0$, only.

Assume $b \geqslant 0$ and let $M = \{j \mid ja \leqslant b\}$.

Since 0 always belongs to M, M is not empty.

We first show that M possesses an upper bound.

Since $b \geqslant 0$ and $a \geqslant 1$, $b \leqslant ba$.

Thus for every $j \in M$
$ja \leqslant b \leqslant ba$, or
$ja \leqslant ba$, or by OM
$j \leqslant b$.

Thus b is an upper bound of M.

By Corollary 2 of 5.4, M has a greatest element k.

Hence
(2) $ka \leqslant b < (k + 1)a$.

Now from (2) we obtain, by subtraction of ka, the result
$0 \leqslant b - ka < (k + 1)a - ka = a$.

Thus if we let $b - ka = \gamma$, we find that
(3) $0 \leqslant \gamma < a$, and
(4) $b - ka = \gamma$, or
(5) $b = ka + \gamma$.

Equations (3) and (5) establish the validity of the theorem.

In familiar language, in equation (1), b is the *dividend*, a is the *divisor*, k is the *quotient*, and γ is the *remainder*.

Corollary 1. *If $b \geqslant 0$, then k and γ can be chosen to be non-negative.*

Proof. This follows from the definition of M in the case of k and from (3) in the case of γ.

Theorem 2. *If a and b are positive integers and p is a prime, then*

(6) $p \mid ab \Rightarrow p \mid a \lor p \mid b$.

Proof. Suppose $a > 0$ and p does not divide a. We set out to show that if $p|ab$, then p must divide b, in which case our theorem is valid. To accomplish this we show that the set

$$S = \{b\,|\,b > 0 \wedge p|ab \wedge p\nmid b\}$$

is empty. (Remember that $p\nmid b$ means p does not divide b.)

To prove S is empty, we shall show that if S is nonempty, a contradiction arises. Assume S nonempty. Because S is a subset of the natural numbers, S has a least element, say b_0.

We first show $b_0 < p$. If this were not the case, then $b_0 - p > 0$, and since $p|ab_0$ and $p|ap$, it follows that $a(b_0 - p)$ is divisible by p while $p\nmid(b_0 - p)$. It then would follow that $b_0 - p \in S$ and $b_0 - p < b_0$. Since b_0 is least in S, we must conclude that $b_0 \leqslant p$. However, $p\nmid b_0$, so $b_0 \neq p$ and hence $b_0 < p$.

Now by Corollary 1, there exist $k \geqslant 0$ and γ such that $0 \leqslant \gamma < b_0$ and

(7) $\qquad\qquad\qquad\qquad p = kb_0 + \gamma.$

Since p is prime, $\gamma \neq 0$ (if $\gamma = 0$, then p could be factored). Multiply both sides of Equation (7) by a to obtain $pa = kab_0 + \gamma a$ and hence

(8) $\qquad\qquad\qquad\qquad pa - kab_0 = \gamma a.$

However, $p|ab_0$, and so $p|(pa - kab_0)$.

Thus $p|\gamma a$,

and since $0 < \gamma < b_0 < p$, we obtain the results $p\nmid\gamma$ and $\gamma > 0$. Therefore $\gamma \in S$ while $\gamma < b_0$ which is the smallest element in S. This contradiction follows from the assumption that S is non-void.

Hence S must be empty.

Therefore if $p\nmid a$ and $p|ab$, then $p|b$.

This proves statement (6).

Corollary 2. *If $p|a_0\,a_1\,a_2\,a_3\,\ldots\,a_n$, then $p|a_i$ for some $i = 0, 1, 2, \ldots, n$.*

Proof. By induction on n.

Theorem 3. *If m is an integer greater than 1, then m has a prime factor.*

Proof. If m is already prime, then $m = m \cdot 1$ and m is the prime factor of m.

If m is not prime then m can be factored into $m_1 n_1 = m$ where $1 < m_1$, $n_1 < m$ (which means $1 < n_1 < m$ and $1 < m_1 < m$). Now if either of m_1 or n_1 is prime, then the theorem holds. If neither is prime, choose one of the numbers m_1 and n_1, say m_1. It can be factored into m_2 and n_2 where

$$1 < m_2, \quad n_2 < m_1 < m.$$

If one of these is prime the proof is complete. Otherwise m_2 can be factored into m_3, n_3 where $1 < m_3, n_3 < m_2 < m_1 < m$ and so on.

Thus if we continue this process, we either (1) arrive at a prime factor or (2) construct for any $k = 1, 2, \ldots$ an m_k such that

$$1 < m_k < m_{k-1} < \ldots < m_1 < m.$$

For some k, however, the process must terminate by OW (that is, because there are at most $m - 2$ natural numbers between 1 and m).

Thus before $k = m - 2$ a prime factor will have been found.

From Theorems 2 and 3 it is easy to prove the **fundamental theorem of arithmetic**.

Theorem 4. *If m is an integer different from zero and ± 1, then m can be written as plus or minus a product of prime numbers and this product is unique except for the order of the factors.*

Proof. If $m > 0$, then by Theorem 3, there is a prime p_1 such that $m = p_1 q_1$.

Similarly, if q_1 is neither prime nor 1, then $q_1 = p_2 q_2$ where p_2 is prime.

Thus $m = p_1 p_2 q_2$.

In a manner similar to that employed in the proof of Theorem 3, this process must terminate (by OW; see exercise 10 of 5.6) with m in the form

$$(9) \qquad\qquad m = p_1 p_2 \cdots p_k$$

where p_i (for $i = 1, 2, \ldots, k$) is prime.

Thus m can be represented as a product of primes.

It remains only to show that these prime factors are unique except for order.

Suppose, in addition to the prime factorization indicated by (9), we have also

(10) $$m = p_1'p_2' \cdots p_l'$$

where p_i' (for $i = 1, ..., l$) is prime.

Combining Equations (9) and (10), we obtain

(11) $$p_1'p_2' \cdots p_l' = p_1p_2 \cdots p_k.$$

Since, for each $i = 1, 2, ..., l$, p_i' divides the left side of (11), it also divides the right.

Thus by Corollary 2, p_i' divides one of the factors $p_1, ..., p_k$. Since they are all prime, p_i' must be equal to one of these factors.

In each case let p_i' be cancelled together with the prime factor on the right side of (11) to which it is equal.

When these cancellations are completed for $i = 1, 2, ..., l$, the left side of (11) equals 1 and the right side, because of the equality, must also be 1.

This indicates that exactly the same primes appeared on the left side of (11) as appeared on the right, that is $k = l$ and only the order of the factors is different.

Thus the theorem is proved for the case $m > 0$.

In case $m < 0$, $-m > 0$, and hence $-m$ possesses a unique factorization

$$-m = p_1p_2 \cdots p_k.$$

Thus, m has the unique factorization

$$m = -p_1p_2 \cdots p_k,$$

and hence the theorem is valid for $m < 0$.

5.6 EXERCISES

1. Show that (a) the integers do not form a group under multiplication,
   (b) the natural numbers do not form a group under addition.
   In both (a) and (b) list the group axioms which are not satisfied.
2. Show that the set $\{0, 1\}$ provided with operation "o" such that
   $$0 \text{ o } 0 = 0, \quad 1 \text{ o } 0 = 0 \text{ o } 1 = 1, \text{ and } 1 \text{ o } 1 = 0$$
   forms a group where 0 plays the part of e.

3. (a) Show that the numbers $\{1, 2, ..., 12\}$ on the face of a clock form a group where $e = 12$ and "o" is identified with $+$ defined in the following natural way

$$m + n = \text{the time } n \text{ hours after } m \text{ o'clock.}$$

For example, $10 + 11 = \text{time 10 hours after 11 o'clock}$
$$= 9.$$

    (b) Can this principle be generalized from 12 to any natural number?

4. Verify that, given the order relation in N, the relation $<$ defined on the integers by statement (1) in 5.4 is actually a total order relation on I according to the axioms of 2.4.

5. Prove that there is always a prime number between p and $p!$ where p is prime and $p \neq 2$.

6. Show that there are more than a finite number of prime numbers. Hint: Use previous exercise.

7. Show that the integers satisfy the *Archimedean* property: If a,b belong to I and $0 < a$, $0 < b$, then there is a natural number n such that $b < na$.

8. Discuss the problem of solving equations of the form $ax = b$ in I.

9. Is $a|b$ a partial order on N? Is it total? (Cf. exercise 8.)

10. (a) Show that $a|b \Rightarrow a \leqslant b$ in N.

    (b) Explain fully the use of OW in the proof of (9) in Theorem 4 of 5.5. (It is sometimes called "the method of infinite descent.")

11. Show that

$$m + n(^-1) = m + (^-1)n$$

holds in N^*.

12. Show that $a + x = b$, $a,b \in N^*$, has a *unique* solution in N^*.

6

Rational Numbers

6.1 THE SET Q

We now study the rational numbers, numbers which most readers have met before. The set Q of all fractions m/n such that $m \in I$ and $n \in I - \{0\}$ is called the set of *rational numbers*. Remember that $I - \{0\}$ designates the set of integers different from zero. Two rational numbers, m/n and p/q, are **equal** if and only if

(1) $$mq = np.$$

For example, $5/3 = 15/9$ because $5 \cdot 9 = 15 \cdot 3$.

Note that the set I of integers becomes a subset of Q when we identify the integer m with the fraction $m/1$. Henceforth, it is assumed that this identification has been made, and so $I \subset Q$.

Note also that the equality of two rational numbers is given by an equation involving integers. Thus, in a certain sense, the rational numbers are defined from the integers. The details of this appear in 7.5.

The sum $m/n + p/q$ of two rational numbers is given by the formula

(2) $$\frac{m}{n} + \frac{p}{q} = \frac{mq + np}{nq},$$

and the product $\dfrac{m}{n} \cdot \dfrac{p}{q}$ is given by

(3) $$\frac{m}{n} \cdot \frac{p}{q} = \frac{mp}{nq}.$$

62

With respect to these operations, Q, like I, satisfies axioms A1 to A4, M1 to M3, and D. In addition Q satisfies M4.

M4   For every $a \neq 0$, there exists a^{-1} such that $a \cdot a^{-1} = a^{-1} \cdot a = 1$.

We name a^{-1} the **multiplicative inverse** of a. The validity of M4 in Q is easy to see. Indeed, $a \in Q$ implies $a = m/n$ where $m \in I$ and $n \in I - \{0\}$. Since $a \neq 0$, $m \neq 0$. Thus $n/m \in Q$, and

$$\frac{n}{m} \cdot \frac{m}{n} = \frac{m}{n} \cdot \frac{n}{m} = \frac{mn}{mn} = \frac{1}{1} = 1.$$

Hence, with $a^{-1} = n/m$, Q satisfies axiom M4.

Note that statements M2, M3, M4 are almost the axioms of a group. The difference lies in the fact that M4 demands that $a \neq 0$. However, if we consider $Q - \{0\}$ provided with the operation "$\cdot$", then by virtue of M2, M3, M4, and the fact that $Q - \{0\}$ is closed under "$\cdot$", we conclude that $Q - \{0\}$ is a group under the operation "$\cdot$".

From 5.3, it is clear that since $Q - \{0\}$ is a group under multiplication, the cancellation law C2 is valid in Q.

By virtue of conditions A1 to A4, M1 to M4, and D, the set Q forms a **field**. Roughly, a field is a set of elements in which it is possible to add, subtract, multiply, and divide. More precisely, a **field** is a set F with operations "$+$" and "$\cdot$" and special elements 0 and 1 ($0 \neq 1$) which satisfy Axioms A1 to A4, M1 to M4, and D. In addition to Q, the rational numbers, there are many other fields. In particular, we encounter the field of real numbers in Chapter 8.

We shall prove two results about fields generally. These are valid for the rational numbers because the latter form a field.

Theorem 1.   *Let F be a field. Then*

$$m \in F \Rightarrow m \cdot 0 = 0 \cdot m = 0.$$

Proof. The proof is identical with the proof of Theorem 2 of 2.2

Theorem 2.   *If $ab = 0$, then either $a = 0$ or $b = 0$.*

Proof.   If $a \neq 0$, then $a^{-1} \in F$ by axiom M4, and by Theorem 1.

$$a^{-1}(ab) = a^{-1}0 = 0.$$

Thus $0 = a^{-1}(ab) = (a^{-1}a)b = 1 \cdot b = b.$

The substance of Theorem 2 is summed up by the remark: In a field there are no **divisors of zero**. The reader should prove that neither N, the natural numbers, nor I, the integers, possess zero divisors. (See Chapter 9 for examples of zero divisors.)

6.2 THE RATIONAL NUMBERS AS AN EXTENSION OF THE INTEGERS

Just as subtraction is not always possible in the natural numbers, division is not always possible in the integers, that is, it is not always possible to solve the equation $ax = b$. However, if we extend the integers to form the rational numbers, then this equation is solvable except when $a = 0$.

Indeed if $a \neq 0$, then $a^{-1} \in Q$. Hence

$$ax = b \Rightarrow a^{-1}(ax) = a^{-1}b,$$

and

$$1 \cdot x = x = a^{-1}b.$$

Thus a solution exists when $a \neq 0$.

Someone might ask, "Why not provide a solution for $ax = b$ when $a = 0$ and $b \neq 0$?" For example, let $b = 1$. This amounts to providing 0 with an inverse 0^{-1}, so that

(1) $$0 \cdot 0^{-1} = 0^{-1} \cdot 0 = 1.$$

If we were to add 0^{-1} to the set of rational numbers and assume that the set so constructed had the properties A1 to A4, M1 to M4, and D, we would find

$$0 + 0 = 0 \qquad\qquad \text{by A3}$$
$$0^{-1}(0 + 0) = 0^{-1}0 + 0^{-1}0 = 0^{-1}0 \qquad \text{by D}$$
and
$$0^{-1}0 + 0^{-1}0 = 1 + 1 = 0^{-1}0 = 1. \qquad \text{by (1)}$$

Thus $1 + 1 = 1$ or, by cancellation, $1 = 0$.

Also by 6.1, Theorem 1,

$$a = a \cdot 1 = a \cdot 0 = 0$$

for all rational numbers a. Thus we would have to conclude that all rational numbers are zero!

6.3 ORDER IN THE RATIONAL NUMBERS

We say that a rational number m/n is **positive** when $m \neq 0$, and m and n agree in sign; that is, when m and n are both positive or both negative integers.

Then,

(1) $\qquad \dfrac{m}{n} < \dfrac{p}{q}$ if and only if $\dfrac{p}{q} - \dfrac{m}{n}$ is positive.

Theorem 1. *The relation "$<$" on Q is a linear ordering.*

Proof. To prove this we must show that O1 to O4 are valid in Q. We leave the proof of O1 and O2 to the reader.

O3: Suppose $\dfrac{m}{n} < \dfrac{p}{q}$ and $\dfrac{p}{q} < \dfrac{r}{s}$. We must show that $\dfrac{m}{n} < \dfrac{r}{s}$.

We choose $m, n, p, q, r,$ and s so that the denominators n, q, s are positive integers. By our assumption and (1), both

$$\frac{p}{q} - \frac{m}{n} = \frac{pn - qm}{qn} \quad \text{and} \quad \frac{r}{s} - \frac{p}{q} = \frac{rq - sp}{sq}$$

are positive. Hence $pn - qm$ and $rq - sp$ must be positive integers. If we multiply the former by s and the latter by n, we obtain two positive integers

$$spn - sqm, \qquad rqn - spn.$$

The sum of these positive integers is $rqn - sqm$, a positive integer.

But $\qquad\qquad\qquad\qquad rqn - sqm = q(rn - sm).$

Hence, we conclude that $rn - sm$ is a positive integer, by OM (Theorem of 2.4).

Thus

$$\frac{r}{s} - \frac{m}{n} = \frac{rn - sm}{sn} \quad \text{is positive.}$$

Then

$$\frac{r}{s} > \frac{m}{n} \quad \text{or} \quad \frac{m}{n} < \frac{r}{s}.$$

O4: To show O4 is valid for Q, we need only prove that if $m/n \not< r/s$ and $m/n \neq r/s$, then $r/s < m/n$. Assume m, n, r, s chosen so that n and s are positive. Now $m/n \not< r/s$ means that $rn - sm$ is not positive.

Also $m/n \neq r/s$ means that $sm \neq rn$, and $rn - sm \neq 0$. Since $rn - sm$ is an integer, it must be negative by O4, and $sm - rn$ must be positive. Therefore $sm - rn$ and ns agree in sign, and $m/n > r/s$.

We leave to the reader the exercise of showing that OA, OM, and OM' (5.4) all hold in Q, that is, the order relation in Q is **compatible with addition and multiplication**. Thus Q is an instance of what is called an **ordered field**, that is, a field with an order relation such that O1 to O4, OA, OM, and OM' are satisfied.

Later when we discuss the real field (Chapter 8), we shall find that the reals also form an ordered field.

In addition to being an ordered field, Q is **densely ordered**, that is Q satisfies the following.

OD.   If $a,b \in Q$ and $a < b$, then there is $c \in Q$ such that $a < c < b$.

Thus between every pair of rational numbers there is a rational number. To show that Q satisfies OD, we need only observe that if $a < b$, then

$$a < \frac{a+b}{2} < b.$$

If we let $c = (a + b)/2$, then OD is satisfied. Hence Q is an ordered field which is dense, that is, a **dense ordered field**.

It should be noted that neither N nor I satisfies OD.

In regard to the integers, which may be represented as in Figure 6–1, it is clear that between any pair of consecutive integers, n and $n + 1$, no third integer exists.

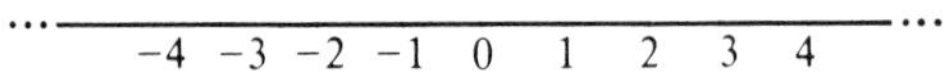

$$\cdots \overline{} \cdots$$
$$\;\;\; -4 \;\; -3 \;\; -2 \;\; -1 \;\;\; 0 \;\;\; 1 \;\;\; 2 \;\;\; 3 \;\;\; 4$$

Figure 6–1

However, between any pair of rationals there is another to "fill the gap." Thus there can be no such thing as two consecutive rational numbers. It appears at first sight therefore that the rationals fill the line. This is not, however, the case, as we shall see in Theorem 5. The rational numbers, although dense, still leave gaps in the line which must be filled by irrational numbers (Chapter 8).

There are some additional properties of the order relation in Q which are useful in the sequel.

Theorem 2.   If $a,b \in Q$ and $0 < a$, $0 < b$, then $a < b \Leftrightarrow a^2 < b^2$.

Proof.   Now $0 < a$, $0 < b$.

      (i) Suppose $a < b$.

         Since $0 < b$, $ab < b^2$.

Since $0 < a,\ a^2 < ab.$

Thus $a^2 < ab < b^2.$

Thus $a^2 < b^2.$

(ii) Suppose $a \not< b.$

Either $a = b$ or $a > b.$

If $a = b,\ a^2 = b^2,$ and $a^2 \not< b^2.$

If $a > b,\ a^2 > b^2,$ and $a^2 \not< b^2.$

Thus if $0 < a$ and $0 < b,\ a < b \Leftrightarrow a^2 < b^2.$

Theorem 3. *If $a,b \in Q$ and $0 < a < b$, then there is an element $k \in Q$ such that $k > 0$ and $a < k^2 b < b$.*

Proof.   First note that

(1) $0 < \dfrac{n-1}{n} < 1,$ for $n = 2, 3, \ldots$

Thus $0 = 0^2 < \left(\dfrac{n-1}{n}\right)^2 < 1^2 = 1,$ by Theorem 2.

Thus $0 < 1 - \left(\dfrac{n-1}{n}\right)^2 < 1.$

Thus $0 < b - \left(\dfrac{n-1}{n}\right)^2 b = \dfrac{n^2 b}{n^2} - \dfrac{(n-1)^2 b}{n^2} = \dfrac{2n-1}{n^2} b.$

Now $\dfrac{2n-1}{n^2} < \dfrac{2n}{n^2} = \dfrac{2}{n}.$

It follows that

(2) $0 < b - \left(\dfrac{n-1}{n}\right)^2 b < \dfrac{2}{n} b$

for $n = 2, 3, \ldots$.

Now $(2/n)b$ can be made smaller than any preassigned positive rational number by making n large enough. (The reader is asked to verify this for himself in Exercise 2 of 6.4.) Thus there is a natural number n_0 such that

(3) $0 < \dfrac{2b}{n_0} < b - a.$

From Expressions (2) and (3) we obtain

$$0 < b - \left(\frac{n_0 - 1}{n_0}\right)^2 b < \frac{2b}{n_0} < b - a,$$

or $\qquad 0 < b - \left(\frac{n_0 - 1}{n_0}\right)^2 b < b - a.$

Thus $\qquad -\left(\frac{n_0 - 1}{n_0}\right)^2 b < -a,$ or

(4) $$a < \left(\frac{n_0 - 1}{n_0}\right)^2 b.$$

Expression (4) together with the remark that

$$\left(\frac{n_0 - 1}{n_0}\right)^2 b < b$$

yields the result

$$a < \left(\frac{n_0 - 1}{n_0}\right)^2 b < b.$$

Let $k = \left(\frac{n_0 - 1}{n_0}\right)$, and we obtain $a < k^2 b < b$.

Theorem 3′. *If a, b are rational and $0 < a < b$, then there is a rational number $k > 0$ such that $a < ak^2 < b$.*

Proof. This follows from Theorem 3 and the fact that if $0 < a < b$, then $0 < 1/b < 1/a$.

Corollary 1. *Let a, b be positive rationals. If $a < b^2$, then there is a $c \in Q$ such that $a < c^2 < b^2$. If $a^2 < b$, then there is a $d \in Q$ such that $a^2 < d^2 < b$.*

Proof. If $a < b^2$, use Theorem 3 to produce k such that $a < k^2 b^2 < b^2$. Then for $c = kb$, $a < c^2 < b^2$.

To treat $a^2 < b$, use Theorem 3′ analogously.

From this corollary and Theorem 4 below, we can show that neither Corollary 1 nor Corollary 3 referred to in 5.4 holds for Q. We shall show that there are subsets of Q which have upper bounds but possess no least upper bound, and subsets with lower bounds having no greatest lower bound.

Theorem 4. *There is no rational number a such that $a^2 = 2$.*

This theorem states in effect that 2 has no rational square root. Since the Pythagoreans showed that $\sqrt{2}$ exists as the length of the hypotenuse of an isosceles right triangle with sides of unit length, from Theorem 4 we conclude that Q does not contain all numbers. It thus becomes necessary to construct a larger system to include, not only Q, but also numbers such as $\sqrt{2}$ and π which are not rational but whose existence is geometrically provable. This construction is developed in Chapter 8 and leads to the set of real numbers. In this book we shall not attempt to show that π is irrational. For a proof of this see *Irrational Numbers*, by I. Niven (John Wiley, 1956).

Proof of Theorem 4.   Suppose $a = m/n$ is a rational number such that

$$a^2 = \left(\frac{m}{n}\right)^2 = 2.$$

Then

(5) $m^2 = 2n^2.$

By the fundamental theorem of arithmetic both m and n have a unique prime factorization. Suppose m contained 2 in its factorization α times, and n contained 2, β times. Then the factorization of m^2 contains 2, 2α times and the factorization of n^2 contains 2, 2β times. Further the factorization of $2n^2$ contains 2, $2\beta + 1$ times.

From Equation (5) the number $m^2 = 2n^2$ possesses two prime factorizations: one with 2α factors of 2 and the other with $2\beta + 1$ factors of 2. This is impossible because for no choice of α, β can $2\alpha = 2\beta + 1$. Therefore the assumption that $a^2 = 2$ must be invalid, and so 2 is not the square of a rational number.

We are now in a position to exhibit a subset of Q which has an upper bound but has no least upper bound.

Theorem 5.   *The set $J = \{a \in Q \,|\, a > 0 \wedge a^2 < 2\}$ has an upper bound but has no least upper bound in Q.*

Proof.   (i) It is clear that 4 is an upper bound of J. Thus the first part of the theorem is proved.

(ii) Assume b is an upper bound of J and $b \in Q$. Then $b \notin J$. Indeed, if $b \in J$, then $b^2 < 2$, and there is a c, by Corollary 1, such that $b^2 < c^2 < 2$. Thus $c \in J$. In that event $b < c$, by Theorem 2, and b is not an upper bound.

An upper bound b of J cannot satisfy the equation $b^2 = 2$, by Theorem 4. Since $b \notin J$ and $b^2 \neq 2$, $b^2 > 2$. Thus any upper bound b of J has the property $2 < b^2$.

By Corollary 1, there is a $c \in Q$ such that $c > 0$ and $2 < c^2 < b^2$. By Theorem 2, $0 < c < b$. Now every $a \in J$ has a square less than 2, so $a < c < b$ for all $a \in J$. Therefore, c is an upper bound of J. Since every upper bound of J lies above another upper bound, J has no least upper bound.

Corollary 2. *The set* $H = \{a \in Q \,|\, a > 0 \wedge a^2 > 2\}$ *is bounded below but has no greatest lower bound.*

Proof. Analogous to the proof of Theorem 5.

Corollary 3. *J has no greatest element and H has no least element.*

The real numbers (to be discussed in Chapter 8) are constructed from the rational numbers by assigning real least upper bounds and greatest lower bounds to sets of rationals (such as H and J) which do not already possess them but do have upper bounds or lower bounds respectively. As a word of caution, despite Theorem 5 rational sets with upper bounds and least upper bounds in Q do exist. For example

$$K = \{a \in Q \,|\, \tfrac{1}{2} < a \leqslant 2\}.$$

Here l.u.b. K is 2 (the greatest element of K), and g.l.b. K is $\tfrac{1}{2}$ (K has no least element).

6.4 EXERCISES

1. Show that for any positive rational number r there is a natural number n such that $0 < 1/n < r$.

2. Use Exercise 1 to show that $\dfrac{2}{n} b$, for natural number n and positive rational b, can be made smaller than any given positive rational c.

3. Give the details of the proof of Theorem 3′ of 6.3.

4. Let F be an arbitrary field. Prove that the following equations are true for $a,b \in F$.
   (a) $a^2 - b^2 = (a - b)(a + b)$.
   (b) $a^3 - b^3 = (a - b)(a^2 + ab + b^2)$.
   (c) $(a + b)^2 = a^2 + ab + ab + b^2$.

5. The group in Exercise 2 of 5.6 can be used to define the addition group of a field. Let $F = \{0, 1\}$ with
$$0 + 0 = 1 + 1 = 0, \qquad 0 + 1 = 1 + 0 = 1,$$
   and
$$0 \cdot 0 = 0 \cdot 1 = 1 \cdot 0 = 0, \qquad 1 \cdot 1 = 1.$$
   Show that F, with addition "+" and multiplication "·" as indicated above, forms a field.

6. Show that Theorem 2 in 6.3 is no longer valid when the conditions "$a > 0$" and "$b > 0$" are removed from its hypothesis.

7. Show that no rational number x exists such that $x^2 = 3$.

8. Show that no rational number y exists such that $y^3 = 4$.

9. Prove Corollary 2 of 6.3.

10. Consider the set $F = \{a + \sqrt{2}\,b \mid a \in Q \text{ and } b \in Q\}$.

    (a) Show that F is closed under multiplication; that is, $(a + \sqrt{2}\,b)(c + \sqrt{2}\,d)$ can be written in the form $e + \sqrt{2}\,f$ where $e \in Q$ and $f \in Q$.

    (b) Show that

$$(a + \sqrt{2}\,b)\left(\frac{a}{a^2 - 2b^2} - \sqrt{2}\,\frac{b}{a^2 - 2b^2}\right) = 1 + \sqrt{2}\cdot 0 = 1.$$

    (c) Using parts (a) and (b), show that F provided with usual operations of "$+$" and "$\cdot$" forms a field with additive identity $0 = 0 + \sqrt{2}\cdot 0$ and multiplicative identity $1 = 1 + \sqrt{2}\cdot 0$.

11. Consider the set $\{0, 1, 2, 3\} = I_4$ where multiplication and addition are defined for I_4 by the following tables:

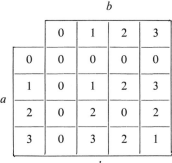

	b			
a		0	1	2	3
	0	0	1	2	3
	1	1	2	3	0
	2	2	3	0	1
	3	3	0	1	2

$a + b$

	b			
a		0	1	2	3
	0	0	0	0	0
	1	0	1	2	3
	2	0	2	0	2
	3	0	3	2	1

$a \cdot b$

    (a) Show that I_4 cannot be a field.

    (b) Which of the field axioms does I_4 satisfy?

12. (a) In an ordered field show that $(-1)^2 > 0$.

    (b) Then show that for any rational number a, $a^2 \geqslant 0$.

    (c) Show that no order relation can be placed on the field of Exercise 5 in order to make it an ordered field.

13. Prove that O1 and O2 of 2.4 hold in Q.

14. Show that OA, OM, and OM' all hold in Q.

7

Relations and Functions

7.1 FUNCTIONS

In 1.1 we said that the concept of "set" was probably the most basic notion in mathematics. The concept we wish to discuss here, **function,** is unquestionably one of the most important concepts in mathematics. As we shall see, relations and operations are also in some sense basically functions.

There are a number of ways to define a function. The following, while not the most elegant or sophisticated, gives a clearer idea of the intuitive idea we are trying to make precise than does any other. We shall examine a more sophisticated approach in 7.6.

Intuitively, a function involves three things; two of them are sets and the third is a **rule of correspondence.** The notion "rule of correspondence" (briefly "rule") will not be precisely defined, just as was the case with "set", but it should become intuitively clear as we proceed. One of the two sets is called the **domain of definition** (briefly **domain**); the other is called the **counter domain.** Let D be the domain and C the counter domain. The rule of correspondence must have the following property: for each $x \in D$ the rule assigns to this x a unique element $y \in C$. There are two key words here. The first is "each"; *every* element in D has some corresponding element from the set C assigned to it by the rule. The second is "unique"; *no* element in D has *more than one* element corresponding to it by the rule.

As a matter of notation, we usually assign some letter, say f, to stand for the rule of correspondence, and for any $x \in D$ we write $f(x)$ (read "f of x" or "f at x") as the name for the corresponding element $y \in C$ under the rule $f: y = f(x)$. (There are many variations to this convention, but we shall ignore them here.)

The following is a precise definition of function.

Definition.   *A **function** is a triple of objects D, C, f such that D is a nonempty set* (*called the **domain of definition***), *C is a nonempty set* (*called the **counter domain***), *and f is a **rule of correspondence** with the property that for any x ∈ D, there is a unique y ∈ C such that y = f(x).*

For each $x \in D$, the element $f(x)$ in C is called the **image** or **functional value** of x (under f).

Figure 7–1 schematically illustrates the function concept. Here, as in the Venn diagrams of Chapter 4, D and C are represented by plane regions. Each point in D corresponds to some (unique) point in C under the rule f. Four such correspondences are shown. The figure is a heuristic device and must never be taken as more than that; for example, it does not show all correspondences, and there is no reason why D and C are necessarily different or disjoint.

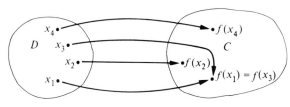

Figure 7–1

Example 1.   Let the domain be $D = \{1, 2, 3\}$ and the counter domain be $C = \{1, 2\}$. Let the rule f be given by $f(1) = 1$, $f(2) = 1$, $f(3) = 2$. This function can be completely illustrated by Figure 7–2, wherein the elements of D and C are listed and the rule of correspondence f is (in this case completely) shown by arrows.

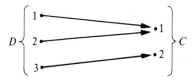

Figure 7–2

All the requirements for a function are satisfied; we have a domain D, a counter domain C and a rule f (written out explicitly) which assigns to each member of D exactly one member of C. Observe that $f(1) = f(2)$ and this *does not violate* the definition of a function.

Since Example 1 involves a finite function the rule can be completely described by a table of the type shown in Figure 7–3.

x	$f(x)$
1	1
2	1
3	2

Figure 7–3

Example 2. Let the domain and counter domain each be I. Let the rule g be defined by $g(x) = 4$. Again all the requirements are satisfied. This Example illustrates a point: nothing was said in the definition to the effect that all the counter domain need be "used up" under the rule of correspondence. In this case, not every element of the counter domain is the image of an element of the domain; only the number 4 is an image. (Such a function is called a **constant** function.) This function can be illustrated by an infinite analogue of Figure 7–2; see Figure 7–4. Its rule can, in a sense, be given by an infinite table as in Figure 7–5.

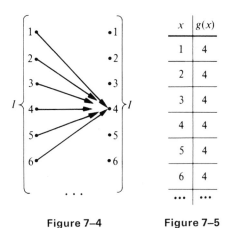

x	$g(x)$
1	4
2	4
3	4
4	4
5	4
6	4
...	...

Figure 7–4   **Figure 7–5**

Example 3. Let the domain and counter domain be I. Let the rule h be given by $h(n) = n^2$. Again the definition is satisfied. Again not all the counter domain is used up.

It is a great convenience to allow a counter domain to be bigger than is actually needed. On the other hand, it is convenient to have a name for the set of all images (or functional values).

Definition 2. *Given a function with rule f, domain D and counter domain C, $\{y \in C \mid y = f(x)$ for some $x \in D\}$ is called the* **range** *of the function.*

The range, of course, is always a subset of the counter domain. In Example 1, the range equals the counter domain. In Example 2 the range is $\{4\}$, a proper subset of the counter domain.

Given a function with domain D, counter domain C, and rule f, it is very common practice to call f the function. This is an ellipsis and is not strictly correct. There is nothing wrong with the practice provided that the domain and counter domain are still understood to constitute essential ingredients of the function. Sometimes one refers to a function f without explicitly mentioning the domain and counter domain. Again there is nothing wrong with this provided these sets are implicitly understood from the context, but it is bad practice if these sets are not unambiguously understood at least implicitly. If the domain is known, the counter domain can always be taken to be the range. We shall not consider two functions to be essentially different if they differ only in having different counter domains, each of which has the range of the function as a subset. (Some authors disagree, but the distinction is pedantic.) There is, however, a suggestive language used to describe the situation.

A function f with domain D is said to be **defined on** D. If f has counter domain C, f is said to **map** or **transform** D **into** C. (Sometimes functions are called **maps, mappings,** or **transformations.**) Alternatively we say f is a function from D to (or into) C. If C happens to be the range of f, we say f maps D **onto** C. In Example 1, f maps D onto C. In Example 2, g maps I into I. Sometimes we say **properly into** in this case, meaning into but not onto. Further, g maps I onto $\{4\}$.

Nothing in the definition of a function implies that the rule of correspondence need be a rule which can be explicitly written down either by listing the functional values (as in Example 1) or by giving some formula (as in Examples 2, 3).

Example 4. If a person throws a pair of dice consecutively 25 times and records the results, he has produced a function from the set $A = \{1, 2, ..., 25\}$ into the set $B = \{2, 3, ..., 12\}$. No formula can be given for the rule of this function, nor can the functional values be listed in advance. If the person repeated the process, he would probably produce a different function from A into B.

Example 5. Let p be a function defined on Q such that $p(x) = 1$ if $x \geqslant 0$, $p(x) = -1$ if $x < 0$. The range of p is $\{1, -1\}$. Any set F such that

$\{1, -1\} \subseteq F$ can be taken as the counter domain. No formula for p is given, at least in the usual sense, but p is well defined.

It is not necessary that the domain and counter domain be sets of numbers; they can be sets of anything whatever. A column in a railroad timetable could be interpreted as a function in which the domain is a set of geographic locations and the range is a set of numbers designating times of the day. A table of postage rates can be similarly viewed as a function. The function of Example 4 could be reinterpreted so that the domain of definition is viewed as a set of "trials," that is the throwing of the dice.

Example 6. Let $\mathscr{P}(x)$ stand for the (open) sentence "x is an even integer". $\mathscr{P}$ is a function whose domain is I and whose counter domain is $\{\mathscr{T}, \mathscr{F}\}$. At least this is one way to look at it. There is an intermediate stage which one can consider, wherein $\mathscr{P}$ maps the set I into the set of all (closed) sentences. Then one can envisage a second truth-assigning function which maps the set of all sentences into the set of truth values. The first point of view is adequate and normally more useful.

7.2   CARTESIAN PRODUCT OF SETS

In addition to the operations on sets discussed in Chapter 4, there is another important operation of a somewhat different kind. This is called the **Cartesian product** of two sets A, B (named after R. Descartes (1596–1650), the inventor of coordinate geometry). The Cartesian product is written $A \times B$ (read "A cross B"). We need first the notion of an **ordered pair** of objects.

Definition 1.   *An **ordered pair** (x, y) is a set containing exactly the elements x, y with the additional property that x is specifically designated to be the first element and y is specifically designated to be the second.*

By 1.3 or Definition 2 of 1.5, the *sets* $\{x, y\}$ and $\{y, x\}$ are equal. However, the *ordered pairs* (x, y) and (y, x) are not in general equal. It follows from Definition 1 above that *the ordered pairs (x, y) and (u, v) are equal if and only if $x = u$ and $y = v$:*

$$(x, y) = (u, v) \Leftrightarrow (x = u) \wedge (y = v).$$

This is really another way of stating Definition 1. It follows that $(1, 3) \neq (3, 1)$.

Definition 2.   *The **Cartesian product**, $A \times B$, of two sets A, B is the set of all ordered pairs (x, y) such that $x \in A$ and $y \in B$.*

$$A \times B = \{(x,y) \mid x \in A, \ y \in B\}.$$

Example 1. Let $A = \{0, 1, 2\}$, $B = \{2, 3\}$. Then $A \times B = \{(0, 2),$ $(0, 3), (1, 2), (1, 3), (2, 2), (2, 3)\}$. Note that A has 3 elements, B has 2 elements, and $A \times B$ has 6 ($=2 \cdot 3$) elements.

Example 2. For A as above, $A \times I = \{(x, y) | x \in A \wedge y \in I\}$. $A \times I$ is a set which contains exactly all those ordered pairs whose first element is $0, 1$, or 2 and whose second element is an integer; for example, $(0, 1)$, $(0, -3)$, $(1, 27)$, $(1, -10^9)$, $(2, 365)$, etc.

The notion of an ordered pair enters into the definition of $A \times B$ in a way other than the one explicitly mentioned. The operator $\times$ is applied to the *ordered pair* A, B. Indeed $B \times A$ is a different thing and may well not equal $A \times B$, that is in general, $A \times B \neq B \times A$. This question also arises in connection with $A \cup B$, $A \cap B$, and $A - B$. See Chapter 4.

Example 1. (continued). $B \times A = \{(2, 0), (2, 1), (2, 2), (3, 0), (3, 1), (3, 2)\}$. $A \times B \neq B \times A$; for example $(0, 2) \in A \times B$, but $(0, 2) \notin B \times A$. Remember $(0, 2) \neq (2, 0)$.

The Cartesian product frequently appears in the form $A \times A$. For example, the reader is no doubt already familiar with $R \times R$ which has as its geometric interpretation all the points in the plane. An ordered pair of real numbers (x, y) uniquely determines a point in plane coordinate geometry; all such ordered pairs form the set $R \times R$. It is in this sense that Descartes invented the Cartesian product.

7.3   RELATIONS

A **relation** on a set A has already been defined in 2.4 as a predicate of two variables which take their values in the set A. We usually name relations by a script letter, say $\mathscr{S}$, followed by the names for the variables in parentheses; thus $\mathscr{S}(x, y)$. $\mathscr{S}$, of course, is simply the name of something which will have been defined in terms of previously known notions. For example we might have: $\mathscr{S}(x, y)$ if and only if $x, y \in N$ and x is the number of distinct prime factors of y. Then $\mathscr{S}(x, y)$ becomes a sentence as soon as x and y are replaced by elements of N, and as such is either true or false. So $\mathscr{S}(1, 2)$ is a true sentence; $\mathscr{S}(2, 2)$ is a false sentence. (Cf. Example 6 of 7.1.)

There is a slightly different way of looking at a relation which allows us to generalize the notion a little. Notice the appearance of the ordered pair "(x, y)" when we speak of the relation "$\mathscr{S}(x, y)$". $\mathscr{S}$ is actually a function whose domain of definition is some set of ordered pairs and whose range of values is the set of truth values $\mathscr{T}$, $\mathscr{F}$. Now there is really no need for the variables x and y to take their values in the same set (although they often do).

Definition. *A* **relation** $\mathcal{R}$ *on the sets* A, B *is a function whose domain is* $A \times B$ *and whose range is* $\{\mathcal{T}, \mathcal{F}\}$.

Example 1.   Let $\mathcal{D}$ be a relation on N, I as follows. $\mathcal{D}(x, y) \Leftrightarrow (y$ is x undirected units distant from 0 on the number line). So $\mathcal{D}$ is a function from $N \times I$ onto $\{\mathcal{T}, \mathcal{F}\}$. $\mathcal{D}(3, 2)$ is false $(= \mathcal{F})$; $\mathcal{D}(2, 2)$ is true $(= \mathcal{T})$; $\mathcal{D}(3, -3)$ is true; and so on.

Example 2.   Let **N** be the set of all subsets of N; that is $\mathbf{N} = \{A \mid A \subseteq N\}$
or
$$A \in \mathbf{N} \Leftrightarrow A \subseteq N.$$

The elements of **N** are subsets of N. For example $N \in \mathbf{N}$, $\{1\} \in \mathbf{N}$, $\{0,1,17\} \in \mathbf{N}$, $\{0, 2, 4, ..., 2n, ...\} \in \mathbf{N}$, $\{x \in N \mid x = a^2 + b^2$ for some $a,b \in N\} \in \mathbf{N}$ and so on. Now consider $N \times \mathbf{N}$. This is a rather complicated looking set; its elements are ordered pairs of the form (m, A) where $m \in N$, $A \subseteq N$ (for example $(17, \{1, 3, 11\}) \in N \times \mathbf{N}$). Now $\in$ of 1.2 is a relation on N, **N**. It is true that we usually write $m \in A$ rather than $\in (m, A)$, but this is simply bowing to convention; we could use the latter if we wished. Applying $\in$ to the ordered pair $(17, \{1, 3, 11\})$ above, we get $17 \in \{1, 3, 11\}$, which is false; thus $\in (17, \{1, 3, 11\}) = \mathcal{F}$.

Example 3.   Let L be the set of all straight lines in the plane. Let $\mathcal{P}$ be the familiar parallel relation on $L \times L$.

$$\mathcal{P}(l_1, l_2) \Leftrightarrow (l_1 \text{ is parallel to } l_2).$$

Sometimes $\mathcal{P}(l_1, l_2)$ is written $l_1 \| l_2$.

Example 4.   We have already seen a number of other relations in 2.4. They were all order relations. The reader should review these now.

It should be emphasized that to call "$\mathcal{R}$" a relation is elliptic in the same way that to call "f" a function is elliptic. When we view a relation, as we did here, as a function, the counter domain is always $\{\mathcal{T}, \mathcal{F}\}$. However, the domain $A \times B$ must always be clear from the context or be explicitly mentioned. Remember further, that a "relation on A" means a relation defined on $A \times A$ in the function sense as defined above.

Example 5. $x \mid y$ is a relation on N (that is, defined on $N \times N$). Here the name or symbol of the relation "$\mid$" appears between the variables. This convention is very common in the case of frequently used relations which have fixed symbols designating them; for example, $<$, $\leqslant$, $\|$, $=$, etc.

In 2.4 we emphasized the difference between a relation on a set A and an operation on a set A. This can be more succinctly stated now as follows: a relation on A is a function from $A \times A$ into $\{\mathcal{T}, \mathcal{F}\}$; an operation on A is a function from $A \times A$ into A.

7.4   GRAPHS OF RELATIONS

If $\mathcal{R}$ is a relation on A, B, then there will, in general, be some ordered pairs $(x, y) \in A \times B$ such that "$\mathcal{R}(x, y)$" is true and some such that "$\mathcal{R}(x, y)$" is false. For example, on I, "$-1 < 2$" is true and "$2 < 0$" is false.

Definition 1.   *The **graph** of a relation $\mathcal{R}$ on A, B is the set of all ordered pairs $(x, y) \in A \times B$ such that "$\mathcal{R}(x, y)$" is true.*

$$\text{Graph of } \mathcal{R} = \{(x, y) \in A \times B \mid \mathcal{R}(x, y)\}.$$

The reader may have the impression that a graph is a kind of picture. However, a graph is a set of ordered pairs. If it can be schematically presented in a pictorial way, this is a great convenience, but it is not always possible.

Example 1.   We shall assume some familiarity with R, the set of real numbers, which is to be discussed in the next chapter. The Cartesian product $R \times R$ has as its geometrical representation all the points in the plane with a rectangular coordinate system imposed on it. The horizontal axis represents the first factor in $R \times R$ and the vertical axis represents the second factor, as in Figure 7–6. Any point in the plane has coordinates $(x, y) \in R \times R$. The

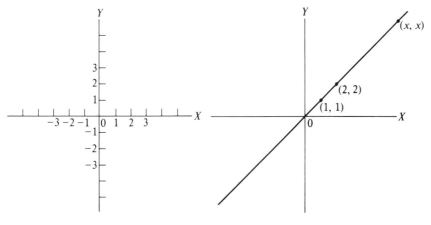

Figure 7–6   Figure 7–7

graph of any relation on R (that is, defined on $R \times R$) will be a subset of $R \times R$; in other words a set of points in the plane. For example, the relation "=" has as its graph the set $\{(x,y) \in R \times R \mid x = y\}$. This set has as its geometrical representation the points on the diagonal straight line in Figure 7–7.

Example 2.   The relation $\mathscr{R}_1$ on R defined by

$$\mathscr{R}_1(x, y) \Leftrightarrow x^2 + y^2 = 4$$

has as its graph $\{(x,y) \in R \times R \mid x^2 + y^2 = 4\}$. This is well known to the reader to be the set of all points on the circle of Figure 7–8.

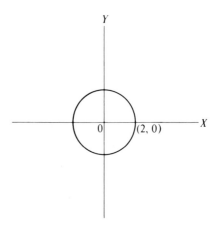

Figure 7–8

Example 3.   Consider relations $\mathscr{R}_2$, $\mathscr{R}_3$, $\mathscr{R}_4$, $\mathscr{R}_5$ on R defined by

$$\mathscr{R}_2(x, y) \Leftrightarrow x^2 + y^2 < 4,$$
$$\mathscr{R}_3(x, y) \Leftrightarrow x^2 + y^2 > 4,$$
$$\mathscr{R}_4(x, y) \Leftrightarrow x^2 + y^2 \leqslant 4,$$
$$\mathscr{R}_5(x, y) \Leftrightarrow x^2 + y^2 \geqslant 4.$$

The graphs of these relations are regions of the plane $R \times R$. The graph of $\mathscr{R}_2$ is the set of all points (that is, ordered pairs) interior to the circle of Figure 7–9, but not on the circle. The graph of $\mathscr{R}_3$ is the set of points exterior to the circle, but not on the circle. The graph of $\mathscr{R}_4$ is the set of all points on the circle or interior to it; in fact Graph $\mathscr{R}_4$ = Graph $\mathscr{R}_1$ ∪ Graph $\mathscr{R}_2$. Similarly, Graph $\mathscr{R}_5$ = Graph $\mathscr{R}_3$ ∪ Graph $\mathscr{R}_1$.

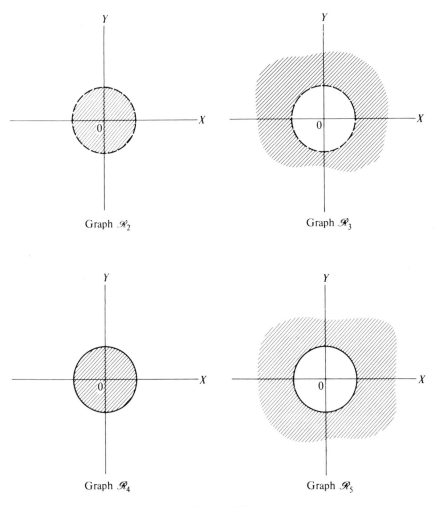

Graph $\mathcal{R}_2$

Graph $\mathcal{R}_3$

Graph $\mathcal{R}_4$

Graph $\mathcal{R}_5$

Figure 7–9

Example 4. Let the relations of Examples 1, 2, 3 be modified so that the rule is unchanged, but the domain is $Q \times Q$ instead of $R \times R$; that is, the relations are on Q instead of R. The graphs of these new relations are not the same as they were for the old ones. For example, $(\sqrt{3}, 1)$ is in the graph of $x^2 + y^2 = 4$ on R, but not in the graph of $x^2 + y^2 = 4$ on Q, because $(\sqrt{3}, 1) \notin Q \times Q$. On the other hand, $(0, 1)$ is in both graphs. If one attempts to represent the graphs of these relations on Q in a pictorial way, as we did in Figures 7–7, 7–8, 7–9, for their counterparts on R, they would *look* the same as Figures 7–7, 7–8, and 7–9. (This is because Q is dense in R, that is, between any two real numbers there is a rational number.) However, they are *not* the same graph.

Example 5. Let the relations of Example 4 be further modified so that the domain is $I \times I$ but the rule remains unchanged. The graph of the relation $=$ on I is $\{..., (-2, -2), (-1, -1), (0, 0), (1, 1), (2, 2), ...\}$. If one wishes to draw a picture, the best one can do is to indicate these isolated points as in Figure 7–10. The graph of $\mathscr{R}_1$ on I is the set of four elements, $\{(0, 2), (2, 0), (-2, 0), (0, -2)\}$. The graph of $\mathscr{R}_2$ on I is the set of five elements, $\{(0, 0), (1, 1), (1, -1), (-1, 1), (-1, -1)\}$. The graph of $\mathscr{R}_4$ on I is the union of these two sets.

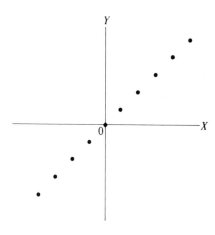

Figure 7–10

Example 6. The graph of the relation $x \,|\, y$ on N is $\{(x,y) \in N \times N \,|\, x \,|\, y\}$. If would be impossible to draw a picture of this graph in any meaningful way.

Example 7. Let us look at an example of a relation on R whose graph cannot possibly be drawn in the plane. The following example is not very useful for other purposes, but it illustrates the point. Let every $x \in R$ be written in its decimal form. (We assume the reader is familiar with this, although it is treated in the next chapter.) Let $x = x_0 . x_1 x_2 ... x_i ...$ where $x_0 \in I$, $x_i \in \{0, 1, 2, ..., 9\}$ for $i > 0$, and similarly write $y = y_0 . y_1 y_2 ... y_i ...$. Let the relation $\mathscr{D}$ be defined by: $\mathscr{D}(x, y) \Leftrightarrow x_1 = y_1$. $\mathscr{D}$ is certainly a relation on R; try to draw its graph! There are many variations on this theme; for example

$$\mathscr{D}_1(x, y) \Leftrightarrow (\text{there is an } i > 0 \text{ such that } x_i = y_i).$$

Example 8. Let A be any fixed subset of $R \times R$. Define a relation $\mathscr{A}$ on R by: $\mathscr{A}(x, y) \Leftrightarrow (x, y) \in A$. Now $\mathscr{A}$ is a perfectly well defined relation, given A. Note that the graph of $\mathscr{A}$ is A.

Example 8 illustrates an important point. Any relation $\mathcal{P}$ on sets E, F uniquely determines the set $\{(x, y) \in E \times F \mid \mathcal{P}(x, y)\}$ (a subset of $E \times F$), which is the graph of $\mathcal{P}$; conversely any set $P \subseteq E \times F$ uniquely determines the relation $(x, y) \in P$ which has P as its graph. From a point of view one level of abstraction higher, so to speak, one could quite correctly say that there is (abstractly) no difference between a relation and its graph. (Remember that a graph is a set, not a picture.) This identification of a relation with its graph is often used as an alternative definition of relation.

Definition 2.　*A relation $\mathcal{R}$ on A, B is a subset of $A \times B$. (A relation $\mathcal{R}$ on A is a subset of $A \times A$.)*

This alternative definition is mathematically elegant, but it has a tendency to miss the point of the idea.

7.5　EQUIVALENCE RELATIONS

Certain relations on a set A enjoy special properties (for example, transitivity). Relations that enjoy a certain particular list of properties occur so frequently that they are given a special name and are subjected to special study. One such is the order relation, which we have already encountered. Another kind, equally as important as the order relation, is the **equivalence relation**.

Definition 1.　*If $\mathcal{R}$ is a relation on a set A, then $\mathcal{R}$ is an **equivalence relation** provided that $\mathcal{R}$ satisfies the following three properties for all $x,y,z \in A$.*

(i)　$\mathcal{R}(x, x)$		*Reflexive Property*
(ii)　$\mathcal{R}(x, y) \Rightarrow \mathcal{R}(y, x)$		*Symmetric Property*
(iii)　$(\mathcal{R}(x, y) \wedge \mathcal{R}(y, z)) \Rightarrow \mathcal{R}(x, z)$		*Transitive Property*

Example 1.　Let A be any set and consider the equality relation. It is certainly true that: $x = x$; $x = y \Rightarrow y = x$; and $(x = y \wedge y = z) \Rightarrow x = z$ for all $x,y,z \in A$. In other words, equality on any set is an equivalence relation. As a matter of fact, equality is that equivalence relation which differs from all others in that it, and it alone, additionally enjoys the Replacement Property discussed in 2.2.

Example 2.　Let M be the set of all mathematical sentences. Consider the relation $\Leftrightarrow$ on M defined in 3.6. For all sentences $\mathcal{P}$, $\mathcal{Q}$, $\mathcal{R}$ the following were established there (or can be deduced easily by truth tables): $\mathcal{P} \Leftrightarrow \mathcal{P}$; $(\mathcal{P} \Leftrightarrow \mathcal{Q}) \Rightarrow (\mathcal{Q} \Leftrightarrow \mathcal{P})$; $[(\mathcal{P} \Leftrightarrow \mathcal{Q}) \wedge (\mathcal{Q} \Leftrightarrow \mathcal{R})] \Rightarrow (\mathcal{P} \Leftrightarrow \mathcal{R})$. Thus logical equivalence is an equivalence relation. (As a matter of fact, logical equivalence is an equivalence relation very much like the equality relation. It does enjoy a property something like the replacement property; in any compound sentence, one component sentence of it can be replaced by a logically equivalent one without changing the truth value of the original compound sentence.)

Example 3.   Let L be the set of all straight lines in the Euclidean plane. Consider the relation $\parallel$, of parallelism (where we must here adopt the convention that a straight line is parallel to itself). It is indeed the case that for any $l,m,n \in L: l \parallel l; l \parallel m \Rightarrow m \parallel l; (l \parallel m \wedge m \parallel n) \Rightarrow l \parallel n$. Thus "$\parallel$" is an equivalence relation. This is a very useful example to illustrate the essential features of an equivalence relation. Choose any line $l \in L$ and consider the set $[l]$ of all lines x parallel to l (viz., $[l] = \{x \in L \mid x \parallel l\}$). The set $[l]$ is a set of lines all of which are parallel to one another, and every line parallel to l is in $[l]$. If $l' \parallel l$ then $l' \in [l]$. Suppose we consider the set $[l'] = \{x \in L \mid x \parallel l'\}$; clearly $[l'] = [l]$. If we consider a line $m \in L$ such that $m \nparallel l$, then the set of lines parallel to m (viz., $[m] = \{x \in L \mid x \parallel m\}$) is a different set; $[m] \neq [l]$. In addition, $[m] \cap [l] = \emptyset$. See Figure 7–11.

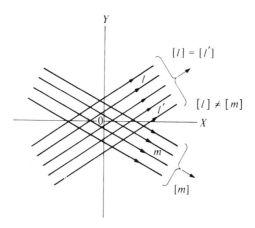

Figure 7–11

The set $[l]$ is called an **equivalence class** determined by l. ("Class" is another name for set.)[1] These equivalence classes represent, or determine, an abstraction. All members of a given equivalence class share a common property; in this case, direction. (We are ignoring a positive or negative sense of direction.) The equivalence classes $[l]$, $[m]$, etc., can be construed abstractly to be directions. In any context in which *direction alone* is considered, the elements of $[l]$ are indistinguishable, and constitute one single mathematical entity. It is said that the equivalence relation "$\parallel$" **partitions** the set L into

---

[1] But see the discussion concluding 10.5.

equivalence classes. Note that the partition of L consists of sets (equivalence classes) $[l]$, $[m]$, etc., with the following properties:

(i) Either $[l] = [m]$ or $[l] \cap [m] = \emptyset$ (depending on whether $l \| m$ or $l \nparallel m$ respectively);

(ii) Every $l \in L$ is in some equivalence class (viz., $[l]$).

The foregoing discussion applies to equivalence relations generally and will be amplified below.

Example 1 (continued). The equivalence classes with respect to = all consist of a single element; that is, for $x \in A$, $[x] = \{x\}$. This indeed describes the essential nature of equality; the only thing equal to a given object is itself. This is quite apart from the fact that the object might have different names; for example, 4, 2^2.

Now let us give the precise definitions of **partition** and **equivalence class**, and after a few more examples we shall consider the situation generally.

Definition 2. *Let A be any set. A **partition** of A is a collection of disjoint subsets of A which covers A in the sense that every element of A is in some set of the partition.*

Definition 3. *Let $\mathcal{R}$ be an equivalence relation on A. Let $a \in A$. The **equivalence class determined by** a, written $[a]$ (or $[a]_{\mathcal{R}}$ if necessary), is defined by:*

$$[a] = \{x \in A \mid \mathcal{R}(x,a)\}.$$

Example 4a. Let the set $A = \{1, 2, 3, 4, 5, 6, 7\}$ be partitioned as follows $\{\{1\}, \{2, 5\}, \{3, 4, 6, 7\}\}$. Let the relation $\mathcal{A}$ be defined on A by $\mathcal{A}(x, y)$ if and only if x and y belong to the same set of the partition; that is, $\{1\}$ or $\{2, 5\}$ or $\{3, 4, 6, 7\}$. Note that these sets do form a partition; they cover A and are disjoint. We leave it to the reader to check that $\mathcal{A}$ is an equivalence relation.

b. Consider the following subsets of A: $A_1 = \{1\}$, $A_2 = \{2, 3, 4\}$, $A_3 = \{4, 5, 6\}$. Then $\{A_1, A_2, A_3\}$ fails to be a partition of A because both requirements are violated: 7 is an element of none of the sets A_i; $A_2 \cap A_3 = \{4\}$. Define a relation $\mathcal{V}$ by: $\mathcal{V}(x, y) \Leftrightarrow$ (for some i, $x \in A_i \lor y \in A_i$, $i = 1, 2, 3$). Now $\mathcal{V}$ is a well-defined relation and it is uniquely determined by the collection of sets $\{A_1, A_2, A_3\}$. Let us see how the fact that this collection fails to be a partition entails the fact that $\mathcal{V}$ is *not* an equivalence relation on A. First, we do not have $\mathcal{V}(x, x)$ for all $x \in A$, because we do not have $\mathcal{V}(7, 7)$; 7 is in none of the given subsets. Second, transitivity fails also because we have $\mathcal{V}(3, 4)$ (since $3,4 \in A_2$) and we have $\mathcal{V}(4, 5)$ (since $4,5 \in A_3$), but we do not have $\mathcal{V}(3, 5)$; this failure results from the fact that 4 is in both A_2 and A_3.

Example 5. Let I be partitioned into two disjoint subsets $E = \{x \in I \mid x$ is even$\}$ and $O = \{x \in I \mid x$ is odd$\}$. (Remember, zero is even.) Then $\{E, O\}$ is a partition of I. Let the relation $\mathscr{E}$ be defined on I by: $\mathscr{E}(x,y) \Leftrightarrow (x,y \in E) \vee (x,y \in O)$ (that is, if and only if both x, y belong to the same set of the partition). Now $\mathscr{E}$ is an equivalence relation; we can say $\mathscr{E}(x, x)$ for any x because every x is in one of the sets and not in both; we have $\mathscr{E}(x, y) \Rightarrow \mathscr{E}(y, x)$ obviously, because to say that x, y belongs to a certain set is the same as saying that y, x belongs to that set; similarly we can see that transitivity holds.

The equivalence relation $\mathscr{E}$ was uniquely determined by the partition $\{E, O\}$. This situation is general, as is its converse, which we saw illustrated in Example 3. We shall state and prove that this is a theorem presently.

Example 6. In Examples 4, 5, there were certain features of finiteness which are not essential. For each $i = 1, 2, 3, \ldots$, let let $C_i = \{(x,y) \in R \times R \mid x^2 + y^2 \leqslant i^2\}$; that is, C_i is the set of all points in the Cartesian plane $R \times R$ which lie on or inside the circle $x^2 + y^2 = i^2$, with center at the origin and radius i units. Let $D_1 = C_1$, $D_2 = C_2 - C_1$, $D_3 = C_3 - C_2$, etc. If $i > 1$, $D_i = \{(x,y) \in R \times R \mid (i - 1)^2 < x^2 + y^2 \leqslant i^2\}$. Then $D_i, i > 1$, are annular regions, (See Figure 7–12) and $\{D_1, D_2, D_3, \ldots\}$ is a partition of $R \times R$. Let $\mathscr{A}$ be a relation on $R \times R$ (not on R) defined by

$$\mathscr{A}[(x, y), (u, v)] \Leftrightarrow [\text{for some } i, (x, y) \in D_i \wedge (u, v) \in D_i].$$

$\mathscr{A}$ is an equivalence relation.

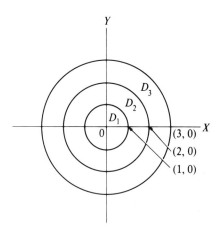

Figure 7–12

Example 7. Before we proceed to the theorem, let us look at some examples of relations on a set which are not equivalence relations and see how they fail to partition the given set.

(a) Consider the relation $<$ on Q. We know $<$ is not an equivalence relation because it is irreflexive and anti-symmetric. Consider the set $[1/2] = \{x \in Q \,|\, x < 1/2\}$ (which would be called an equivalence class if $<$ were an equivalence relation), and the set $[3/2] = \{x \in Q \,|\, x < 3/2\}$. Certainly $[1/2] \cap [3/2] \neq \varnothing$.

(b) Consider the perpendicularity relation on the set L of Example 3. This is not an equivalence relation because it is neither reflexive nor transitive. We leave it to the reader to check that $\perp$ does not partition L. Note that if we form the set $[l] = \{x \in L \,|\, x \perp l\}$, then $l \notin [l]$. This situation never occurs with equivalence classes.

Theorem. (i) *Let $\mathscr{R}$ be an equivalence relation on a set A. For each $a \in A$ form the equivalence class $[a] = \{x \in A \,|\, \mathscr{R}(x, a)\}$. Then the set of all equivalence classes $\{[a] \,|\, a \in A\}$ is a partition of the set A.*

(ii) *Let $\{A_i\}$ be a collection of subsets of a set A which partitions A. Define a relation $\mathscr{R}$ on A by $\mathscr{R}(x,y) \Leftrightarrow ($ for some $i, x \in A_i \wedge y \in A_i)$. Then $\mathscr{R}$ is an equivalence relation on A.*

Proof. (i) The set of equivalence classes covers A because for any $a \in A$ we have $a \in [a]$ by the reflexive property, $\mathscr{R}(a, a)$. Now choose two equivalence classes $[a]$, $[b]$. We wish to show that either $[a] = [b]$ or $[a] \cap [b] = \varnothing$. So suppose that $[a] \cap [b] \neq \varnothing$; if we can show that $[a] = [b]$ we are finished. To this end, pick any $x \in [a] \cap [b]$. Then $x \in [a]$ and $x \in [b]$. That is, $\mathscr{R}(x, a)$ and $\mathscr{R}(x, b)$. But $\mathscr{R}(x, a)$ implies $\mathscr{R}(a, x)$ by symmetry, and $\mathscr{R}(a, x)$ and $\mathscr{R}(x, b)$ together imply $\mathscr{R}(a, b)$ by transitivity. Now we can show that $[a] \subseteq [b]$. Pick any $y \in [a]$; that is, $\mathscr{R}(y, a)$. But this with $\mathscr{R}(a, b)$ and transitivity gives $\mathscr{R}(y, b)$; that is, $y \in [b]$. So we have shown $[a] \subseteq [b]$. Similarly, we can show $[b] \subseteq [a]$ and hence that $[a] = [b]$.

(ii) We need to show that $\mathscr{R}$ has the required properties. Choose any $x, y, z \in A$.

(a) Since the partition covers A, $x \in A_i$ for some i. So $\mathscr{R}(x, x)$. (It is the covering property that ensures reflexiveness.)

(b) If $x, y \in A_i$ for some i, then $y, x \in A_i$ for the same i; that is, $\mathscr{R}(x, y) \Rightarrow \mathscr{R}(y, x)$.

(c) Suppose $\mathscr{R}(x, y)$, that is, $x, y \in A_i$ for some i, and $\mathscr{R}(y, z)$, that is, $y, z \in A_j$ for some j. Then $y \in A_i$ and $y \in A_j$. Since the partitioning sets are disjoint $A_i = A_j$. So $x, z \in A_i$; that is $\mathscr{R}(x, z)$. (So it is the disjointness property that ensures transitivity.)

Equivalence relations on a set and the resulting partitioning of the set into equivalence classes provides one of the major tools whereby new mathematical structures are defined from old ones. This device can be (and usually is) used to define the integers from the natural numbers. We did not do it this way above because the device which we did use has some pedagogical value through its similarity with the definition of the complex numbers from the real numbers. In the following example, we use the former method to define the rational numbers from the integers. The same method, suitably modified, could be used to define the integers from the natural numbers.

Example 8. This example has already been sketched briefly without reference to equivalence relations in 6.1. Let $Q' = I \times (I - \{0\}) = \{(x,y) \mid x \in I \land y \in I \land y \neq 0\}$. Intuitively, we think of (x, y) as the fraction or rational number x/y, but since this latter is being defined here, we must not beg the question and employ rational numbers as if they already existed. We simply keep in mind what we want to accomplish, and try to develop our definitions in the right way to achieve this. By the definition of ordered pair, in 7.2, $(2, 3) \neq (4, 6)$ and $(2, 3) \neq (1, 2)$. On the other hand, we do want $(2, 3)$ and $(4, 6)$ to be "equal" in this new number system that we are defining, but not $(2, 3)$ and $(1, 2)$. We define a relation on Q' as follows: $(x, y) \sim (u, v) \Leftrightarrow xv = yu$. Note that $xv = yu$ is a statement about integers only, and is therefore admissible. Observe that $(2, 3) \sim (4, 6)$ and $(2, 3) \nsim (1, 2)$. Now $\sim$ is an equivalence relation on Q'. (The reader should now prove this.)

Since $\sim$ is an equivalence relation, it partitions Q' into equivalence classes; for example, $[(u, v)] = \{(x, y) \mid (x, y) \sim (u, v)\} = \{(x, y) \mid xv = yu\}$. (We shall write $[u, v]$ instead of $[(u, v)]$ from now on.) The set Q is defined to be the set of all these equivalence classes; $Q = \{[u, v] \mid u \in I, v \in I - \{0\}\}$. For example, $[1, 4] \in Q$ and $[1, 4] = \{(1, 4), (-1, -4), (-2, -8), (3, 12), (-3, -12), \ldots\}$. It is the equivalence class $[1, 4]$ itself which is the rational number (usually written $1/4$) *by definition*. Note that $[1, 4]$ contains all the ordered pairs which are names for the same rational number; for example, $-11/-44$. In fact, $[-11, -44] = [1, 4]$, because $(-11, -44) \sim (1, 4)$; thus the equivalence class itself has different names in the same way. We now let the set Q of rational numbers be equivalence classes of ordered pairs of integers. It remains to define addition and multiplication in Q, as well as order, which we omit for the present. Addition and multiplication of rational numbers were considered in a less precise manner in 6.1. Temporarily, we use symbols $\oplus$, $\otimes$ for these operations in Q to distinguish them from their counterparts $+$, $\cdot$ in I. The operations $\oplus$ and $\otimes$ must be *defined* in terms of the operations $+$ and $\cdot$.

(1) $$[m, n] \oplus [p, q] = [m \cdot q + n \cdot p, n \cdot q].$$

(2) $$[m, n] \otimes [p, q] = [m \cdot p, n \cdot q].$$

Observe that these definitions are based on what we want to happen. (We cannot always be so cavalier in mathematics; what we invent must be consistent; this is another matter beyond the scope of this discussion.) For example, $[2, 3] \oplus [2, 5] = [2 \cdot 5 + 3 \cdot 2, 3 \cdot 5] = [16, 15]$ and $[-1, 3] \otimes$

$[3, 5] = [-3, 15] = [-1, 5]$ (since $(-3, 15) \sim (-1, 5)$). (Cf. 6.1.) There is one flaw in our process so far; (1) and (2) appear to be all right in the sense that they provide us with the behavior we expect of the rational numbers, but since we cannot rely on our knowledge of the intuitively understood rational numbers beyond heuristics, we must be more careful. Formula (1) defines $\oplus$ as an operation on equivalence classes, but the definition on the right side depends only on the particular members of these classes, viz., (m, n) and (p, q). Formula (1) would be nonsense unless whenever $(m', n') \sim (m, n)$ and $(p', q') \sim (p, q)$ we also had $(m'q' + n'p', n'q') \sim (mq + np, nq)$. (Remember $(m', n') \sim (m, n)$ means the same as $(m', n') \in [m, n]$.) In other words, the operation is not well defined unless it is provably independent of the members (or names) of the equivalence classes used in the definition. This is indeed the case for both $\oplus$ and $\otimes$, and is left to the reader to check (Exercise 10).

A number of jobs remain:

(i) We must show that Q provided with the operations $\oplus$ and $\otimes$ as an algebraic system forms a field. This is treated in Exercise 11.

(i) We must define the correct order relation $\prec$ on Q (in terms of $<$ on I) and show that the resulting system is an ordered field. (See Exercise 12.)

(iii) Finally, we must show that Q is an extension of I in the right sense. This is discussed below in Example 3 of 7.9. After all this has been done, we can drop the special symbols $[m, n]$, $\oplus$, $\otimes$, in favor of m/n, $+$, $\cdot$, and carry on with the set of rationals properly defined as an extension of I.

7.6   FUNCTIONS (continued)

In 7.1 we introduced the function concept as basic and defined the relation concept in terms of it. On the other hand, a function can be viewed as a relation with certain additional features. This, of course, would be circular if used as a definition, and hence unacceptable. We can avoid this by taking Definition 2 of 7.4 as the definition of relation. In this section, we shall consider the connection between relation and function, starting from Definition 2 of 7.4. The exact nature of this connection appears in Theorem 1 below.

Consider a function f that maps a set A into a set B. If we use the usual functional notation $y = f(x)$, recall that this means that $y \in B$ is the *unique* element associated with $x \in A$ under the correspondence f. The function under consideration is a certain mathematical object; "$y = f(x)$" is not the function. This latter is a predicate of two variables x, y (for fixed f). What predicate it is depends on what the function is; whether the resulting sentence $b = f(a)$ is true or false depends on what the particular elements a, b are. It is only a small step now to define a relation $\mathscr{R}$ on A, B as follows: $\mathscr{R}(x, y) \Leftrightarrow y = f(x)$.

The reader should not be confused by the appearance of "x" to the left of "y" on one side and vice versa on the other; this is only a convention. Now this relation $\mathscr{R}$ has a special property; for each $x \in A$ there is a unique $y \in B$ such that $\mathscr{R}(x, y)$. Note that it is possible to have a $w \in B$ such that $w \neq f(x)$ for all $x \in A$ (if f maps A properly into B); in this case $\mathscr{R}(x, w)$ is false for all $x \in A$ by the definition of $\mathscr{R}$.

Example 1. Let $A = \{x \in R \mid -2 \leqslant x \leqslant 2\}$. Let f be defined by $f(x) = -\sqrt{(4 - x^2)}$. Then f is a function mapping A (properly) into R. The range of f is $B = \{y \in R \mid -2 \leqslant y \leqslant 0\}$. If we define $\mathscr{R}$ by $\mathscr{R}(x, y) \Leftrightarrow y = f(x)$, then the relation $\mathscr{R}$ on A, B can also be described as $(x^2 + y^2 = 4 \wedge y \leqslant 0)$. Note that the graph of $\mathscr{R}$ is the set of points on the semicircle in Figure 7–13. Observe how this graph pictorially illustrates the fact that corresponding to each x there is a unique y.

Example 2. Let $\mathscr{S}$ be the relation on R defined by $\mathscr{S}(x, y) \Leftrightarrow x^2 + y^2 = 4$. The graph of $\mathscr{S}$ is illustrated in Figure 7–14. Observe that $\mathscr{S}$ differs from $\mathscr{R}$ of Example 1 in that y is not restricted by $y \leqslant 0$. In this case there is not a unique y corresponding to each x and we cannot say that "y is a function of x". (Nor can we say that "x is a function of y", either here or in Example 1, because corresponding to each y, there is not a unique x.)

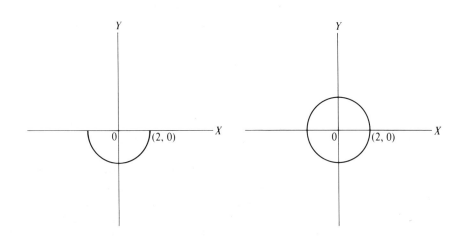

Figure 7–13   Figure 7–14

Now consider any relation $\mathscr{R}$ on given sets A, B. Suppose $\mathscr{R}$ has the property that for each $x \in A$ there is a unique $y \in B$ such that $\mathscr{R}(x, y)$. *Then we can define a function f mapping A into B by defining $f(x)$ equal to the unique y such that $\mathscr{R}(x, y)$.* Thus we have proved the following theorem.

Theorem 1. (alternative definition of function). *A function that maps a set A into a set B is a relation $\mathscr{R}$ on A, B with the property that for each $x \in A$ there is a unique $y \in B$ such that $\mathscr{R}(x, y)$.*

Example 1. (continued). If we begin with the relation $\mathscr{R}(x, y) \Leftrightarrow (x^2 + y^2 = 4 \land y \leqslant 0)$ where we view $\mathscr{R}$ as a relation on A, R, then $\mathscr{R}$ has the following property. For each $x \in A$ there is a unique $y \in R$ such that $\mathscr{R}(x, y)$, and hence $\mathscr{R}$ is a function mapping A into R. The rule for this function is $y = -\sqrt{(4 - x^2)}$. This can also be viewed as a function mapping A onto B, the range of the function.

Nothing much need be said now about the graph of a function; it is simply its graph as a relation, which has already been discussed. If f is a function mapping A into B, then the graph of f is

$$\{(x, y) \in A \times B \,|\, y = f(x)\}.$$

The graph is always a subset of $A \times B$ with the following properties:

(i) for every $x \in A$ there is a $y \in B$ such that (x, y) belongs to the graph; and

(ii) if (x, y_1) and (x, y_2) belong to the graph of the function, then $y_1 = y_2$ (that is, there are no elements of the graph with the same first member and different second members).

Conversely, of course, if we select a subset of $A \times B$ with these properties, then this set uniquely determines a function. Abstractly then, just as was the case for relations as discussed in 7.4, functions can be identified with their graphs. In analogy with Definition 2 of 7.4 we can state the following theorem.

Theorem 2. (alternative definition of function). *A function mapping a set A into a set B is a subset C of A × B with the properties:*

(i) *for each $x \in A$ there is a $y \in B$ such that $(x, y) \in C$, and*

(ii) *if $(x, y_1) \in C, (x, y_2) \in C$, then $y_1 = y_2$.*

The preceding "ordered pair" definition of function has considerable conceptual merit (although possibly at a pedagogical price). If the reader will refer to Example 1 and Figure 7–3 of 7.1, he will recall that when the domain is finite, the function can be "described" by a table. This table *is* the function, and *it is a set of ordered pairs* (viz., in this case, $\{(1, 1), (2, 1), (3, 2)\}$) with the appropriate uniqueness condition (ii) of Theorem 2. When we speak of the "rule" of a function, there is a natural tendency to think of this rule as being described, or describable, by some mathematical-linguistic expression. The fact is, that there are not enough mathematical-linguistic expressions to go around. It is impossible to establish a one-to-one correspondence (cf. 7.8 below) between these expressions and all functions from (say) N into N; there will always be functions left over. This is a technical matter beyond the scope of this book, but it is a fact. On the other hand, the definition given by Theorem 2 is the obvious natural generalization into the infinite of the finite table in 7.1.

7.7   SEQUENCES

There is a special type of function which is both very common and very useful in mathematics. This particular type of function, called an *infinite sequence* or, in this book, simply a **sequence**, is often treated as though it were not a function, but it is, and it is best understood if this is made very clear.

When we write

(1) $1, 1/2, 1/3, 1/4, \ldots, 1/n, \ldots,$

the reader understands quite well what is meant. This is a sequence whose first *member* or *term* is 1, whose second member is $1/2$, and generally whose nth member is $1/n$. The sequence does not terminate; after each term there is a next, as is the case with the natural numbers.

The members of a sequence need not be all different. The following are sequences.

(2) $1, 1/2, 1, -1/4, 1, 1/6, 1, -1/8, \ldots .$
(3) $1, 1/2, 1/2, 1/3, 1/3, 1/3, 1/4, 1/4, 1/4, 1/4, \ldots .$

The reader will have no difficulty in extending these sequences as far as he likes; the initial information given is sufficient. However, a rule for the general, or nth term is not as simple to find for (2) and (3) as it is for (1).

In talking about a sequence in general, it is common practice to use a letter with a subscript to represent a term of the sequence, thus

(4a) $a_1, a_2, a_3, \ldots, a_n, \ldots .$

This sequence is often abbreviated as $\{a_n\}$. Each a_n is some number (integral, rational, real, complex). Occasionally the numbering of the terms starts at 0, thus

(4b) $b_0, b_1, b_2, \ldots, b_n, \ldots .$

Again this sequence is abbreviated as $\{b_n\}$, but it might be necessary to specify that $n \geqslant 0$ (rather than $n > 0$).

Definition. *A **sequence** is a function whose domain of definition is N or N − {0} and whose range is a set of numbers.*

The fact that we did not use the usual functional notation above is immaterial; we could, and will often, write sequences as, for example,

(4c) $$f(0), f(1), f(2), ..., f(n), ...$$

or as $\{f(n)\}$ where f is the correspondence which assigns to each natural number some unique number (in I, Q, R, etc. as the case may be).

Sequences (1), (2), (3), are sequences of rationals; that is, functions from $N - \{0\}$ into Q. If we define $p(n) = 1/n$, then the sequence (1) can be described by the function p, whose domain is $N - \{0\}$ and whose counter domain is Q. It can also be written as $\{p(n)\}$, $n > 0$, or as $\{1/n\}$, $n > 0$. When we write $\{1/n\}$, $n > 0$ we are, in fact, thinking of a sequence as a set, but the set in question is actually the range of the function.

Define q by

$$q(n) = \begin{cases} (-1)^{n/2} \cdot (-1/n) & \text{if } 2|n, \\ 1 & \text{otherwise,} \end{cases}$$

where $n \in (N - \{0\})$. Then the sequence (2) is the sequence $\{q(n)\}$.

When one thinks of a sequence as a set, one often says for example, that "1 is a member of the sequence," or "2 is not a term of the sequence," or "$-1/4$ belongs to the sequence." One often writes: $1 \in \{q(n)\}$, $2 \notin \{q(n)\}$, $-1/4 \in \{q(n)\}$. There is nothing wrong with this, but it does not tell the whole story. It is true that $1 \in \{q(n)\}$, but more than that, $1 = q(1) = q(3) = q(5) = \dots$. Remember, the set in question is the range of the function; and the sequence, properly considered, is the function itself.

New sequences can be defined from old ones. Let $r(n) = q(2n - 1)$. Then $\{r(n)\}$, $n > 0$ is the sequence

(5) $$1, 1, 1, ..., 1, ...$$

This too is a sequence, and it differs greatly from the set $\{1\}$.

In summary then, a sequence is a set of numbers, not necessarily all different, indexed by N (or by $N - \{0\}$); as such it is a function from N (or from $N - \{0\}$) into some set of numbers.

7.8 COMPOSITE FUNCTIONS AND INVERSE FUNCTIONS

Consider a function f that maps a set A into a set B. To know what function f is, it is necessary to know, for every x in A, which element of B is the specific $f(x)$. Expressed differently, a particular function is defined when we state, by some rule, list (if possible), or other means, what the image is for every x in the domain of the function. On the other hand, the function itself must be thought of as a single mathematical object, quite apart from its functional values or images. When we speak of a function f mapping A into B, f is actually a name for the rule of correspondence and it is only elliptically that we call f the function. Allowing ourselves to do this for the moment, there are some further distinctions which are important. The function f, as we said, is a mathematical object in its own right. On the other hand, "$f(x)$" is not the function, but the name of some number. The variable "x" is a term standing for an arbitrary member of the domain A, and "$f(x)$" is a term standing for the image of x in B. The term "$f(x)$" is sometimes called the **ambiguous value of the function at** x. The value $f(x)$ is ambiguous to the extent that x is, since x stands for an arbitrary, not a specific, object in the domain. If $\sqrt{2}$ is in the domain, then "$f(\sqrt{2})$" names a specific, though possibly unknown, number, not an ambiguous one. Incidentally, "$\sqrt{2}$" itself is a term naming a specific real number. Many mathematics books use the term "$f(x)$" as the name of the function as well as using it correctly as the name of a number. When terms are combined by relation symbols, one obtains a sentence, possibly open; for example, "$y = f(x)$" is an open sentence (predicate).

Consider now the set **A** of all functions that map a set A into itself. Remember that the elements of **A** are functions (from A into A). There is an important operation on **A** which we shall designate by the symbol o. Since o is an operation on **A**, for any $f, g \in \mathbf{A}$, $f \circ g \in A$. To define o, we have to define the *function* $f \circ g$ for all $f, g \in \mathbf{A}$. To define the function $f \circ g$ we have to define the value $(f \circ g)(x)$ for all $x \in A$. This, of course, assumes that we already know f and g; that is, we already know the values $f(x)$, $g(x)$ for all $x \in A$.

Definition 1. *Let g be a function mapping A into B, and f be a function mapping B into C. Then $f \circ g$, called the* **composite function** *of f, g, is the function mapping A into C defined by*

$$(f \circ g)(x) = f[g(x)].$$

Note that $x \in A \Rightarrow g(x) \in B$, and $g(x) \in B \Rightarrow f[g(x)] \in C$, or $(f \circ g)(x) \in C$.

Diagrammatically, $f \circ g$ can be visualized as in Figure 7–15 (cf. Figure 7–1 of 7.1).

Of course, if f and g are functions that map a set A into itself, so that, in the definition, $B = A$, then C is also identical with A, and the composite function $f \circ g$ also maps A into itself.

Example 1. Let $A = \{1, 2, 3, 4\}$. Let f be defined by $f(1) = f(2) = 3$, $f(3) = f(4) = 1$. Let g be defined by $g(1) = 2$, $g(2) = 3$, $g(3) = 4$, $g(4) = 1$. Then $f \circ g$ is the function mapping A into itself given by $(f \circ g)(1) = f[g(1)] = f(2) = 3$, $(f \circ g)(2) = f[g(2)] = f(3) = 1$, etc. The function

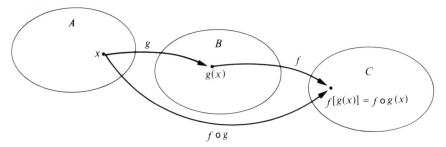

Figure 7–15

$g \circ f$ is given by $(g \circ f)(1) = g[f(1)] = g(3) = 4$. Note that $(g \circ f)(1) \neq (f \circ g)(1)$ and hence $g \circ f \neq f \circ g$ (that is, $g \circ f$ and $f \circ g$ are *different* functions). This example proves that o is in general non-commutative.

In the definition of course, there is no special relationship among A, B, C. They may be disjoint or not; they may be equal or not; etc.

Theorem 1. (a) $f \circ g \neq g \circ f$ *in general.*

(b) $f \circ (g \circ h) = (f \circ g) \circ h$ *whenever these expressions are defined.*

Proof. (a) See Example 1.

(b) Exercise 13.

Example 2. Let A be a set of people in some sample (for example, all the students in a particular classroom). Let g be a function on A which assigns to each $x \in A$ his (or her) height in inches to the nearest inch. Then g maps A into I. Let f map I into I as follows: $f(n) = 0$ if $n \leqslant 60$, $f(n) = 1$ if $n > 60$. Then $f \circ g$ maps A into I in such a way that $(f \circ g)(x) = 0$ if and only if x is not over 5 feet tall to the nearest inch and $(f \circ g)(x) = 1$ if and only if x is over 5 feet tall. Note that $g \circ f$ has no meaning (is not defined), because $f(x)$ for $x \in I$ is 0 or 1, but g is a function defined on a set of people, and $g(0)$ or $g(1)$ has no meaning.

A simple but special function mapping a set A into itself is the function i, called the **identity function**, where $i(x) = x$ for all $x \in A$. Of course, the function is determined by the rule i and the set A; there is a different identity function for each set A. Sometimes we write i_A to signify the identity function on A.

Theorem 2. *Let f be a function mapping A into B. Then $f \circ i = f$ where i is the identity function on A, and $i \circ f = f$ where i is the identity function on B.*

Proof. $(f \circ i)(x) = f[i(x)] = f(x)$ for all $x \in A$. So $f \circ i = f$. Similarly, $i \circ f = f$.

The identity functions behave with respect to o in much the same way as the real number 1 (the multiplicative identity) behaves with respect to multiplication. This partial analogy can be carried a bit further.

Recall that when a function f maps A into B, then for each $x \in A$ there is a unique $y \in B$ such that $y = f(x)$, but there is nothing in the definition to prohibit two different elements $x_1, x_2 \in A$ from having the same image. That is, $f(x_1) = f(x_2)$ is possible although $x_1 \neq x_2$. (Check through the Examples for instances of this.) If $y = f(x)$, "x" is sometimes called the **pre-image** of y (under f). So the definition of a function carries the stipulation that images are unique, but pre-images need not be. The class of functions with the property that pre-images *are* unique is an especially important one. These functions are called **one-to-one** or **one-one**.

Definition 2.   *If f is a function mapping A into B such that* $f(x_1) = f(x_2) \Rightarrow$ $x_1 = x_2$ *for all* $x_1, x_2 \in A$, *then f is called a* **one-one** *function.*

The function of Example 1 of 7.1 is not one-one. But the identity function on $D = \{1, 2, 3\}$ is one-one. See Figure 7–16 and compare it with Figure 7–2 of 7.1.

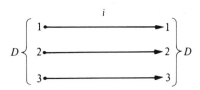

Figure 7–16

Example 3.   Let $f(x) = x^2$. Then f as a function mapping N (properly) into itself is one-one, whereas f as a function mapping I into N is not one-one $((-2)^2 = 2^2 = 4)$.

When f maps A into B, the rule of correspondence can be viewed in reverse, so to speak. We can look at any $y \in B$ and see what, if any, $x \in A$ it corresponds to under f; that is, given $y \in B$, find every $x \in A$ if any, such that $y = f(x)$. If f is not onto B, there will be elements $y \in B$ which correspond to no $x \in A$, that is $y = f(x)$ for no $x \in A$. So let us restrict our attention to the case: f maps A onto B. Now if f is not one-one, there will be elements $y \in B$ which correspond to more than one $x \in A$. In this case, the reverse correspondence from B back *onto* A is not a function from B onto A. Then let us restrict our attention to the case: f maps A onto B and is one-one. In this case, and only in this case, we have: for *each* $y \in B$ there is a *unique* $x \in A$ such that $y = f(x)$. Now this is exactly the definition of a function where x is the image and y is the pre-image.

We have shown: *In the case that f maps A one-one onto B, and only in that case, the reverse correspondence from B onto A is also a function.*

It should be clear that this function which maps B onto A is also one-one. This function is called the **inverse** of f, and is written f^{-1}.

Definition 3.   *If f is a function mapping A one-one onto B, then the* **inverse** *of f, written f^{-1}, is the function that maps B one-one onto A so that* $f^{-1}(y) = x \Leftrightarrow y = f(x)$.

If f is not one-one and onto, then it has no inverse. Clearly, $i^{-1} = i$; see Figure 7–16 for the finite case.

Example 4.   Let $N_m = \{1, 2, 3, \ldots, m\}$. There are $m! = m(m - 1)$ $(m - 2) \cdots 3 \cdot 2 \cdot 1$ different functions one-one (including i) from N_m onto N_m. Such functions are called permutations of N_m. The reader should prove this result for himself, and write out all $m!$ permutations of N_m for $m = 1, 2, 3, 4$. On the other hand there are m^m different functions from N_m into N_m.

Example 5.   Let L be the set of straight lines in the plane as discussed in Example 3 of 7.5 and let **L** be the set of equivalence classes defined there. For any $l \in L$, let

$$\phi(l) = \text{(the angle } l \text{ makes with the } x\text{-axis)}$$

(where the angle in question is measured in degrees in the usual counter-clockwise direction with $\phi(l) = 0$ if l is parallel to the x-axis). So $0° \leqslant \phi(l) < 180°$ for all $l \in L$. In this way, ϕ is a function from L onto the set $B = \{\text{angles } \alpha \,|\, 0° \leqslant \alpha < 180°\}$. Clearly ϕ is not one-one; we can have $l_1 \neq l_2$ but $l_1 \| l_2$ and thence $\phi(l_1) = \phi(l_2)$. On the other hand, ϕ can be defined on **L** as follows: $\phi([l]) = \phi(l)$. Note that ϕ is well defined on **L** because $\phi(l_1) = \phi(l)$ for every $l_1 \in [l]$. Furthermore, ϕ *is* one-one from **L** onto B. This is just another way of saying what we said in Example 3 of 7.5 about each element of **L** being characterized by its direction.

The following theorem should be obvious.

Theorem 3.   *If f maps A one-one onto B, then $f \circ f^{-1} = i_B$ and $f^{-1} \circ f = i_A$.*

If f maps A one-one onto B, we sometimes say that f **establishes a one-one correspondence** between A and B. If there is such an f, we say A and B are in one-one correspondence; A and B are said to be **equivalent** or **equipollent**. Equivalence of sets generalizes to infinite sets the notion of "same number of elements".

Example 6.   (a) The sets $A = \{1, 2, 3, 4\}$ and $B = \{\frac{1}{2}, \frac{1}{4}, \frac{1}{8}, \frac{1}{16}\}$ are equipollent. There exist 4! one-one functions from A onto B. One such is $f(1) = \frac{1}{2}, f(2) = \frac{1}{4}, f(3) = \frac{1}{8}, f(4) = \frac{1}{16}$. The reader should construct some others.

(b) On the other hand, there is *no* one-one correspondence from A onto $\{\frac{1}{2}, \frac{1}{4}, \frac{1}{8}\}$, or onto $\{1, 2, 3\}$.

Example 7.   The function $h(x) = 2x$ is a one-one function from I onto E ($=$ the even integers); $h^{-1}(x) = x/2$. Yet E is a *proper* subset of I. So I can be put into one-one correspondence with a proper subset of itself. This is a characterizing property of infinite sets.

7.9   ISOMORPHISMS

Some sets have certain algebraic structures imposed on them. We shall use the word "structure" to denote a system consisting of a set together with certain associated operations and relations, and properties of these operations and relations. For example, the set Q of rational numbers possesses two operations $+, \cdot$ and an order relation $<$ and, as we have seen, Q is an ordered field with respect to these. Now, it is possible that there is another set Q^* with operations $+^*$ and * and an order relation $<^*$, which behaves exactly like $Q, +, \cdot, <$. We should like to make precise what we mean by saying that $Q^*, +^*, {}^*, <^*$ and $Q, +, \cdot, <$ behave exactly alike. Intuitively, we wish to be able to say that any true statement about $Q^*, +^*, {}^*, <^*$ is also true of $Q, +, \cdot, <$ and vice versa. This would be possible if we possessed a mechanism such that any true statement about $Q, +, \cdot, <$ could be "translated" into the *corresponding* true statement about $Q^*, +^*, {}^*, <^*$ and vice versa. Further, this would be possible if there were a one-to-one correspondence ϕ mapping Q onto Q^* such that

(1) $$\phi(q_1 + q_2) = \phi(q_1) +^* \phi(q_2)$$
(2) $$\phi(q_1 \cdot q_2) = \phi(q_1) * \phi(q_2)$$
(3) $$q_1 < q_2 \Leftrightarrow \phi(q_1) <^* \phi(q_2).$$

We often sum up the results of (1), (2), (3) by saying ϕ preserves the operations and the order relation. Suppose an equation

(4) $$q_1 + q_2 = q_3$$

is valid in Q; then by (1)

$$\phi(q_3) = \phi(q_1 + q_2) = \phi(q_1) +^* \phi(q_2)$$

or

(5) $$\phi(q_1) +^* \phi(q_2) = \phi(q_3).$$

Therefore the equation (5) corresponding to (4) under ϕ is valid in Q^*. Indeed, note that $\phi(q_1)$, $\phi(q_2)$, $\phi(q_3)$ are the elements in Q^* corresponding to q_1, q_2, q_3 in Q; $+^*$ is the operation corresponding to $+$; and (5), the equation corresponding to (4), is valid in Q^* if and only if (4) is valid in Q. The situation is analogous for $\cdot$ and $<$. Thus, using ϕ, we can translate true statements about Q, $+$, $\cdot$, $<$ into true statements about Q^*, $+^*$, *, $<^*$. Since it is easy to show that for $r_1, r_2, r_3 \in Q^*$

(1') $$\phi^{-1}(r_1 +^* r_2) = \phi^{-1}(r_1) + \phi^{-1}(r_2),$$

(2') $$\phi^{-1}(r_1 {}^* r_2) = \phi^{-1}(r_1) \cdot \phi^{-1}(r_2),$$

(3') $$r_1 <^* r_2 \Leftrightarrow \phi^{-1}(r_1) < \phi^{-1}(r_2),$$

we can show that true statements about Q^*, $+^*$, *, $<^*$ can be translated into true statements about Q, $+$, $\cdot$, $<$ using ϕ^{-1}. Thus any statement $\mathscr{P}$ about Q, $+$, $\cdot$, $<$ involving equality, elements q_1, q_2, q_3, ... of Q, and $+$, $\cdot$, $<$ can be translated into a statement $\mathscr{P}^*$ involving the corresponding elements $\phi(q_1)$, $\phi(q_2)$, ... of Q^*, and $+^*$, *, $<^*$, and in addition $\mathscr{P}$ is true if and only if $\mathscr{P}^*$ is true. We say that Q, $+$, $\cdot$, $<$ and Q^*, $+^*$, *, $<^*$ are isomorphic algebraic structures.

The following definition will render this concept more precise. For simplicity's sake let us consider algebraic systems with but one operation and one relation. The extension of the definition to cover systems with more than one operation and/or more than one relation should be obvious.

Definition. *Let A be a set with operation $*_A$ and relation $\mathscr{R}_A$ defined on it. Let B be a set with an operation $*_B$ and relation $\mathscr{R}_B$. A, $*_A$, $\mathscr{R}_A$ is* **isomorphic** *to B, $*_B$, $\mathscr{R}_B$ if there is a one-to-one correspondence ϕ mapping A onto B such that for $a_1, a_2 \in A$*

(6) $$\phi(a_1 *_A a_2) = \phi(a_1) *_B \phi(a_2),$$

(7) $$\mathscr{R}_A(a_1, a_2) \Leftrightarrow \mathscr{R}_B[\phi(a_1), \phi(a_2)].$$

Such a ϕ is called an **isomorphism** of A, $*_A$, $\mathscr{R}_A$ onto B, $*_B$, $\mathscr{R}_B$. The reader should verify that for such a ϕ, the mapping ϕ^{-1} is an isomorphism of B, $*_B$, $\mathscr{R}_B$ onto A, $*_A$, $\mathscr{R}_A$.

In case the system contains no relation, then (7) is deleted from this definition, and, on the other hand, if the system has no operations (6) is deleted.

Thus two groups $G, \cdot$ and $H, *$ are isomorphic if there is a one-to-one correspondence ϕ mapping G onto H such that for $g_1, g_2 \in G$,

$$\phi(g_1 \cdot g_2) = \phi(g_1) * \phi(g_2),$$

and two partially ordered systems $S, <$ and $T, \prec$ are isomorphic provided there is a one-to-one correspondence ϕ mapping S onto T such that for $s_1, s_2 \in S$,

$$s_1 < s_2 \Leftrightarrow \phi(s_1) \prec \phi(s_2).$$

Example 1. As an example of an isomorphism of groups, consider the following: The set R together with the operation $+$ forms a group which is commutative, that is, $a, b \in R \Rightarrow a + b = b + a$. Furthermore, R^+ and $\cdot$ also form a commutative group. These two groups are isomorphic in the sense that $\phi(r) = 10^r$ is a one-to-one correspondence from R onto R^+ such that

$$\phi(r_1 + r_2) = 10^{r_1 + r_2} = 10^{r_1} \cdot 10^{r_2}$$
$$= \phi(r_1) \cdot \phi(r_2).$$

The inverse $\phi^{-1}(r) = \log_{10} r$ of ϕ is readily available in any table of base 10 logarithms. Through use of this isomorphism ϕ, its inverse, ϕ^{-1}, and log tables, multiplications and divisions can be reduced to additions and subtractions.

Example 2. A less useful isomorphism exists between the ordered sets N with the natural order $<$, and $E = \{n \in N \mid 2|n\}$, (the even numbers) provided with the order relation $<$ which is already defined on $N \supset E$. This isomorphism ϕ is of the form $\phi(n) = 2n$ for $n \in N$. It clearly carries N onto E, and since

$$m < n \Leftrightarrow 2m < 2n$$

for all $m \in N$ and $n \in N$,

$$m < n \Leftrightarrow \phi(m) < \phi(n)$$

for all $m \in N$ and $n \in N$. Therefore we have established that ϕ is an isomorphism of the ordered set $N, <$ onto the ordered set $E, <$.

Example 3. In this example we clear up a point discussed in Example 8 of 7.5. It was mentioned without proof that Q, as defined there, was an extension of I. It is natural to ask how this could be, since in no direct sense is I a subset of Q (as defined therein) because Q is actually a set of equivalence classes of ordered pairs of integers. Q is an extension of I in the following sense: there is a subset J of Q which, when provided with the operations $+, \cdot$ and relation $<$ defined in J by virtue of its role as a subset of Q, is isomorphic to $I, +, \cdot, <$.

The J we have in mind is $J = \{[m, n] \in Q \,|\, n = 1\} = \{[1, 1], [-1, 1], \ldots\}$. Remember that because we are using the operations and relation of Q on J we have for $[m, 1] \in J$ and $[n, 1] \in J$

(8)
$$[m, 1] + [n, 1] = [m \cdot 1 + n \cdot 1, 1 \cdot 1]$$
$$= [m + n, 1],$$

(9)
$$[m, 1] \cdot [n, 1] = [m \cdot n, 1 \cdot 1] = [m \cdot n, 1], \qquad \text{and}$$

(10)
$$[m, 1] < [n, 1] \Leftrightarrow m \cdot 1 < n \cdot 1$$
$$\Leftrightarrow m < n.$$

Now, let the mapping ϕ be defined on J into I as follows: $\phi[m, 1] = m$. The definition of ϕ is not ambiguous because $[m, 1] = [n, 1]$ can occur only if $m \cdot 1 = n \cdot 1$ or $m = n$. It is easy to verify that ϕ is a one-to-one correspondence from J onto I. Now we must test the interaction of ϕ with operations $+$, $\cdot$, and relation $<$.

By (8),

(11)
$$\phi([m, 1] + [n, 1]) = \phi([m + n, 1])$$
$$= m + n$$
$$= \phi[m, 1] + \phi[n, 1]$$

for any $[m, 1]$ and $[n, 1]$ in J;
By (9),
$$\phi([m, 1] \cdot [n, 1]) = \phi[m \cdot n, 1] = m \cdot n$$
$$= \phi[m, 1] \cdot \phi[n, 1]$$

for any $[m, 1]$, $[n, 1]$ in J; and finally by (10), $\phi[m, 1] < \phi[n, 1]$ in I if and only if $[m, 1] < [n, 1]$ in J. Thus according to the definition the algebraic systems

$$J, +, \cdot, < \quad \text{and} \quad I, +, \cdot, <$$

are isomorphic and are identical in behavior—only the names are changed. This is the sense in which Q is an extension of I.

7.10   ISOMORPHISMS AND ORDERED FIELDS

In this section we show that every ordered field F is an extension of the rational field in the sense that there is a subset Q^* of F such that if $+$, $\cdot$ are the operations and $<$ the order relation of F, then Q^* provided with these is isomorphic to Q.

Let F be an ordered field. In general, we say a subset $F_1 \subseteq F$ is an **ordered subfield** of F if F_1 itself forms an ordered field with respect to the operations and order relation as they are defined in F. Thus, in terms of this definition we are setting out to show that every ordered field F contains an ordered sub-field Q^* which is isomorphic to the ordered field of rational numbers. When there is no possibility of confusion we say Q^* is isomorphic to Q instead of saying the subfield $Q^*, +, \cdot, <$ of $F, +, \cdot, <$ is isomorphic to $Q, +, \cdot, <$; that is we omit the lists of respective operations and relations.

To accomplish our goal we must deepen our knowledge of ordered fields. First, by definition we have $0 < 1$ where 0 is the additive identity and 1 the multiplicative identity of F and so adding 1 to both sides

$$0 < 0 + 1 < 1 + 1.$$

By induction we show that

$$0 < \underbrace{1 + \cdots + 1}_{n \text{ terms}}$$

for any $n \in N - \{0\}$. For the sake of brevity we form the following definition.

Definition. *For any $a \in F$, let*

(1) $$na = \underbrace{a + a + \cdots + a}_{n \text{ terms}}$$

and

(2) $$0 = 0 \cdot a$$

where n, and the 0 on the left side of (2), are natural numbers.

Note that $1 \cdot a = a$, $n \cdot 0 = \underbrace{0 + \cdots + 0}_{n \text{ terms}} = 0$, and $n \cdot 1 = \underbrace{1 + \cdots + 1}_{n \text{ terms}}$.

Thus we have $0 < n \cdot 1$ for $n \in N - \{0\}$.

Lemma 1. *If F is an ordered field, then for every $m,n \in N$ such that $m \neq n$, $m \cdot 1 \neq n \cdot 1$.*

Proof. $m \cdot 1 = n \cdot 1$ means that

(3) $$\underbrace{1 + \cdots + 1}_{m \text{ terms}} = \underbrace{1 + \cdots + 1}_{n \text{ terms}}.$$

Now suppose $n > m$. Then since the cancellation law for addition is valid in F, we can cancel m terms from both sides of (3) to leave

$$0 = \underbrace{1 + \cdots + 1}_{n - m \text{ terms}} = (n - m) \cdot 1.$$

However, $n - m \in N$ and $n - m \neq 0$, so

$$1 + \cdots + 1 = (n - m) \cdot 1 > 0.$$

Hence $0 < 0$, a contradiction. Therefore $n \not> m$. Similarly we can show $m \not> n$. Thus $m = n$.

Theorem 1. *If $N^* = \{a \in F \,|\, a = n \cdot 1 \text{ for } n \in N\}$, then N^*, provided with the operations and order relations imposed upon it as a subset of F, is isomorphic to N, the set of natural numbers.*

   Proof.   Let us define the mapping ϕ_0 by setting

(4) $$\phi_0(n \cdot 1) = n.$$

Now ϕ_0 is a function because $n \cdot 1 = m \cdot 1 \Rightarrow n = m$ by Lemma 1. It is much easier to see that ϕ_0 is one-to-one because $\phi_0(a) = \phi_0(b) = n \Rightarrow a = n \cdot 1$ and $b = n \cdot 1$ and so no two different elements of N^* are carried into the same element of N. The reader should verify that ϕ_0 maps N^* onto N.

   Now all that remains is to show that operations and the order relation are preserved by ϕ_0, that is, that equations analogous to (1), (2), and (3) of 7.9 are valid.

   To do this we first note that

(5)
$$
\begin{cases}
m \cdot 1 + n \cdot 1 = \underbrace{1 + \cdots + 1}_{m \text{ terms}} + \underbrace{1 + \cdots + 1}_{n \text{ terms}} \\[2mm]
\qquad = \underbrace{1 + \cdots + 1}_{m + n \text{ terms}} \\[2mm]
\qquad = (m + n)1,
\end{cases}
$$

and

(6)
$$
\begin{cases}
(m \cdot 1) \cdot (n \cdot 1) = \underbrace{(1 + \cdots + 1)}_{m \text{ terms}} \underbrace{(1 + \cdots + 1)}_{n \text{ terms}} \\[2mm]
\qquad = \overbrace{\underbrace{(1 + \cdots + 1)}_{m \text{ terms}} + \underbrace{(1 + \cdots + 1)}_{m \text{ terms}} + \cdots + \underbrace{(1 + \cdots + 1)}_{m \text{ terms}}}^{n \text{ terms}} \\[2mm]
\qquad = \underbrace{1 + \cdots + 1}_{m \cdot n \text{ terms}} = (m \cdot n) \cdot 1.
\end{cases}
$$

In addition we note that

$$n \cdot 1 < m \cdot 1 \Leftrightarrow \underbrace{1 + \cdots + 1}_{n \text{ terms}} < \underbrace{1 + \cdots + 1}_{m \text{ terms}}.$$

Since in an ordered field $a < b$ if and only if $a + c < b + c$ for $a, b, c \in F$, there are fewer terms on the left in the expression

$$\underbrace{1 + \cdots + 1}_{n \text{ terms}} < \underbrace{1 + \cdots + 1}_{m \text{ terms}}$$

than on the right, which is to say $n < m$. Thus

(7) $$n \cdot 1 < m \cdot 1 \Leftrightarrow n < m$$

is valid. Now by equations (4)–(7)

(8) $$\begin{cases} \phi_0(n \cdot 1 + m \cdot 1) = \phi_0[(n + m) \cdot 1] \\ \qquad = n + m \\ \qquad = \phi_0(n \cdot 1) + \phi_0(m \cdot 1), \end{cases}$$

(9) $$\begin{cases} \phi_0[(n \cdot 1)(m \cdot 1)] = \phi_0[(n \cdot m) \cdot 1] \\ \qquad = n \cdot m \\ \qquad = \phi_0(n \cdot 1) \cdot \phi_0(m \cdot 1) \end{cases}$$

and

(10) $$n \cdot 1 < m \cdot 1 \Leftrightarrow n < m \Leftrightarrow \phi_0(n \cdot 1) < \phi_0(m \cdot 1).$$

Therefore ϕ_0 carries N^* onto N in a one-to-one fashion and preserves the operations and order relation. Therefore N^* and N are isomorphic.

Now define $(-n) \cdot 1 = a$ for that $a \in F$ with the property $a + n \cdot 1 = 0$, that is, $(-n) \cdot 1 = -(n \cdot 1)$.

Lemma 2. $n(-1) = (-n) \cdot 1$ for $n \in N$.

Proof. Since $n(-1) = (-1) + (-1) + \cdots + (-1)$ (n terms),
$$n(-1) + n \cdot 1 = [(-1) + \cdots + (-1)] + (1 + \cdots + 1),$$
and by commutativity and associativity
$$n(-1) + n \cdot 1 = [(-1) + 1] + \cdots + [(-1) + 1]$$
$$= 0 + \cdots + 0 = 0.$$
Therefore
$$n(-1) + n \cdot 1 = 0 = (-n) \cdot 1 + n \cdot 1$$

and so by cancellation (or the uniqueness of the additive inverse) $(-n) \cdot 1 = n(-1)$.

Now let us define the subset

$$I^* = \{a \in F \mid a = n \cdot 1 \text{ for some } n \in I\}$$

of F and show it is isomorphic to the integers.

Theorem 2. *The set I^* given the operations and order relation defined in F is isomorphic to the integers I, $+$, $\cdot$, $<$.*

Proof. We only sketch the proof. First we can use Lemma 2 to show that

(11) $$(m + n) \cdot 1 = m \cdot 1 + n \cdot 1,$$

(12) $$(m \cdot n) \cdot 1 = (m \cdot 1)(n \cdot 1),$$

(13) $$n \cdot 1 < m \cdot 1 \Leftrightarrow n < m$$

for any pair of integers m, n. Then we can show that for any integers m, n, $m \cdot 1 = n \cdot 1 \Rightarrow m = n$. From this it follows that the mapping $\phi_1(m \cdot 1) = m$ is a one-to-one mapping of I^* onto I by an argument analogous to that used in Theorem 1. In addition, just as Equations (3), (9), (10) follow directly from (5), (6), (7) respectively, the equations

(14) $$\phi_1(m \cdot 1 + n \cdot 1) = \phi_1(m \cdot 1) + \phi_1(n \cdot 1),$$

(15) $$\phi_1[(m \cdot 1)(n \cdot 1)] = \phi_1(m \cdot 1) \cdot \phi_1(n \cdot 1),$$

(16) $$n \cdot 1 < m \cdot 1 \Leftrightarrow \phi_1(n \cdot 1) < \phi_1(m \cdot 1)$$

follow directly from (11), (12), (13), and therefore the isomorphism is established.

Now we are ready to define Q^*, the subset of F isomorphic to Q. First define

$$\frac{1}{n} \cdot 1 = (n \cdot 1)^{-1} \text{ and } \left(-\frac{1}{n}\right) \cdot 1 = [(-n) \cdot 1]^{-1}$$

for $n \in N - \{0\}$ where a^{-1} stands for the multiplicative inverse of $a \in F$.

Lemma 3. (a) $\left(-\dfrac{1}{n}\right) \cdot 1 = [n \cdot (-1)]^{-1}$

(b) $\left(-\dfrac{1}{n}\right) \cdot 1 = -\left(\dfrac{1}{n} \cdot 1\right)$

for $n \in N - \{0\}$.

Proof. (a) By Lemma 2, $(-n) \cdot 1 = n(-1)$ and since the multiplicative inverse is unique, $[(-n) \cdot 1]^{-1} = [n \cdot (-1)]^{-1}$ for $n \in N - \{0\}$. Since the left side is

$$\left(-\frac{1}{n}\right) \cdot 1, \text{ part (a) is proved.}$$

(b) $\left(-\dfrac{1}{n}\right) \cdot 1 = [n(-1)]^{-1}$ by part (a).

Now $n(-1) = (-1) + \cdots + (-1) \quad (n \text{ terms})$
$$= -1(1 + \cdots + 1)$$
$$= -1(n \cdot 1)$$

Now it is easy to show that for any field $F, a, b \in F \Rightarrow (ab)^{-1} = a^{-1} b^{-1}$. Therefore in our case $[n(-1)]^{-1} = (-1)^{-1} (n \cdot 1)^{-1}$. Since $(-1)(-1) = 1$,

$$(-1)^{-1} = -1.$$

Thus
$$[n(-1)]^{-1} = (-1)(n \cdot 1)^{-1} = (-1)\left(\frac{1}{n} \cdot 1\right) = -\left(\frac{1}{n} \cdot 1\right).$$

Now let
$$\left(\frac{m}{n} \cdot 1\right) = m\left(\frac{1}{n} \cdot 1\right)$$

for $m \in I$, and then define

$$Q^* = \left\{\frac{m}{n} \cdot 1 \,\middle|\, m \in I \text{ and } n \in N - \{0\}\right\}$$

Theorem 3. *The subset Q^* is an ordered subfield of F.*

The proof of this theorem, which involves verifying that each of the axioms of an ordered field is satisfied by Q^*, is left to the reader. With the aid of the lemmas it is not a difficult task to accomplish.

Theorem 4. *Q^* is isomorphic to Q.*

Proof. Let $\phi\left(\dfrac{m}{n} \cdot 1\right) = \dfrac{m}{n}$. Suppose $\dfrac{m}{n} \cdot 1 = \dfrac{m'}{n'} \cdot 1$.

Then $m\left(\dfrac{1}{n} \cdot 1\right) = m'\left(\dfrac{1}{n'} \cdot 1\right)$ and so

(17) $$m(n \cdot 1)^{-1} = m'(n' \cdot 1)^{-1}.$$

If we multiply both sides of (17) by $(n \cdot 1)(n' \cdot 1)$ we have $m(n' \cdot 1) = m'(n \cdot 1)$.

(18)   Hence, $m \cdot n' \cdot 1 = m' \cdot n \cdot 1$.

Since ϕ_1 is an isomorphism,

$$m \cdot n' = \phi_1(m \cdot n' \cdot 1) = \phi_1(m' \cdot n \cdot 1) = m' \cdot n$$

and so $m \cdot n' = m' \cdot n$ which is just the property which ensures $m/n = m'/n'$.

Thus every element of Q^* has exactly one representative in the form

$$\frac{m}{n} \cdot 1.$$

Therefore ϕ is a one-to-one mapping from Q^* into Q. It is clear that ϕ is onto Q as well, and so ϕ is a one-to-one correspondence from Q^* onto Q.

We will show that ϕ preserves "$+$" and will leave to the reader the proof that ϕ preserves "$\cdot$" and "$<$".

First note that

$$\frac{m}{n} \cdot 1 + \frac{m'}{n'} \cdot 1 = m(n \cdot 1)^{-1} + m'(n' \cdot 1)^{-1}$$

$$= m(n' \cdot 1)(n' \cdot 1)^{-1}(n \cdot 1)^{-1} + m'(n \cdot 1)(n \cdot 1)^{-1}(n' \cdot 1)^{-1}$$
$$= [(mn' \cdot 1) + (m'n \cdot 1)](n \cdot 1)^{-1} \cdot (n' \cdot 1)^{-1}$$
$$= [(mn' + m'n) \cdot 1][(n \cdot 1)(n' \cdot 1)]^{-1}$$
$$= [(mn' + m'n) \cdot 1](nn' \cdot 1)^{-1}$$
$$= \underbrace{(1 + \cdots + 1)}_{mn' + m'n \text{ terms}}(nn' \cdot 1)^{-1}$$

$$= \underbrace{(nn' \cdot 1)^{-1} + \cdots + (nn' \cdot 1)^{-1}}_{mn' + m'n \text{ terms}} \quad \text{by the distributive law.}$$

Therefore

$$\frac{m}{n} \cdot 1 + \frac{m'}{n'} \cdot 1 = (mn' + m'n)(nn' \cdot 1)^{-1}$$

$$= \left(\frac{mn' + m'n}{nn'}\right) \cdot 1,$$

and so

$$\phi\left(\frac{m}{n} \cdot 1 + \frac{m'}{n'} \cdot 1\right) = \phi\left(\frac{mn' + m'n}{nn'} \cdot 1\right)$$

$$= \frac{mn' + m'n}{nn'}$$

$$= \frac{m}{n} + \frac{m'}{n'}$$

$$= \phi\left(\frac{m}{n} \cdot 1\right) + \phi\left(\frac{m'}{n'} \cdot 1\right).$$

Thus addition is preserved.

Hence the isomorphism of Q^* and Q is proved after, of course, the reader has shown that " $\cdot$ ", "$<$" are preserved by ϕ.

7.11 EXERCISES

1. (a) Explain why the relation $A = \{(x, y)|y = \sqrt{x},\ x,\ y \in R\}$ is a function while $B = \{(x, y)|y^2 = x,\ x,\ y \in R\}$ is *not* a function.

   (b) If for fixed $k \in R$, $g = \{(x, kx)|x \in R\}$, under what conditions on k does $g(x_1 \cdot x_2) = g(x_1)\, g(x_2)$?

2. In each of the following, state whether the mapping is into or onto I if the domain of the function is I.

   (a) $f: x \rightarrow |x|$ (d) $f: x \rightarrow |x|^2$

   (b) $f: x \rightarrow \cos x$ (e) $f: x \rightarrow x^2 + x + 1$

   (c) $f: x \rightarrow 3x$ (f) $f: x \rightarrow 2^x$

3. (a) Determine which of the following are the defining equations of functions with domain R.

   (b) For those which *do* represent functions determine which are one-to-one and obtain the inverse function.

   (c) For those functions which are *not* one-to-one, restrict the domain so that they become one-to-one.

   (i) $f(x) = x^3$

   (ii) $f(x) = x(x + 3)(x - 2)$

   (iii) $f(x) = |x|$

   (iv) $f(x) = \sin x,\ -\pi/2 \leqslant x \leqslant \pi/2$

   (v) $y = 1/(x^2 - 1)$

   (vi) $K(x) = x$

4. For each of the following, form $f \circ g$ and $g \circ f$. In each case state the domain of $f \circ g$ and of $g \circ f$.

   (a) $f(x) = x,\ g(x) = 1/x^2$

   (b) $f(x) = 1/x^2,\ g(x) = 1/x^2$

   (c) $f(x) = \sqrt{(x^2 + 1)},\ g(x) = \sqrt{x}$

   (d) $f(x) = \cos x,\ g(x) = 2x$

   (e) $f(x) = x^3,\ g(x) = x^{1/3}$

5. If $[P_1, P_2, ..., P_n]$ and $[Q_1, Q_2, ..., Q_m]$ are two partitions of a set T, a new partition of T can be obtained from the class of all subsets of T of the form $P_i \cap Q_j$. This new partition is called the *cross-partition* of the original two partitions. If $T = \{1, 2, 3, 4, 5, 6\}$ find the cross-partitions of the partitions

   (a) $[\{1, 2, 3\}, \{4, 5, 6\}], [\{1, 3, 4\}, \{2, 5, 6\}]$

   (b) $[\{1\}, \{2, 3\}, \{5, 6, 4\}], [\{1, 6\}, \{2, 4, 5\}, \{3\}]$

6. A partition $[P_1, P_2, \cdots, P_n]$ is a *refinement* of the partition $[Q_1, Q_2, \cdots, Q_m]$ if every P_i is a subset of some Q_j. Show that a cross-partition of two partitions is a refinement of each of the partitions from which the cross-partition was formed.

7. Let n, m be integers. Let F be an ordered field with multiplicative identity 1. If $m \cdot 1$ and $n \cdot 1$ are defined as in 7.10, then show that the following conditions hold.

   (a) $m \cdot 1 = n \cdot 1 \Leftrightarrow n = m$.

   (b) $m \cdot 1 + n \cdot 1 = (m + n) \cdot 1$. Note that you must consider more than one case.

   (c) $(m \cdot 1)(n \cdot 1) = (m \cdot n) \cdot 1$.

8. Let F be a field and let $a, b, c \in F$. Show that

   (a) If $a + b = a + c = 0$, then $b = c$, thus proving that the additive inverse of $a \in F$ is unique.

   (b) If $ab = ac = 1$, then $b = c$, thus proving that the multiplicative inverse of $a \in F$ is unique.

   (c) $(ab)^{-1} = a^{-1} b^{-1}$.

   (d) $(-1)^{-1} = -1$.

9. If F_1 is a subfield of F, show that the additive identity 0 and multiplicative identity 1 of F_1 are the additive and multiplicative identities of F.

10. Show that the operations $\oplus$, $\otimes$ of Example 8 of 7.5 are well defined. That is, show that if $[m, n] = [m', n']$ and $[p, q] = [p', q']$, then

$$[m, n] \oplus [p, q] = [m', n'] \oplus [p', q']$$

and

$$[m, n] \otimes [p, q] = [m', n'] \otimes [p', q'].$$

11. Show that $Q, \oplus, \otimes$ is a field.

12. Define the natural order relation (call it $\prec$ for now) on Q, $\oplus$, $\otimes$, in terms of the order relation $<$ on I. Show that $Q, \oplus, \otimes, \prec$ is an ordered field. (Hint: If m, n, p, q are integers, $n \neq 0$, $q \neq 0$, under what conditions on m, n, p, q is the fraction m/n less than the fraction p/q?)

13. Suppose f, g, and h are three functions with domains and counter domains such that $(f \circ g) \circ h$ is defined. Show that

   (a) $f \circ (g \circ h)$ is defined, and

   (b) $f \circ (g \circ h) = (f \circ g) \circ h$.

14. Prove that there are $m!$ permutations of $\{1, 2, 3, ..., m\}$.

15. Prove that there are m^m different functions from $\{1, 2, ..., m\}$ into itself.

16. Find a one-to-one function from N onto $N \times N$ in order to show that these sets are equipollent.

8

Real Numbers

8.1 IRRATIONAL NUMBERS

In Chapter 6 the rational numbers were defined and their properties discussed. From a naive point of view, since the rationals are dense, one might be tempted to think that rational numbers would be adequate for all purposes in mathematics. However, this is not the case. It was shown in fact in Theorem 4 of 6.3 that $\sqrt{2}$ is not a rational number. Thus we are forced to define new numbers, the **irrationals**. Before we discuss the formal definition we shall investigate in more detail the reasons for wanting to work with irrational numbers.

The irrationals arose originally through attempts to solve algebraic equations. In the case of $\sqrt{2}$ we wish to find an x satisfying the equation $x^2 - 20 =$. If we try to solve other algebraic equations we encounter other irrational numbers. For example the equation $x^2 + 8x + 5 = 0$ has as its roots the two numbers $-4 + \sqrt{11}$, $-4 - \sqrt{11}$. After we have proved the theorem below we shall show that these numbers are irrational. It is helpful to attempt to represent the irrationals by points on a line as was done in Chapter 6, where we represented rationals in this way. For example in Figure 8–1 we take L to be any line, and O (the origin) is any point on L. Using ruler and compasses we make OA one unit and erect AB also one unit and perpendicular to L. With center O and radius OB we make a circle to cut L in C. It is easily seen that OC is $\sqrt{2}$ units long. Since O is the origin, the point C represents the irrational number $\sqrt{2}$.

To construct $\sqrt{11}$ we form a right triangle with sides OC of length $\sqrt{2}$ and CD of length 3; then the hypotenuse is $\sqrt{11}$ units. In Figure 8–2, E has the coordinate $\sqrt{11}$; the points F with position $-4 + \sqrt{11}$ and G with position $-4 - \sqrt{11}$ are readily found.

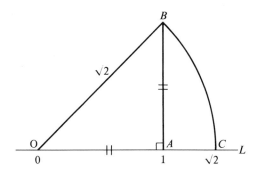

Figure 8–1   The geometrical construction of $\sqrt{2}$

Investigations of the type just considered show us that the rational number line has "gaps" in it. In other words, although all the rationals are represented by points on the line, there are points on the line, such as C in Figure 8–1 or E, F, G in Figure 8–2, that have no rational numbers corresponding to them. The following theorem enables us to prove that many of the numbers that occur as roots of polynomial equations are irrational.

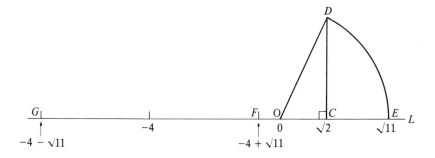

Figure 8–2   The geometrical construction of $-4 + \sqrt{11}$, $-4 - \sqrt{11}$

Theorem 1.   *Let* $x^n + a_1 x^{n-1} + \cdots + a_n = 0$ *be a polynomial equation of degree n, where* $a_i \in I$. *The rational roots of this equation, if there are any, are all integers.*

Proof. Suppose that $x = a/b$ is a root of the equation where $a, b \in I$, $b \neq 0$, and a, b are relatively prime. After substituting and clearing of fractions, we have $a^n + a_1 a^{n-1} b + \cdots + a_n b^n = 0$.

Hence
$$a^n = -a_1 a^{n-1} b - \cdots - a_n b^n.$$

Since b divides the right-hand side, we see that b is a divisor of a^n. Thus every prime factor of b divides a^n, and hence divides a. But a and b have no common divisor; therefore $b = \pm 1$ and $x = \pm a$. This proves the theorem.

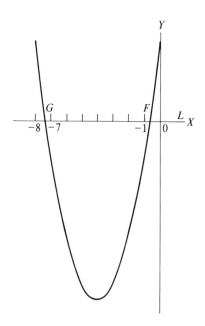

Figure 8–3   Graph of $y = x^2 + 8x + 5$

Example 1.   To illustrate the use of this theorem we shall prove that the roots of $x^2 + 8x + 5 = 0$ are irrational.

Since $x^2 + 8x + 5 = (x + 4)^2 - 11$ and $11 < 4^2$, if x is a root of the equation, $-4 < x + 4 < 4$, and $-8 < x < 0$. On substitution, no integer between -8 and 0 satisfies the equation. Therefore, by Theorem 1, the roots $-4 + \sqrt{11}$, $-4 - \sqrt{11}$ are irrational.

Example 2.   We shall now prove that $\sqrt[3]{7}$ is irrational. We see that if $x \geq 2$, then $x^3 - 7 > 0$, and if $x \leq 1$, then $x^3 - 7 < 0$. Therefore $x^3 - 7 = 0$ for some x in the interval $1 < x < 2$. By Theorem 1 it is clear that the solution $\sqrt[3]{7}$ must be irrational. It is worth noting that there is no construction using ruler and compasses for the point representing $\sqrt[3]{7}$ on the number line. This fact is related to the unsolvability of the ancient Greek problem of duplicating the cube; the proof that there is no construction is given in the book[1] by Courant and Robbins. However, despite this result, there is still a point on the number line corresponding to $\sqrt[3]{7}$—it is "near" the rational number 1.913. By carrying out a certain arithmetical process we can obtain an approximation to $\sqrt[3]{7}$ that is as accurate as we desire. Thus we can locate the poisition of the point on the number line representing $\sqrt[3]{7}$ to any degree of accuracy. In 8.5 we shall have much more to say about this procedure of approximating an irrational by rationals.

In order to make up for the evident deficiencies in the rationals, such as the presence of gaps in the rational number line, we might try to extend the rationals by adjoining all the roots of polynomial equations of the form $x^n + a_1 x^{n-1} + \cdots + a_n = 0$, $a_i \in I$. However, this procedure would produce major problems. The first difficulty we would encounter is illustrated by the equation $x^2 + 1 = 0$ or, as a slightly more complicated example, $x^2 + 4x + 7 = 0$. It is clear that these equations can have no rational roots since, for all rational x, $x^2 + 1 > 0$, and $x^2 + 4x + 7 = (x+2)^2 + 3 > 0$. The absence of rational roots for these equations follows directly from the property of the rationals that the square of any rational number is positive or zero (see 6.4 Exercise 12(b)). To produce the extended number system we could invent a number i (say) such that $i^2 = -1$; then the roots of $x^2 + 1 = 0$ would be $+i$, $-i$, and the roots of $x^2 + 4x + 7 = 0$ would be $-2 \pm i\sqrt{3}$. The numbers so formed are called **complex numbers** which are discussed in Courant and Robbins. However, this extended number system would fail to satisfy the basic property of the rationals that the square of *any* element is positive or zero (see 6.4 Exercise 12(b)). For example, $i^2 = -1$. In other words the set of complex numbers is not an **ordered** field. If we wish to preserve this property in the extended number system, we must reject such complex numbers as elements of our extended number line.

Even if we reject equations which have complex roots, we still have problems. It can be shown[2] that the irrational number π, for example, cannot be the root of any equation of the form

$$x^n + a_1 x^{n-1} + \ldots + a_n = 0, \qquad a_i \in Q.$$

After we have adjoined all non-complex roots of equations of this type, there will still remain gaps on the number line corresponding to irrational numbers like π. Irrational numbers of this type are called **transcendental**. It is clear that a different approach is needed to build up what is called the **real number system**. The real numbers will have the property that every point on the line

---

[1] *What is Mathematics?* by Courant and Robbins, Oxford, 1948.

[2] *Irrational Numbers*, by I. Niven, John Wiley, 1956.

will be represented by a real number; in other words there will be no gaps. The additional property needed to extend the rationals to the reals is discussed in the next section.

8.2 COMPLETE ORDERED FIELDS

The property which essentially fills in the gaps in the number line is **completeness**, as we shall see in the sequel.

In 6.3 it was shown that the rationals Q form an ordered field, and in 7.10 it was shown that every ordered field contains a subfield Q^* which is isomorphic to the ordered field Q. Since Q^* behaves exactly like Q, we can think of Q^* as if it were Q, and say that every ordered field contains Q as a subfield.

We shall assume that the real number system forms a **complete ordered field**, where completeness (of order) is defined as follows.

OC   *An ordered field is (order)* **complete** *if and only if every nonempty subset of the field which has an upper bound has a least upper bound (in the ordered field). (Thus we are assuming that the set R of real numbers is an ordered field like Q, contains Q as a subfield, but unlike Q, is a complete ordered field.)*

With assumption OC, the set $J = \{a \in Q \mid a > 0 \wedge a^2 < 2\}$ of Theorem 4 of 6.3 *does have* a l.u.b. (but not a rational one); we call it $\sqrt{2}$.

The real number π is the l.u.b. of the set of the lengths of the perimeters of regular polygons inscribed in a circle of diameter 1. (This is sometimes taken as the definition of π.)

We are, of course, skirting the issue a bit when we simply postulate the nature of R. To be mathematically correct, we should now prove that (i) there does exist a complete ordered field, and (ii) any two complete ordered fields are isomorphic. This can be done, but we shall not do it here.

Historically, this matter was first handled by constructing R from Q. (Q was constructed from I in 7.5, and I was constructed from N in 5.2.) J. W. R. Dedekind (1831–1916) defined a system of "cuts" in the rational number line, in which each cut corresponds to a real (rational or irrational) number. In his set of cuts he defined $+$, $\times$, $<$ and showed that the resulting system was a complete ordered field.

A **Dedekind cut** is defined to be an ordered pair (A, B) of subsets of Q such that:

(i) $A \cup B = Q$,

(ii) $A \cap B = \emptyset$,

(iii) $(a \in A \wedge b \in B) \Rightarrow a < b$,

(iv) B has no least element.

Let $P = \{a \in Q \mid a \leqslant 2\}$,

$\quad\quad T = \{a \in Q \mid a > 2\}$,

then (P, T) is a Dedekind cut which corresponds to, or *is*, the real number 2.

Let $S = \{a \in Q \mid a \leqslant 0\} \cup J$, where $J = \{a \in Q \mid a > 0 \wedge a^2 < 2\}$,

$\quad\quad H = \{a \in Q \mid a > 0 \wedge a^2 > 2\}$,

then (S, H) is a Dedekind cut (see 6.3); it is the real number $\sqrt{2}$.

The reader can consider for himself how to define $+$, $\times$, $<$ for cuts, and how to prove (what we shall henceforth assume) that the resulting system is a complete ordered field with a subfield isomorphic to Q. See *Principles of Mathematical Analysis*, by W. Rudin, McGraw–Hill, 1953, for details.

8.3   PROPERTIES OF THE REAL NUMBERS

The assumption that the real numbers form a complete ordered field will now be used to deduce some important properties of the real numbers. An immediate deduction from OC is the following Theorem.

Theorem 1. *Every nonempty set of real numbers that has a lower bound has a greatest lower bound.*

Proof.   Let A be a set of real numbers with a lower bound a; that is, for all $x \in A$, $x \geqslant a$. Now consider the set $-A = \{x \in R \mid x \leqslant -a \text{ for some } a \in A\}$. This is a nonempty set with an upper bound $-a$. By OC the set $-A$ must have a l.u.b. say b. Hence the set A has a g.l.b., $-b$.

Lemma 1. *The set I has no upper bound and no lower bound.*

Proof.   Assume I has an upper bound; then, since I is nonempty, by OC the set must have a l.u.b., say, b. Now $b - 1$ cannot be an upper bound of I. Hence there exists an integer n such that $n > b - 1$, and so $n + 1 > b$. But $n + 1 \in I$ and this contradicts the assumption that b is a l.u.b. of I. Thus I has no upper bound. Using Theorem 1 it can be proved that I has no lower bound.

Theorem 2. *(Archimedean Property of R). If x is a positive real number and y is an arbitrary real number, then there exists a positive integer n such that $nx > y$.*

Proof.   There must exist a positive integer n such that $n > y/x$, for if this were not so, y/x would be an upper bound for the set of integers, and this is not possible by Lemma 1. Hence there must exist a positive integer n such that $nx > y$.

Theorem 3. *If x is an arbitrary real number, there exists an integer n such that $n \leqslant x < n + 1$.*

Proof. We can find integers n_1, n_2 such that $n_1 < x < n_2$. For if there is no n_2 such that $n_2 > x$, x is an upper bound for I, and this is impossible by Lemma 1. Similarly, if there is no n_1 such that $n_1 < x$, x is a lower bound for I.

Consider the set of integers $\leqslant x$. This set is nonempty (n_1 is a member) and it has an upper bound, n_2. Consequently the set has as l.u.b. a unique integer n, where $n \leqslant x$. Now $n + 1 > x$ otherwise n would not be a l.u.b. This proves the result.

This integer n such that $n \leqslant x < n + 1$ is called the **greatest integer in** x, and it is written $[x]$. Thus we can write the result of Theorem 3, as $[x] \leqslant x < [x] + 1$.

Examples: $[2\frac{1}{2}] = 2$, $[-\frac{28}{3}] = -10$, $[\sqrt{2}] = 1$,

$\qquad\qquad$ $[75] = 75$, $[+2.715] = 2$, $[-2.715] = -3$.

Theorem 4. *Let x and y be real numbers such that $x < y$. Then there exists a rational number r such that $x < r < y$.*

Proof. Let $y - x = z$; z is a real number, $z > 0$. Now choose a positive integer n such that $nz > 1$. This is always possible by Theorem 2. Then $ny - nx = nz > 1$. Thus $ny - 1 > nx$. Let m be the integer $m = 1 + [nx]$. By Theorem 3, $nx < [nx] + 1 = m$ or $nx < m$. Also, $ny - 1 > nx \geqslant [nx] = m - 1$ by Theorem 3, that is, $ny - 1 > m - 1$ or $ny > m$. Thus $nx < m < ny$, and $x < m/n < y$. Since $r = m/n$ is a rational number, this establishes the result.

Corollary 1. *Between any two real numbers there are infinitely many rational numbers.*

Proof. If x, y are two real numbers such that $x < y$, we have shown that there exists a rational number r with the property $x < r < y$. But r itself is also real, so there exists a rational number s such that $r < s < y$. This process can be repeated indefinitely and the result follows. This result also implies that a rational number can be found as close as desired to any given real number.

Corollary 2. *Between any two real numbers there are infinitely many irrational numbers.*

Proof. Let x, y be real numbers, such that $x < y$. Let r be any rational number such that $x < r < y$. Now $\sqrt{2}/n$, $n \in I$, is irrational (Exercise 4(a)

of 8.7) and $r + \sqrt{2}/n$, $n \in I$, is irrational (Exercise 4(b) of 8.7). By Theorem 2, we can choose $n_1 \in I$ such that for $n \geqslant n_1$, $n > \sqrt{2}/(y - r)$. Thus for all $n \geqslant n_1$, $r + \sqrt{2}/n < y$ and $x < r < r + \sqrt{2}/n < y$. This proves the result.

The result of Theorem 4 can be expressed by saying that the rationals are **dense** in the reals. This use of the word dense should be distinguished from the use in the phrase **dense ordered.** It was shown in 6.3 that the rationals are dense ordered. By Theorem 4, the reals are also dense ordered, since between any two reals there is another real. Because Corollary 1 establishes that between any two irrationals there is an unlimited number of rationals, the reader might think that there are "as many" rationals as irrationals. Contrary to this expectation there are, in a certain sense, more irrationals than rationals. For details see Courant and Robbins page 81.

The following concept is fundamental in the study of limits of sequences discussed in the next section. For any given real number x we define an associated real number $|x|$ which is read as "mod x" or "the absolute value of x".

Definition.   *Let x be a real number. Then*

$$|x| = \begin{cases} x, & \text{if } x \geqslant 0, \\ -x, & \text{if } x < 0. \end{cases}$$

Examples: $|7| = 7$, $|-5| = 5$, $|-2^3| = 2^3$.

If we represent real numbers by points on a line, then for any given real number x, $|x|$ is simply the distance from the point x to the origin (see Figure 8–4).

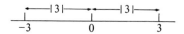

Figure 8–4 $|3|$ is the distance of 3, or -3, from 0

In the following two theorems we shall establish two important properties of $|x|$.

Theorem 5.   *If $a \geqslant 0$ is a real number then $|x| \leqslant a \Leftrightarrow -a \leqslant x \leqslant a$.*

Proof.   Suppose $|x| \leqslant a$, then $-a \leqslant -|x|$. Now $x = -|x|$ or $x = |x|$. If $x = -|x|$, then $-a \leqslant x$. If $x = |x|$, then $x \leqslant a$. Hence, $-a \leqslant x \leqslant a$. Conversely, suppose $-a \leqslant x \leqslant a$. If $x = |x|$, then $|x| \leqslant a$; if $x = -|x|$, then $-a \leqslant -|x|$, or $|x| \leqslant a$; and the proof is complete.

The geometric interpretation of this theorem is given in Figure 8–5.

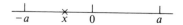

Figure 8–5 $|x| \leqslant a \Leftrightarrow -a \leqslant x \leqslant a$

Theorem 6.   *For any two real numbers x and y,* $|x + y| \leqslant |x| + |y|$.

Proof. By Theorem 5, $-|x| \leqslant x \leqslant |x|$ and $-|y| \leqslant y \leqslant |y|$. Now by Exercise 19 of 8.7, $-|x| - |y| \leqslant x + y \leqslant |x| + |y|$. Hence, by Theorem 5, $|x + y| \leqslant |x| + |y|$.

The inequality established in Theorem 6 is called the **triangle inequality**.

8.4   SEQUENCES OF RATIONALS

Sequences were defined in 7.7. We shall be concerned here with sequences of rationals, $\{a_n\}$. (In Chapter 11 we shall discuss sequences of real numbers.)
There are two ways of illustrating sequences pictorially which are sometimes helpful. One way, illustrated in Figure 8–6, is to plot the members, or terms, of the sequence on the number line. Another way is to plot the graph of the sequence (as a function) in the usual way; that is, plot the points (n, a_n), $n \in N$, in the plane. (See Figure 8–7.) Of course, in neither case is it possible to plot all the points.

Figure 8–6   Representation of the sequence $\{1/(n +1)\}$ by points on the line

A sequence $\{a_n\}$ is **monotone nondecreasing** if $a_{n+1} \geqslant a_n$ for all $n \in N$, and **monotone nonincreasing** if $a_{n+1} \leqslant a_n$ for all $n \in N$. If $a_{n+1} > a_n$, the sequence is called **monotone increasing**, and if $a_{n+1} < a_n$ for all n, it is **monotone decreasing**. It is clear that if a sequence is monotone increasing, it is also monotone nondecreasing. If we wish to refer to a sequence which may be either monotone nonincreasing or monotone nondecreasing, we shall call it simply a monotone sequence.

Example 1. The sequence $\{1, \frac{1}{2}, \frac{1}{3}, ...\}$ of Figures 8–6, 8–7 is monotone decreasing since $1/(n+2) < 1/(n+1)$ for all $n \in N$. Since $1/(n+2) \leqslant 1/(n+1)$, it is also monotone nonincreasing.

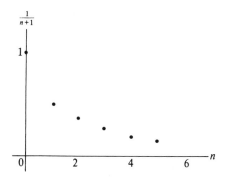

Figure 8–7   Graph of the sequence $\{1/(n+1)\}$

Example 2. The sequence $\{n^2\} = \{0, 1, 4, 9, 16, \ldots\}$ is obviously monotone increasing (as well as being monotone nondecreasing).

Example 3. The sequence $\{1, 1, 2, 2, 3, 3, \ldots\}$ is monotone nondecreasing but not monotone increasing.

Example 4. The sequence $\{2, 2, 2, 2, \ldots\}$ is monotone nonincreasing and monotone nondecreasing. Any constant sequence will possess both properties simultaneously. No other sequences will.

Example 5. The sequence $\{(-1)^n n\} = \{0, -1, 2, -3, 4, \ldots\}$ is not monotone in any sense.

It is useful at this stage to recall Definition 2 given in 2.6 for the upper and lower bounds of a set and apply this definition to sequences.

Definition 1. *A sequence $\{a_n\}$ has an upper bound a if $a_n \leqslant a$ for all $n \in N$, and a lower bound b if $a_n \geqslant b$ for all $n \in N$.*

Of course the upper (or lower) bound need not exist for a particular sequence nor need the upper (or lower) bound, if it does exist, be a member of the sequence. Thus the sequence in Example 1 has a lower bound 0 which is not a member of the sequence, and an upper bound 1 which is a member of the sequence. The sequence in Example 5 has neither an upper nor a lower bound.

It can be seen that the successive members of the sequence $\{1, \frac{1}{2}, \frac{1}{3}, \ldots\}$ are getting progressively smaller. We say that the **limit** of this sequence is 0, and that the sequence is **convergent**. *By definition, a convergent sequence has a limit, and any sequence which has a limit is* **convergent**. It should not be thought that all sequences are convergent. For example the sequence of Example 2 has no limit and is thus not convergent. The same remark can be made of the sequences in Examples 3 and 5. *Any sequence which is not convergent is called* **divergent**. Another example of a divergent sequence is $\{(-1)^n\}$. The successive elements of this sequence are $+1$ and -1, and since the successive members are not getting closer and closer to a single number, the sequence is divergent. The sequence of Example 4 is convergent, and the limit is 2.

We shall give the precise definition of the limit of a sequence later, but roughly when we say that the limit of a sequence $\{a_n\}$ is a number a, we mean that simply by taking n sufficiently large we can make a_n as close to a as we please. Thus, in Example 1, we observe that $1/(n+1)$ becomes as close to 0 as we please when we take n sufficiently large; consequently we say that the limit is 0. In Example 4, the limit is 2 since this is the only number we approach (we are actually there!) as we go further out in the sequence. More precisely, we shall say that the limit of a sequence $\{a_n\}$ is a if we can make a_n as close as we wish to a merely by taking n sufficiently large. In order to save writing, we shall replace the statement "the limit of the sequence $\{a_n\}$ is a" by "$\lim_n a_n = a$".

In what follows we consider two kinds of limit, an *o*-limit (order-limit) which is closely related to the concepts of l.u.b. and g.l.b. and the usual definition of limit (with no prefix; see Definition 3). Ultimately we show that these two limit concepts are identical.

Definition 2. *A monotone nondecreasing sequence with an upper bound is said to be* **o-convergent** *and its l.u.b. is the* **o-limit of the sequence**. *A monotone nonincreasing sequence with a lower bound is also said to be o-convergent and its g.l.b. is the* **o-limit of the sequence**. The limits exist by OC.

From an intuitive point of view, it should be clear that in either of these cases the definition agrees with the remark made above that a sequence $\{a_n\}$ has an *o*-limit a if a_n can be made as close as desired to a provided n is chosen sufficiently large. A more usual definition of limit is given in the following.

Definition 3. *A sequence $\{a_n\}$ is* **convergent** *and* $\lim_n a_n = a$ *if, for every* $\varepsilon > 0$, *no matter how small, we can find an integer n_0 such that* $|a_n - a| < \varepsilon$ *for all $n > n_0$.*

We shall prove below that if a sequence is *o*-convergent according to Definition 2, it is also convergent according to Definition 3. It is clear, however, that the definition of convergence in Definition 3 covers a wider range of possibilities than does the *o*-convergence of Definition 2; using Definition 3 we can establish that some arbitrary (that is, non-monotone) sequences are convergent, whereas Definition 2 applies only to monotone sequences. For example the sequence $\{a_n\} = \{1, \frac{1}{2}, \frac{1}{3}, \frac{1}{4}, \frac{1}{3}, \frac{1}{4}, \frac{1}{5}, \frac{1}{6}, \frac{1}{5}, ...\}$ (formed from the sequence $\{1, \frac{1}{2}, \frac{1}{3}, \frac{1}{4}, ...\}$ by moving ahead four terms then moving back one term successively) is not monotone in any sense. However, it is possible to show, by Definition 3, that this sequence is convergent and the limit is 0.

We can extend Definition 2 to handle this case in the following way. Consider the two sequences $\{b_n\} = \{\frac{1}{2}, \frac{1}{3}, \frac{1}{4}, \frac{1}{5}, \frac{1}{6}, ...\}$ and $\{c_n\} = \{1, \frac{1}{2}, \frac{1}{3}, \frac{1}{3}, \frac{1}{3}, \frac{1}{4}, \frac{1}{5}, \frac{1}{5}, \frac{1}{5}, \frac{1}{6}, \frac{1}{7}, \frac{1}{7}, \frac{1}{7}, \frac{1}{8}, ...\}$. The sequence $\{b_n\}$ is monotone decreasing with o-limit 0 (that is, $o\text{-}\lim_n b_n = 0$) by Definition 2; and the sequence $\{c_n\}$ is monotone nonincreasing with o-limit 0 (that is, $o\text{-}\lim_n c_n = 0$) by Definition 2. Further, for all n, $b_n \leqslant a_n \leqslant c_n$. We shall use this example to form the following definition, which is an extension of Definition 2.

Definition 4. *An arbitrary sequence $\{a_n\}$ is said to be o-convergent and has o-limit a if there exist two sequences $\{b_n\}$ and $\{c_n\}$ which are respectively monotone nondecreasing and monotone nonincreasing such that*

$$b_n \leqslant a_n \leqslant c_n$$

for all $n \in N$ and

$$o\text{-}\lim_n b_n = o\text{-}\lim_n c_n = a.$$

We shall prove later in this section that for an arbitrary sequence $\{a_n\}$ of real numbers, if either side exists, then

$$\lim_n a_n = o\text{-}\lim_n a_n.$$

Example 6. The sequence $\{1, \frac{1}{2}, \frac{1}{3}, \frac{1}{4}, \frac{1}{3}, \frac{1}{4}, \frac{1}{5}, \frac{1}{6}, \frac{1}{5}, ...\}$ converges to 0 according to Definition 4.

Definition 3 is useful only if we have found (possibly by guesswork) an alleged limit a. Then, by using Definition 3, we can establish that the sequence is in fact convergent to the limit a. A theoretically more useful condition for convergence called Cauchy's Convergence Criterion, will be discussed in Section 11.3. This states that the sequence $\{a_n\}$ is convergent if, corresponding to an arbitrary positive number ε, there is an integer, n_0, such that $|a_n - a_m| < \varepsilon$ for all $m, n > n_0$. This enables one to determine whether a sequence is convergent without knowing the limit.

By Corollary 1 of 8.3, the "ε" of Definition 3 can be restricted to be rational if we wish. This is so also for Cauchy's Convergence Criterion. Hence, using the C.C.C. and sequences of rationals only, no reference is made to arbitrary real numbers. But the limit, if it exists, may be irrational. This suggests the possibility of defining, or constructing, the real numbers from the rationals in terms of sequences of rationals which converge according to Cauchy's criterion. This is developed further at the end of 8.5.

There is a graphical interpretation of Definition 3 that may give the reader a firmer intuitive appreciation of its meaning. As a particular example, we shall examine graphically the limit of the sequence $\{a_n\}$ where $a_n = 1 + (-1)^n/n$. It is intuitively clear that $\lim_n a_n = 1$. Suppose we are given a small positive number ε, say $\varepsilon = 0.1$. On the graph of $\{a_n\}$ draw a horizontal strip of width 2ε ($=0.2$) with the center line parallel to the N axis and 1 unit above it (Figure 8–8). We observe that if we go far enough out in the sequence (that is, choose n sufficiently large) all the terms in the sequence fall within the strip. In the case that $\varepsilon = 0.1$, we choose $n_0 = 10$. If the given ε is smaller (say $\varepsilon = 0.001$), then the strip will be narrower, but by going out sufficiently far in the sequence we are assured that all the terms fall within the strip. If $\varepsilon = 0.001$ we must go beyond the 1000th term in order to have all the terms within the strip. Since, no matter how narrow the strip is, we can always find an n_0 such that for $n > n_0$ all terms of the sequence fall in the strip, we say that $\lim_n a_n = 1$ for this sequence.

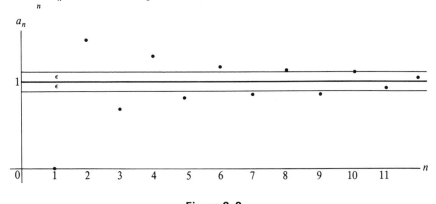

Figure 8–8

In the case of the sequence $\{a_n\}$ where $a_n = (-1)^n$, it is readily seen that there is no arbitrarily narrow horizontal strip such that all but a finite number of terms fall within the strip. Thus this sequence, by Definition 3, is divergent.

Example 7. Let us consider an important example called the geometric sequence. This is given by $\{a_n\}$ where

$$a_n = 1 + r + r^2 + \cdots + r^n, \text{ and } n \in N.$$

Now it is well known that

$$1 + r + r^2 + \cdots + r^n = \frac{1 - r^{n+1}}{1 - r}, \quad r \neq 1.$$

This result can be established by multiplication. Thus

$$a_n = \frac{1 - r^{n+1}}{1 - r} = \frac{1}{1 - r} - \frac{r^{n+1}}{1 - r}.$$

If $0 \leqslant r < 1$, the sequence $\{a_n\}$ is monotone increasing with l.u.b. $1/(1 - r)$. Thus, by Definition 2, the sequence is o-convergent and the o-limit is $1/(1 - r)$. If $-1 < r < 0$, Definition 2 no longer applies since the sequence is not monotone. However, let us form the two sequences

$$\{b_n\} = \{0,\ 1 + r,\ 1 + r + r^2 + r^3,\ 1 + r + r^2 + r^3 + r^4 + r^5, \ldots\}$$

$$\{c_n\} = \{1,\ 1 + r + r^2,\ 1 + r + r^2 + r^3 + r^4,$$
$$1 + r + r^2 + r^3 + r^4 + r^5 + r^6, \ldots\}.$$

By inspection, the sequence $\{b_n\}$ is monotone increasing with limit $1/(1 - r)$, and the sequence $\{c_n\}$ is monotone decreasing also with limit $1/(1 - r)$. Further we see that $b_n \leqslant a_n \leqslant c_n$. Thus, by Definition 4, the sequence $\{a_n\}$ converges and has limit $1/(1 - r)$. We write

$$o\text{-}\lim_{n} (1 + r + r^2 + \cdots + r^n) = \frac{1}{1 - r}.$$

If $|r| > 1$, the sequence diverges.

Example 8. Consider the sequence $\{a_n\}$ where

$$a_n = \frac{1 + 2 + 3 + \cdots + n}{n^2}, \quad n \in N.$$

It has been proved in 2.5 that

$$1 + 2 + 3 + \cdots + n = \frac{n(n + 1)}{2}.$$

Thus

$$a_n = \frac{n(n + 1)}{2n^2} = \frac{1}{2}\left(1 + \frac{1}{n}\right).$$

By Definition 2, since $\{a_n\}$ is monotonic decreasing with a lower bound, $o\text{-}\lim_{n} a_n = \frac{1}{2}$, the g.l.b. of $\{a_n\}$.

Example 9. The decimal expansion of a number is a concept that can be related to the notion of the limit of a sequence. For example consider $2.\overline{3}$; this stands for $2.333 \ldots =$

$$2 + \frac{3}{10} + \frac{3}{10^2} + \frac{3}{10^3} + \cdots .$$

In other words $2.\overline{3}$ is the limit of the sequence $\{a_n\}$ where

$$a_n = 2 + \frac{3}{10} + \frac{3}{10^2} + \cdots + \frac{3}{10^n}.$$

Now we can write

$$a_n = 2 + \frac{3}{10}\left(1 + \frac{1}{10} + \frac{1}{10^2} + \cdots + \frac{1}{10^{n-1}}\right)$$

and from Example 7,

$$\text{o-lim}_n \, a_n = 2 + \frac{3}{10}\left(\frac{1}{1 - 1/10}\right) = 2 + \frac{3}{9} = 2\tfrac{1}{3}.$$

A slightly more complicated example is $6.\overline{142857}$, which stands for the limit of the sequence $\{b_n\}$ where

$$b_n = 6 + \frac{142857}{10^6} + \frac{142857}{10^{12}} + \cdots + \frac{142857}{10^{6n}}$$

$$= 6 + \frac{142857}{10^6}\left(1 + \frac{1}{10^6} + \cdots + \frac{1}{10^{6(n-1)}}\right).$$

Now

$$\text{o-lim}_n \, b_n = 6 + \frac{142857}{10^6}\left(\frac{1}{1 - 1/10^6}\right) \text{ (from Example 7),}$$

$$= 6 + \frac{142857}{999999} = 6\tfrac{1}{7}.$$

We shall now prove some theorems which will be useful in the sequel.

Lemma 1. *Let $\{a_n\}$ be a monotone nondecreasing sequence with the least upper bound a. For any $\varepsilon > 0$ there exists an n_0 such that*:

$$n > n_0 \Rightarrow a_n > a - \varepsilon.$$

Proof. Let $\varepsilon > 0$ be a fixed arbitrary positive number. Since a is the l.u.b. of a_n, then $a - \varepsilon$ is not an upper bound. Hence, there is some value of n, say n_0, for which $a_{n_0} > a - \varepsilon$. But for all n, $a_{n+1} \geqslant a_n$; hence,

$$n \geqslant n_0 \Rightarrow a_n > a - \varepsilon.$$

Theorem 1. *If a_n be a sequence of real numbers, then a_n is o-convergent if and only if it is convergent and*

$$\text{o-lim}_n \, a_n = \lim_n \, a_n.$$

Proof. We divide the proof into two parts: (a) The proof that an o-convergent sequence $\{a_n\}$ is also convergent and if o-$\lim_n a_n = a$, then $\lim_n a_n = a$, and (b) the converse of (a).

(a) Lemma 1 can be readily extended by the reader to the case of a monotone nonincreasing sequence with a lower bound. Now, since $\{a_n\}$ o-converges to a, by Definition 4 there exist two sequences $\{p_n\}$ monotone nondecreasing with limit a, and $\{q_n\}$ monotone nonincreasing with limit a, such that $p_n \leqslant a_n \leqslant q_n$ for all n. The sequences $\{|p_n - a|\}$, $\{|q_n - a|\}$ are both monotone nonincreasing with limit 0.

Hence, by the extension of Lemma 1, for any given $\varepsilon > 0$, we can choose an integer n_1, such that $|p_n - a| < \varepsilon$ for $n \geqslant n_1$, and an integer n_2 such that $|q_n - a| < \varepsilon$ for $n \geqslant n_2$. Thus for $n > \max(n_1, n_2)$, both $|p_n - a| < \varepsilon$ and $|q_n - a| < \varepsilon$. Now for each n, we will have either $|a_n - a| \leqslant |p_n - a|$ or $|a_n - a| \leqslant |q_n - a|$ (see Figure 8–9 and Figure 8–10).

Figure 8–9   Case in which $|a_n - a| \leqslant |q_n - a|$

Figure 8–10   Case in which $|a_n - a| \leqslant |p_n - a|$

Hence for $n > n_0$, $|a_n - a| < \varepsilon$ where $n_0 = \max(n_1, n_2)$. This establishes that o-convergence implies convergence to the same limit.

(b) Assume $\{a_n\}$ converges (Definition 3) to a. Thus for every $\varepsilon > 0$, there is an n_ε such that

$$n > n_\varepsilon \Rightarrow |a_n - a| < \varepsilon.$$

In particular for $\varepsilon = 1/n_0$, for each $n_0 \in N - \{0\}$, there is an m_{n_0} such that

$$n > m_{n_0} \Rightarrow |a_n - a| < \frac{1}{n_0},$$

that is,

$$n > m_{n_0} \Rightarrow a - \frac{1}{n_0} < a_n < a + \frac{1}{n_0}.$$

Now consider the sequence $\{m_{n_0} + 1\}$ and use it to construct the following two monotone sequences $\{p_n\}$ and $\{q_n\}$. Let

$$p_i = \max\{|a_1|, ..., |a_{m_1}|, |a| + 1\} \text{ and } q_i = -p_i \text{ for } i = 1, 2, ..., m_1;$$

for $m_1 + 1 \leqslant i \leqslant m_2$ let $p_i = a + 1$ and $q_i = a - 1$; for $m_2 + 1 \leqslant i \leqslant m_3$ let $p_i = a + \frac{1}{2}$ and $q_i = a - \frac{1}{2}$; and so on. It is clear that the resulting sequences $\{p_n\}$ and $\{q_n\}$ are respectively monotone nonincreasing and nondecreasing, and that

$$q_n \leqslant a_n \leqslant p_n$$

for all $n = 1, 2, \ldots$. In addition, g.l.b. $\{p_n\} = a$ and l.u.b. $\{q_n\} = a$; thus $o\text{-}\lim_n a_n = a$ by Definition 4.

Now that Theorem 1 has been proved, we know that o-convergence and convergence are the same thing in the sense that every sequence $\{a_n\}$ is o-convergent if and only if it is convergent and

$$\lim_n a_n = o\text{-}\lim_n a_n$$

if either side exists. Thus we no longer need to distinguish between the two kinds of convergence, and so the Definitions 2, 3, and 4 can be used interchangeably in arguments concerning limits.

It is now natural to enquire if the following hold.

(1) $\lim_n a_n + \lim_n b_n = \lim_n (a_n + b_n)$.

(2) $c \lim_n a_n = \lim_n (ca_n)$.

(3) $(\lim_n a_n)(\lim_n b_n) = \lim_n (a_n b_n)$.

(4) $\lim_n a_n / \lim_n b_n = \lim_n (a_n / b_n)$ (provided, of course, that $\lim_n b_n \neq 0$).

In fact they do, provided the limits on the left sides exist. We shall prove the first two equations here. The remaining two are Exercises 16, 17 of 8.7.

Lemma 2. *If $\{a_n\}$ and $\{b_n\}$ are monotone nondecreasing sequences with upper bounds, then the sequence $\{c_n\}$ such that $c_n = a_n + b_n$ is convergent and*

$$\lim_n c_n = \lim_n a_n + \lim_n b_n.$$

Proof. Since $\{a_n\}$ is a monotone nondecreasing sequence with an upper bound, by Definition 2 it is convergent to its l.u.b. that is,

$$\lim_n a_n = a \quad \text{(say)}.$$

Similarly $\{b_n\}$ is convergent and

$$\lim_n b_n = b.$$

Now $c_{n+1} = a_{n+1} + b_{n+1} \geqslant a_n + b_n = c_n$, and $\{c_n\}$ is a monotone non-decreasing sequence.

Since $c_n = a_n + b_n \leqslant a + b$ for all n, $a + b$ is an upper bound, and by Definition 2, $\{c_n\}$ converges. Not only is $a + b$ an upper bound of $\{c_n\}$, but it is the least upper bound. To prove this, let us suppose that for some $\varepsilon > 0$ $a + b - \varepsilon$ is the l.u.b. Now for all n sufficiently large, $a_n > a - \varepsilon/2$ and $b_n > b - \varepsilon/2$ by Lemma 1.

So $c_n = a_n + b_n > a + b - \varepsilon$. (In making this last statement we are actually making use of the fact that the real numbers a and b (not necessarily rational although all a_n, b_n, c_n are rational) are elements of an ordered field, and the result follows by applying Exercise 19 of 8.7.)

But if $c_n > a + b - \varepsilon$ for all n sufficiently large, this contradicts our supposition that $a + b - \varepsilon$ is the l.u.b. Hence

$$\lim_n c_n = a + b = \lim_n a_n + \lim_n b_n.$$

Corollary. *If $\{a_n\}$ and $\{b_n\}$ are monotone nonincreasing sequences with lower bounds, then the sequence $\{c_n\}$ such that $c_n = a_n + b_n$ is convergent and*

$$\lim_n c_n = \lim_n a_n + \lim_n b_n.$$

Proof. The proof is almost identical to the proof of Lemma 2 and will be left to the reader.

Theorem 2. *If $\{a_n\}$ and $\{b_n\}$ are two convergent sequences, then the sequence $\{c_n\}$, where $c_n = a_n + b_n$, is convergent and*

$$\lim_n c_n = \lim_n a_n + \lim_n b_n.$$

Proof. Since $\{a_n\}$ converges, by Definition 3 there exist two sequences $\{p_n\}$ and $\{q_n\}$, $\{p_n\}$ monotone nondecreasing, $\{q_n\}$ monotone nonincreasing, both with the same limit and such that $p_n \leqslant a_n \leqslant q_n$ for all n.

Similarly there exist two sequences $\{r_n\}$ monotone nondecreasing and $\{s_n\}$ monotone nonincreasing both with the same limit such that $r_n \leqslant b_n \leqslant s_n$ for all n.

Now $p_n + r_n \leqslant c_n \leqslant q_n + s_n$ and since $\{p_n + r_n\}$ is a monotone non-decreasing sequence and $\{q_n + s_n\}$ is monotone nonincreasing, and both sequences have the same limit (by Lemma 2), it follows from Definition 3 that $\{c_n\}$ is convergent and

$$\lim_n c_n = \lim_n a_n + \lim_n b_n.$$

Note: This result can obviously be extended to the case where $\{c_n\}$ is the algebraic sum of any finite number of sequences. For example, if $c_n = a_n + b_n - d_n$ where $\{a_n\}$, $\{b_n\}$, $\{d_n\}$ are convergent then

$$\lim_n c_n = \lim_n a_n + \lim_n b_n - \lim_n d_n.$$

Theorem 3. *If $\{a_n\}$ is a monotone nondecreasing sequence with an upper bound and $c > 0$, then*

$$\lim_n (ca_n) = c \lim_n a_n.$$

Proof. Since $\{a_n\}$ is a monotone nondecreasing sequence with an upper bound, $\{a_n\}$ converges, by Definition 2, and

$$\lim_n a_n = a \quad \text{(say)}.$$

Now the sequence $\{b_n\}$ where $b_n = ca_n$ is monotone nondecreasing with an upper bound hence it is convergent. Now

$$\lim_n b_n = ca,$$

for suppose that $ca - \varepsilon$ $(\varepsilon > 0)$ is the l.u.b. of $\{b_n\}$.

Then for all n sufficiently large, $a_n > a - \varepsilon/c$ and $b_n = ca_n > ca - \varepsilon$ contrary to the assumption. Thus the result follows.

Corollary. *If $\{a_n\}$ is a convergent sequence and c is any real number, then*

$$\lim_n (ca_n) = c \lim_n a_n.$$

The proof is left to the reader.

8.5 REPRESENTATION OF REAL NUMBERS BY SEQUENCES OF RATIONAL NUMBERS

If we were asked to find an approximate value for $\sqrt{2}$, we might proceed in the following way. A guess at an approximate value is 1.4. But $1.4^2 = 1.96$, which is too small. If we try 1.5, we find that $1.5^2 = 2.25$, which is too large.

Since $1.41^2 = 1.9881$ and $1.42^2 = 2.0164$, the results are again respectively too small and too large. If we continue in this way we obtain two sequences $\{a_n\} = \{1.4, 1.41, 1.414, 1.4142, 1.41421, ...\}$ and $\{b_n\} = \{1.5, 1.42, 1.415, 1.4143, ...\}$. The sequence $\{a_n\}$ is monotone increasing, and the sequence $\{b_n\}$ is monotone decreasing. Furthermore, the sequence $\{a_n\}$ has an upper bound since

$$a_n \leqslant 1 + \frac{9}{10} + \frac{9}{10^2} + \cdots + \frac{9}{10^n} = 1 + \frac{9}{10}\left[\frac{1 - (1/10)^n}{1 - 1/10}\right]$$

$$= 1 + \frac{9}{10}\left[\frac{1}{1 - 1/10} - \frac{1}{10^n(1 - 1/10)}\right] < 1 + \frac{9}{10} \cdot \frac{1}{(1 - 1/10)}$$

$$= 1 + 1 = 2.$$

Thus, by Definition 2 of 8.4, the sequence $\{a_n\}$ is convergent and

$$\lim_n a_n = a$$

where a is the l.u.b. of the sequence. Now $b_n = a_n + 1/10^n$, and by (1) of 8.4 and Exercise 7(b) of 8.7

$$\lim_n b_n = \lim_n a_n + 0 = a.$$

We shall now prove that $a^2 = 2$, using the following lemma.

Lemma 1. *If $\{a_n\}$ is a monotone increasing (decreasing) sequence with an upper (lower) bound and*

$$\lim_n a_n = a, \text{ then } \lim_n a_n{}^2 = a^2.$$

Proof. We shall restrict the proof to the case $a > 0$ and $\{a_n\}$ increasing, and leave the other cases to the reader. Now $a > 0$ implies that $a_n > 0$ for all sufficiently large n. It follows that $\{a_n{}^2\}$ is ultimately monotone increasing; since it clearly has an upper bound a^2, it converges to its l.u.b. and l.u.b. $\{a_n{}^2\} \leqslant a^2$.

We wish to show that l.u.b. $\{a_n{}^2\} = a^2$.

Suppose, on the contrary that l.u.b. $\{a_n{}^2\} < a^2$ and let $\varepsilon > 0$ be chosen so that

(1)   l.u.b. $\{a_n{}^2\} = a^2 - \varepsilon$.

By Lemma 1 of 8.4, $a_n > a - \varepsilon/2a$ for all sufficiently large n. Hence

$$a_n{}^2 > a^2 - \varepsilon + \varepsilon^2/4a^2 > a^2 - \varepsilon$$

for all sufficiently large n. But this contradicts (1), and so the proof is complete.

We now return to an approximation to $\sqrt{2}$ by means of the sequences $\{a_n\}$ and $\{b_n\} = \{a_n + 1/10^n\}$. Proceeding as in our proof that $a_n^2 < 2$, we see that $b_n^2 \geqslant 2$ for all n. But $b_n = a_n + 1/10^n$ so

$$b_n^2 = a_n^2 + \frac{2a_n}{10^n} + \frac{1}{10^{2n}}.$$

Therefore for all n,

$$a_n^2 + \frac{2a_n}{10^n} + \frac{1}{10^{2n}} \geqslant 2,$$

and so

$$0 \leqslant 2 - a_n^2 \leqslant \frac{2a_n}{10^n} + \frac{1}{10^{2n}} \leqslant \frac{4}{10^n} + \frac{1}{10^{2n}}.$$

Therefore by the results of Exercise 10 below,

$$0 \leqslant \lim_n (2 - a_n^2) \leqslant \lim_n \frac{4}{10^n} + \lim_n \frac{1}{10^{2n}} = 0.$$

Thus $\lim_n a_n^2 = 2$; hence $a^2 = 2$ and $a = \sqrt{2}$.

Figure 8–11   Nest of intervals for $\sqrt{2}$

We have formed two sequences $\{a_n\}$ and $\{b_n\}$ each of which has the property that the limit of the sequence is $\sqrt{2}$. By pairing successive terms of the two sequences we can form a new kind of sequence

$$\{(a_0, b_0), (a_1, b_1), (a_2, b_2), \ldots, (a_n, b_n), \ldots\}.$$

Any member of the sequence, such as (a_n, b_n), can be thought of as determining an interval from a_n to b_n on the number line. The length of the interval is $b_n - a_n$ and each interval in the sequence is contained in its predecessor. The sequence of lengths of the intervals is called a **nest of intervals** and is an important concept in certain developments[3] of the real number system. Pictorially, the nest of intervals for $\sqrt{2}$ can be shown as in Figure 8–11.

[3] *Theory and Application of Infinite Series,* by K. Knopp, Blackie and Son Ltd.

It is not difficult to construct other sequences whose limit is $\sqrt{2}$. For example consider the sequence $\{a_n\}$ where $a_{n+1} = \frac{1}{2}(a_n + 2/a_n)$, $a_0 = 2$. In this case, the sequence is given by what is known as a recursion formula which gives the value of a_{n+1} in terms of a_n for every $n \in N$. The sequence is $\{2, \frac{3}{2}, \frac{17}{12}, \ldots\}$. Now this sequence has a lower bound, for

$$a_{n+1}{}^2 = \frac{1}{4}\left(a_n{}^2 + \frac{4}{a_n{}^2} + 4\right) = \frac{a_n{}^4 - 4a_n{}^2 + 4 + 8a_n{}^2}{4a_n{}^2}$$

$$= \frac{a_n{}^4 - 4a_n{}^2 + 4}{4a_n{}^2} + 2 = \frac{(a_n{}^2 - 2)^2}{4a_n{}^2} + 2 > 2.$$

The sequence is monotone decreasing, for

$$a_{n+1} = \frac{a_n}{2} + \frac{1}{a_n} = a_n - \frac{a_n{}^2 - 2}{2a_n} < a_n.$$

Thus, by Definition 2 of 8.4, the sequence has a limit, which we shall call a. Now $a_n{}^2 = 2a_n a_{n+1} - 2$. Hence, by (1), (2), (3) of 8.4 (and the obvious fact that $\lim_n a_{n+1} = a$),

$$\lim_n(a_n{}^2) = 2\left(\lim_n a_n\right)\lim_n(a_{n+1}) - \lim_n 2$$

$$= 2a^2 - 2.$$

But

$$\lim_n(a_n{}^2) = \lim_n(a_n \cdot a_n) = \left(\lim_n a_n\right)\left(\lim_n a_n\right) = a^2$$

by (3) of 8.4. Hence, $a^2 = 2a^2 - 2$; that is, $a^2 = 2$. We have shown that

$$\lim_n a_n = \sqrt{2}.$$

We are now in a position to *characterize* $\sqrt{2}$ by means of sequences. Let $\{x_n\}$ be a convergent sequence of rational numbers such that

$$\lim_n x_n{}^2 = 2.$$

The sequence $\{x_n\}$ could be one of the sequences in the previous examples, or some other. We shall say that any sequence $\{y_n\}$ of rational numbers such that

$$\lim_n(|y_n - x_n|) = 0$$

is **equivalent** to the sequence $\{x_n\}$. If the sequence $\{y_n\}$ is equivalent to the sequence $\{x_n\}$, we write symbolically $\{y_n\} \sim \{x_n\}$. *The set of all such sequences*

$\{y_n\}$ *equivalent to a given sequence* $\{x_n\}$ *forms an equivalence class.* (The idea of an equivalence class has been discussed in 7.5.) To prove this result we need the following Lemmas.

Lemma 2. *Consider a sequence* $\{a_n\}$. *Then*

$$\lim_n |a_n| = 0$$

if and only if $\lim_n a_n$ *exists and is equal to zero.*

    Proof. By Definition 3 of 8.4

$$\lim_n a_n = a$$

if and only if, corresponding to an arbitrary $\varepsilon > 0$, there exists an integer $n_0 \in N$ such that $|a_n - a| < \varepsilon$ when $n \geqslant n_0$. If $a = 0$ this means that $|a_n - 0| = |a_n| < \varepsilon$ when $n \geqslant n_0$. But this, by definition, is precisely the condition that holds if and only if $\lim_n |a_n|$ exists and is equal to 0.

Lemma 3. *Let* $\{a_n\}$ *be a convergent sequence. Then any sequence* $\{x_n\}$ *such that*

$$\lim_n |x_n - a_n| = 0$$

is convergent to the same limit as the sequence $\{a_n\}$.

    Proof. Let $x_n - a_n = b_n$, that is, $x_n = a_n + b_n$, and assume that

$$\lim_n a_n = a \quad \text{(say)}.$$

Thus

$$\lim_n |b_n| = 0$$

and, by Lemma 2,

$$\lim_n b_n = 0.$$

Hence, by (1) of 8.4 $\lim_n x_n$ exists and

$$\lim_n x_n = \lim_n a_n + \lim_n b_n = a.$$

Theorem 1.   *Let $\{a_n\}$ be a convergent sequence. The set of all sequences $\{x_n\}$ such that*

$$\lim_n |x_n - a_n| = 0$$

forms an equivalence class.

   Proof.   We see by Lemma 2 that the sequences $\{x_n\}$ are all convergent and

$$\lim_n x_n = \lim_n a_n.$$

To establish the result we need to prove:

   (a) $\{x_n\} \sim \{x_n\}$;

   (b) If $\{x_n\} \sim \{y_n\}$, then $\{y_n\} \sim \{x_n\}$;

   (c) If $\{x_n\} \sim \{y_n\}$ and $\{y_n\} \sim \{z_n\}$, then $\{x_n\} \sim \{z_n\}$.   (Cf. 7.5.)

The first two are trivial. To prove (c), assume

$$\lim_n |x_n - y_n| = 0 \text{ and } \lim_n |y_n - z_n| = 0.$$

By Lemma 2,

$$\lim_n (x_n - y_n) = 0 \text{ and } \lim_n (y_n - z_n) = 0.$$

But $x_n - z_n = (x_n - y_n) + (y_n - z_n)$. So (1) of 8.4 gives

$$\lim_n (x_n - z_n) = 0,$$

and Lemma 2 yields

$$\lim_n |x_n - z_n| = 0.$$

This completes the proof of the Theorem.

   In particular, if

$$\lim_n a_n = \sqrt{2},$$

then the equivalence class $[\{a_n\}]$ of sequences equivalent to $\{a_n\}$ can be used to "define" $\sqrt{2}$. (This will be discussed in more detail below.) Recall from 7.5 that:

$$\{b_n\} \sim \{a_n\} \Leftrightarrow [\{b_n\}] = [\{a_n\}].$$

Furthermore:

$$\{b_n\} \sim \{a_n\} \Leftrightarrow \lim_n b_n = \lim_n a_n \ (= \sqrt{2}).$$

We can extend the ideas developed above, concerning the irrational number $\sqrt{2}$, to any irrational number. Consider for example the sequences

$$\{c_n\} = \left\{1, 1 - \frac{1}{3} + \frac{1}{5}, 1 - \frac{1}{3} + \frac{1}{5} - \frac{1}{7} + \frac{1}{9}, \ldots\right\}$$

and

$$\{d_n\} = \left\{1 - \frac{1}{3}, 1 - \frac{1}{3} + \frac{1}{5} - \frac{1}{7}, \ldots\right\}$$

It is readily seen that $\{c_n\}$ is monotone decreasing and $\{d_n\}$ is monotone increasing. We also have

$$d_2 = 1 - \frac{1}{3} + \frac{1}{5} - \frac{1}{7} > 1 - \frac{1}{3} + \frac{1}{5} - \frac{1}{5} = \frac{2}{3}$$

$$d_3 = 1 - \frac{1}{3} + \frac{1}{5} - \frac{1}{7} + \frac{1}{9} - \frac{1}{11} > 1 - \frac{1}{3} + \frac{1}{5} - \frac{1}{5} + \frac{1}{9} - \frac{1}{9} = \frac{2}{3}$$

and, in general,

$$d_n = 1 - \frac{1}{3} + \frac{1}{5} - \frac{1}{7} + \cdots - \frac{1}{4n-1} > 1 - \frac{1}{3} + \frac{1}{5} - \frac{1}{5} + \frac{1}{9} - \frac{1}{9} + \cdots$$

$$\cdots + \frac{1}{4n-1} - \frac{1}{4n-1} = \frac{2}{3}.$$

Similarly,

$$c_n = 1 - \frac{1}{3} + \frac{1}{5} - \frac{1}{7} + \cdots + \frac{1}{4n-3} < 1 - \frac{1}{3} + \frac{1}{3} - \frac{1}{7} + \cdots$$

$$\cdots - \frac{1}{4n-3} + \frac{1}{4n-3} = 1.$$

Thus

$$\frac{2}{3} < d_n \leqslant c_n < 1.$$

Since $\{c_n\}$ is monotone decreasing with a lower bound, by Definition 2 of 8.4 $\{c_n\}$ converges. Similarly $\{d_n\}$ converges since $\{d_n\}$ is monotone increasing with an upper bound.

Now

$$c_n - \frac{1}{4n-1} = d_n,$$

and by (1), (2) of 8.4,

$$\lim_{n} c_n - \lim_{n} \frac{1}{4n-1} = \lim_{n} d_n.$$

But

$$\lim_{n} \frac{1}{4n-1} = 0$$

(Exercise 7(e) of 8.7) and

$$\lim_{n} c_n = \lim_{n} d_n.$$

It was proved by Leibniz in 1673 that the common limit of the two sequences $\{c_n\}$ and $\{d_n\}$ is $\pi/4$. A further result,[4] which we shall not demonstrate here, is the proof that π (or $\pi/4$) is irrational. In this connection, we should remark that for a given sequence it is not obvious by inspection whether the limit is rational or irrational and there are many sequences for which the rationality of the limit, despite centuries of intensive investigation, is still an open question.

It is relatively easy, once the methods have been developed, to deduce other sequences whose limit is π. Recently one of these sequences was used to calculate π to 100,000 decimal places (for most purposes the first four or five are all that are needed). We can now represent π as the equivalence class of sequences converging to this same limit.

The preceding considerations concerning the numbers $\sqrt{2}$ and π lead us to the notion of representing *any* real number by an equivalence class of convergent sequences of rationals. If we find any sequence of rationals that is convergent to the particular real number, then all sequences in this equivalence class are convergent to the same limit. The importance of this idea of representing a real number by an equivalence class is that we can prove properties of the real number by using any member of the equivalence class as a representative of the real number. As a rather simple example of this procedure we shall prove that $\sqrt{2} \cdot \sqrt{3} = \sqrt{6}$.

We have shown that the sequence $\{b_n\}$, where $b_{n+1} = \frac{1}{2}(b_n + 2/b_n)$, $b_0 = 2$, converges to $\sqrt{2}$. Similarly (Exercise 18 of 8.7) the sequence $\{r_n\}$ where $r_{n+1} = \frac{1}{2}(r_n + 3/r_n)$, $r_0 = 3$, converges to $\sqrt{3}$. By (3) of 8.4,

$$\sqrt{2} \cdot \sqrt{3} = \left(\lim_{n} b_n\right)\left(\lim_{n} r_n\right) = \lim_{n}(b_n r_n) = d \text{ (say)}.$$

Now

$$2b_n b_{n+1} = b_n^2 + 2, \qquad 2r_n r_{n+1} = r_n^2 + 3.$$

Hence

$$4b_n r_n b_{n+1} r_{n+1} = b_n^2 r_n^2 + 2r_n^2 + 3b_n^2 + 6.$$

---

[4] The proof (given first by Lambert in 1761) will be found in *Irrational Numbers*, by I. Niven, John Wiley, 1956.

Now

$$\lim_n b_n{}^2 = 2, \quad \text{and} \quad \lim_n r_n{}^2 = 3.$$

Hence, by (1), (2), (3) of 8.4, $4d^2 = d^2 + 6 + 6 + 6$. Thus $3d^2 = 18$, $d^2 = 6$, $d = \sqrt{6}$.

The proof that $\sqrt{a} \cdot \sqrt{b} = \sqrt{(ab)}$ holds generally in R can be deduced from Theorem 2 below, which, in effect, proves that if we combine real numbers by the usual algebraic operations, the properties so deduced are independent of the particular sequences chosen as the representatives of the numbers.

Theorem 2. *If $\{a_n\} \sim \{p_n\}$ and $\{b_n\} \sim \{q_n\}$, then*

(a) $\{a_n + b_n\} \sim \{p_n + q_n\}$,

(b) $\{a_nb_n\} \sim \{p_nq_n\}$,

(c) $\{a_n/b_n\} \sim \{p_n/q_n\}$ (assuming $\lim_n \{b_n\} \neq 0$, $\lim_n \{q_n\} \neq 0$).

Proof. The hypothesis of the Theorem is $\{a_n\} \sim \{p_n\}$ and $\{b_n\} \sim \{q_n\}$. In other words,

$$\lim_n |a_n - p_n| = 0 \text{ and } \lim_n |b_n - q_n| = 0.$$

By Lemma 2,

(2) $$\lim_n (a_n - p_n) = 0 \text{ and } \lim_n (b_n - q_n) = 0.$$

We shall prove (b) and leave the rest to the reader. First note that

$$\{a_nb_n - p_nq_n\} = \{a_n(b_n - q_n) + q_n(a_n - p_n)\}.$$

Hence, by (1), (3) of 8.4,

$$\lim_n(a_nb_n - p_nq_n) = \lim_n a_n \lim_n(b_n - q_n) + \lim_n q_n \lim_n(a_n - p_n).$$

Whence using (2),

$$\lim_n(a_nb_n - p_nq_n) = 0,$$

and by Lemma 2,

$$\lim_n |a_nb_n - p_nq_n| = 0.$$

This proves $\{a_nb_n\} \sim \{p_nq_n\}$.

The question arises: how do the rational numbers—these, of course, form part of the real number system—fit into this picture. As an example, the rational number 3/5 can be represented by the sequence $\{3/5, 3/5, 3/5, \ldots\}$

or by the sequence $\{3/5 + 1/n\}, n \in N$. In fact, the set of all sequences converging to $3/5$ will form an equivalence class, and this class serves to determine the *real number* $3/5$. We can identify the equivalence class with the rational number. Much the same sort of situation arises in the construction of
the rational numbers when we identify the integer 2 with the equivalence
class of rational numbers m/n equivalent to the rational number $2/1$ (cf.
Example 8 of 7.5). The representation of real numbers by equivalence
classes of sequences of rationals is due to A. L. Cauchy and G. Cantor. It
is an alternative construction of the reals from the rationals to the method
used by Dedekind (8.2).

The rational numbers form a **subfield** of the field of real numbers (cf.
7.10), that is, a subset of R with the property that the elements of the subset
form a field under the operations of addition and multiplication in R. The
identification of the equivalence classes of rational sequences converging to
rational numbers with the rational numbers themselves is an example of an
isomorphism (cf. 7.9, 7.10).

The development of the real number system can now be used to throw
light on the questions raised in 8.1. For example, are there solutions of
$x^2 + 8x + 5 = 0$ in the real number system? (Example 1 and Figure 8–3
of 8.1). This question, and similar questions, are answered by the following
theorem due to Bolzano; this will be proved in 12.9. The theorem is a consequence of OC.

Theorem 3. *Let the polynomial f, defined by $f(x) = x^n + a_1 x^{n-1} + \cdots +
a_n$, be such that for real numbers a and b, $f(a) < 0$ and $f(b) > 0$. Then there
exists a real number $c, a < c < b$, such that $f(c) = 0$* (see Figure 8–12).

It should be noted that if we were working with just rational numbers
Theorem 3 would not be true; consider for example the graph of $y = x^2 +
8x + 5$ discussed in 8.1. This emphasizes the fundamental importance of OC.

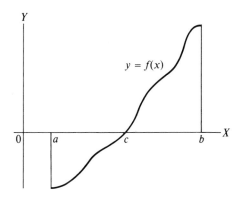

Figure 8–12   Graph of $y = f(x)$

As we have remarked in 8.4, we could have defined convergence of a sequence of rationals by Cauchy's Convergence Criterion (11.3). Then two sequences $\{a_n\}$ and $\{b_n\}$ belong to the same equivalence class, and consequently define the same real number, if

$$\lim_n |a_n - b_n| = 0.$$

Many of the theorems in the preceding sections of this chapter would then become definitions. For example, Theorem 2 of 8.4 would be replaced by the following definition for the sum of two real numbers. If $\{a_n\}$, $\{b_n\}$ are two sequences of rationals defining real numbers a and b then the real number $a + b$ is defined to be the class of sequences equivalent to the sequence $\{a_n + b_n\}$. The difficult part of this approach is to show that convergent sequences of *reals* always have *real* limits. It turns out that this is equivalent to establishing that the condition OC is satisfied (that the ordered field is complete). Although we have only sketched the theory very briefly in 8.2, the Dedekind cut method of constructing the reals is considerably briefer than the Cantor–Cauchy sequence approach. However, the complete ordered fields obtained by these two approaches are isomorphic.

8.6   REPRESENTATION OF RATIONALS BY DECIMALS

As we have seen in Example 9, of 8.4, the decimal representation of a real number is a particular case of the representation of the number by a sequence. Any real number can be represented by a decimal, say, $d_0 + 0.d_1 d_2 d_3 \ldots$, where $d_0 \in I$ and d_i, $i = 1, 2, 3, \ldots$ are chosen from the integers 0, 1, 2, ..., 9. The expression $d_0 + 0.d_1 d_2 d_3 \ldots$ really stands for the limit of the sequence

$$\left\{ d_0 + \left(\frac{d_1}{10} + \frac{d_2}{10^2} + \cdots + \frac{d_n}{10^n} \right) \right\}.$$

Now any sequence of this form is convergent, for it is monotone nondecreasing and has the upper bound

$$d_0 + \left(\frac{9}{10} + \frac{9}{10^2} + \cdots \right) = d_0 + 1;$$

thus by Definition 2 of 8.4, it is convergent.

Since the rationals form a subset of the reals, we should be able to represent them by decimals. In Example 9 of 8.4 we found $2.\overline{3} = 2\frac{1}{3}$ and $6.\overline{142857} = 6\frac{1}{7}$. From experience, the reader may have observed that rational numbers always appear to have terminating (for example, $\frac{1}{2} = 0.5$), or periodic (that is, infinitely repeating) decimal representations. This observation can be established as a theorem.

Theorem 1. *A number is rational if and only if its decimal representation is terminating or periodic.*

Proof. If a decimal is terminating, then it represents a rational number by definition.

Now suppose that the number c has a periodic decimal, that is, suppose $c = d_0 + 0.d_1d_2 \ldots d_p\overline{p_1p_2 \ldots p_q}$. Let $p_1 \ldots p_q = d$. Then, as in Example 9 of 8.4,

$$c = d_0 + \frac{d_1}{10} + \frac{d_2}{10^2} + \cdots + \frac{d_p}{10^p} + \frac{1}{10^{p+q}}\left(d + \frac{d}{10^q} + \frac{d}{10^{2q}} + \cdots\right)$$

$$= d_0 + \frac{d_1}{10} + \frac{d_2}{10^2} + \cdots + \frac{d_p}{10^p} + \frac{d}{10^{p+q}}\left(\frac{1}{1-1/10^q}\right)$$

$$= d_0 + \frac{d_1}{10} + \frac{d_2}{10^2} + \cdots + \frac{d_p}{10^p} + \frac{d}{10^p(10^q-1)},$$

and this number is rational.

Conversely, suppose that c is rational. We shall show that it can be expressed either as a terminating decimal or as a periodic or infinitely repeating decimal. Let $c = a/b$ be a positive rational number in lowest terms, so that a and b are positive integers and are prime to each other. By successive applications of Theorem 1 of 5.5 we divide a by b, at each step getting a quotient and a partial remainder which becomes the dividend for the next step. Since the partial remainder for each step is always less than the divisor b, the only possible partial remainders are $0, 1, 2, \ldots, (b-1)$. Therefore after at most b steps we either obtain the remainder 0, in which case the division terminates and a/b is represented by a terminating decimal, or we obtain a partial remainder that we have already had at an earlier step; in this case the rest of the division proceeds as it did after that earlier step, and a/b is represented by an infinitely repeating decimal. If c is a non-positive rational a proof can easily be constructed.

8.7   EXERCISES

(All sequences in these exercises are sequences of rational numbers.)

1. Prove that $1 + \sqrt{2}$ is irrational.
2. Prove that $\sqrt{3} + \sqrt{7}$ is irrational.
3. Prove that the roots of $x^2 + bx + c = 0$ are rational if $b = k + c/k$, where b, c, k are rational.

4. (a) Prove that $\sqrt{2}/n, n \in N - \{0\}$ is irrational.
    (b) Prove that $r + \sqrt{2}/n, n \in N - \{0\}, r \in Q$, is irrational.
    (c) Prove Corollary 2 of 8.3 another way. (Hint: What can you say about $(x + r)/2$?)

5. If a is rational and b is irrational prove that $a + b, a - b, ab, a/b, b/a$ are irrational.

6. Prove: (a) $|xy| = |x|\,|y|$,
            (b) $|x| = 0$ if and only if $x = 0$,
            (c) $\big||x| - |y|\big| \leqslant |x - y|$,
            (d) $|x - z| \leqslant |x - y| + |y - z|$.

7. Prove that

    (a) $\lim\limits_{n} \left(\dfrac{1}{n^2} + \dfrac{2}{n^2} + \cdots + \dfrac{n}{n^2} \right) = \dfrac{1}{2}$,

    (b) $\lim\limits_{n} \left(\dfrac{1}{10^n} \right) = 0$,

    (c) $\lim\limits_{n} (\sqrt[n]{n}) = 1$,

    (d) $\lim\limits_{n} \left(\dfrac{1}{n^a} \right) = 0, \qquad a > 0$.

    (e) $\lim\limits_{n} \left(\dfrac{1}{an + b} \right) = 0$.

8. (a) Prove that the two sequences $\{a_n\}$ and $\{b_n\}$ where $a_n = (1+1/n)^n$, $b_n = (1+1/n)^{n+1}$ are respectively monotone increasing and monotone decreasing.
    (b) Prove that these sequences converge to a number e where $2 < e < 3$.
    (c) Show that, approximately, $e = 2.718$.

9. If $\{a_n\}$ is a convergent sequence and c is any real number, prove that
$$\lim\limits_{n} ca_n = c \lim\limits_{n} a_n.$$

10. If $a_n \leqslant b_n$ for all n and $\lim\limits_{n} a_n = a$, $\lim\limits_{n} b_n = b$, prove that $a \leqslant b$.

11. Prove that $\lim\limits_{n} \left(\dfrac{1}{1 \times 2} + \dfrac{1}{2 \times 3} + \cdots + \dfrac{1}{n(n+1)} \right) = 1$.

$\left(\text{Hint: Let } \dfrac{1}{n(n+1)} = \dfrac{1}{n} - \dfrac{1}{n+1} \right).$

12. Prove that $\lim\limits_{n} \left(\dfrac{1}{2^2 - 1} + \dfrac{1}{3^2 - 1} + \cdots + \dfrac{1}{n^2 - 1} \right) = \dfrac{3}{4}$.

13. Express as fractions
    (a) $0.\overline{123}$,
    (b) $0.5\overline{1}$,
    (c) $0.3\overline{45}$.

14. (a) Prove that the two sequences $\{a_n\}$ and $\{b_n\}$ where

$$a_n = 1 - \frac{1}{2} + \frac{1}{3} - \frac{1}{4} + \cdots + \frac{1}{2n+1}$$

$$b_n = 1 - \frac{1}{2} + \frac{1}{3} - \frac{1}{4} + \cdots - \frac{1}{2n}$$

are monotone decreasing with a lower bound, and monotone increasing with an upper bound, respectively.

(b) Prove that these sequences converge to a number which is approximately 0.693.

15. If $\{a_n\}$ and $\{b_n\}$ are monotone nonincreasing sequences with lower bounds, prove that the sequence $\{c_n\}$ where $c_n = a_n + b_n$ is convergent and $\lim_n c_n = \lim_n a_n + \lim_n b_n$.

16. If $\{a_n\}$ and $\{b_n\}$ are convergent sequences, prove that $\{c_n\}$, where $c_n = a_n b_n$, is convergent, and $\lim_n c_n = \left(\lim_n a_n\right)\left(\lim_n b_n\right)$.

17. If $\{a_n\}$ and $\{b_n\}$ are convergent sequences, prove that $\{c_n\}$, where

$$c_n = \frac{a_n}{b_n},$$

is convergent and

$$\lim_n c_n = \frac{\lim_n a_n}{\lim_n b_n}$$

(assume $\lim_n b_n \neq 0$ and *hence* $b_n \neq 0$ for all sufficiently large n—show this).

18. Prove that the sequence $\{a_n\}$ where

$$a_{n+1} = \frac{1}{2}\left(a_n + \frac{a}{a_n}\right),$$

$a_0 = a > 0$ converges to $\sqrt{a}$, for a rational.

19. Let a, b, c, d be elements of an ordered field F. If $a < b$ and $c < d$ prove that $a + c < b + d$.

20. (a) Prove that the limit of the sequence $\{a_n\}$ where

$$a_{n+1} = \frac{m-1}{m} a_n + \frac{a}{m a_n^{m-1}},$$

$a > 0$, $a_0 = a$ and $a \in Q$, is the number $a^{1/m}$, $m \in N - \{0\}$.

(b) Prove that $a^{1/p} \cdot a^{1/q} = a^{(p+q)/pq}$, $a > 0$, $p, q \in N - \{0\}$.

(c) Prove that $a^r \cdot a^s = a^{r+s}$ where $a > 0$, and $r, s \in Q$.

21. (a) Prove that the limit of the sequence $\{a_n\}$, where

$$a_{n+1} = \frac{3}{4} a_n + \frac{1}{20 a_n} + \frac{3}{5 a_n + 25} + \frac{3}{5 a_n - 25},$$

$a_0 = 4$, is the number $\sqrt{3} + \sqrt{2}$.

(b) Using the same sequence as in (a) but with $a_0 = 0.3$ prove that the limit of the sequence is $\sqrt{3} - \sqrt{2}$.

9

The Algebra of Real Functions

9.1 COMBINING FUNCTIONS

In Chapter 7 the basic properties of functions were defined and additional properties developed. In this chapter we continue this development by giving definitions of operations for combining functions in various ways. We shall be concerned almost exclusively with **real-valued functions**, a concept defined below in 9.2. It will be recalled that in 6.1 the eight properties A1, A2, A3, A4, M1, M2, M3, and D were postulated, and that these properties enabled one to form new rational numbers from any given finite set of rationals. In much the same way we shall define methods for operating on real functions. It turns out, as we shall see in 9.8, that the resulting mathematical system is *not* a field (as the rational numbers were shown to be in 6.1), but is another system called a **commutative ring**.

9.2 REAL FUNCTIONS

We make the following definition.

Definition 1. *Let f be a function mapping the domain D into the counter domain C. Then:*

(a) *If $D \subseteq R$, f is called a **function of a real variable**.*

(b) *If $C \subseteq R$, f is called a **real-valued function**.*

(c) *If $D \subseteq R$ and $C \subseteq R$, f is called a **real valued function of a real variable** or simply a **real function**.*

142

Example 1. Let C be the set whose elements are the first word on each page of a book, and D be the set of page numbers. Then if $x \in D$, $f(x)$ will denote the first word on page x. This is an example of a function of a real variable.

Example 2. Let C be the set whose elements are the weights to the nearest pound of students in a school and D be the set of students. If $x \in D$ is a particular student, define $f(x)$ to be the weight of student x. This is an example of a real-valued function.

Example 3. Let $C = R$ and $D = \{x \in R \mid x > 0\}$. For $x \in D$, define $f(x)$ to be $\log_{10} x$. This is an example of a real function.

If A, B, C are sets, then $A \times B \times C$, **the Cartesian product** of A, B, C (cf. 7.2), is the set of all ordered triples of the form (a, b, c) where $a \in A$, $b \in B$, $c \in C$. (By convention, $(A \times B) \times C = A \times (B \times C)$; that is we identify pairs of the form $((a, b), c)$ or $(a, (b, c))$ with triples of the form (a, b, c). In general, $A \times B \times C$, $A \times C \times B$, etc. are different.) The Cartesian product with more factors is defined similarly. We abbreviate $A \times A \times \ldots \times A$, with n factors, as A^n. The set R^n consists of all ordered n-tuples of real numbers: it can be interpreted geometrically as the set of all points in n-dimensional Euclidean space.

Functions of several variables can be defined using this generalized Cartesian product. In particular, Definition 1 can be extended in the following way.

Definition 2. *Let f be a function mapping D into C. If $D \subseteq R^n$, f is called a function of n real variables.*

Example 4. Let C be the set $\{1868, 1869, \ldots, 1964\}$ and D be the set $\{1, 2, \ldots, 12\}$. If $x \in C$ and $y \in D$, we define $f(x, y)$ to be the Prime Minister of Canada in year x and month y (or on the first day of month y). Thus $f(1868, 1)$ is Sir John A. Macdonald and $f(1964, 12)$ is L. B. Pearson. The function so defined is an example of a function of two real variables.

Example 5. Suppose the position of a point in a room can be specified by three real numbers (the distances from two adjoining walls and the floor). Then the air temperature in the room over a period of time will be a function of four real variables, the three space coordinates and time.

The reader will have no difficulty in devising other examples.

9.3 EQUALITY OF FUNCTIONS

A basic notion in the algebra of functions is the definition of equality.

Definition 1. *Let f and g be two real-valued functions. We say that f = g if and only if f and g both have the same domain D and for each $x \in D$, $f(x) = g(x)$.*

This definition applies to functions generally, not just to real-valued functions. It can properly be regarded as a **definition** of the **notation** $f = g$. In substance it is really a theorem, and is implied by two definitions given in Chapter 7. In Theorem 2 of 7.6 a function is (alternatively) defined as a set of ordered pairs (x, y) such that for each $x \in D$ there is a unique $y \in C$. It follows from Definition 1 of 7.2 that two ordered pairs are equal if and only if the two first elements are equal and the two second elements are equal. Thus these two definitions imply Definition 1 above.

Definition 1 is a meaningful definition in the sense that equality has already been defined for real numbers (recall that $f(x)$ and $g(x)$ are both real numbers).

Definition 2. *The function s where*

$$s(x) = \begin{cases} 1 \text{ if } x \geqslant 0 \\ 0 \text{ if } x < 0 \end{cases}, \quad D = R$$

*will be called the **unit step function**. See Figure 9–1.*

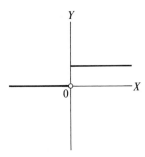

Figure 9–1   The unit step function

Example 1. Let $g(x) = 1$ and $D = R$. Then $s \neq g$ since $s(x) \neq g(x)$ if $x < 0$.

Example 2. Let $f(x) = \frac{1}{2}(x + |x|)$, $D = R$, and $g(x) = x\,s(x)$, (see Figure 9–2). By Definition 1 it is readily seen that $f = g$.

Example 3. Let $f(x) = x^2$, $D = \{x \in R \mid x \geqslant -1\}$; $g(x) = x^2$, $D = \{x \in R \mid x < 0\}$. By Definition 1, $f \neq g$ since the domains of these two functions are different.

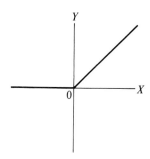

Figure 9–2 The function $g: y = x\,s(x)$, $D = R$

9.4 ADDITION OF FUNCTIONS

Let f be a function that maps a set A into a set B. In 7.8 we discussed the important distinction between "f" as the name of the function and "$f(x)$" as the name of an element in B.

There is another method of naming (rules of) functions which is often more convenient than assigning letters of the alphabet.

For example, instead of saying, "Let f be the function defined on R by the rule $f(x) = x^2$", we shall often say, "Consider the function $x \rightarrow x^2$ defined on R". The expression "$x \rightarrow x^2$" is the *name* of the rule; the variable x is bound by the arrow. In contrast, "x^2" is the name of a number; the variable x is free for substitution.

Mnemonically, $x \rightarrow x^2$ means that x is mapped onto x^2 for any $x \in R$.

Using this device, the variable x in the term $f(x)$ can be bound by the arrow to give the name of the function: $x \rightarrow f(x)$. It makes perfectly good sense to write: $f = (x \rightarrow f(x))$.

Definition 1. *Let f be a real-valued function with domain D_f and let g be a real-valued function with domain D_g ($D_f \cap D_g \neq \emptyset$). Then $f + g$ is a real-valued function with domain $D_{f+g} = D_f \cap D_g$ and the rule of correspondence is $x \rightarrow f(x) + g(x)$ where $x \in D_{f+g}$.*

It should be clear that $f + g$ is indeed a function, for by Definition 1 of 7.1, all we need to show is that corresponding to each number of D_{f+g} there is a unique number in the counter domain. This, however, follows from the closure property of addition for real numbers.

Example 1.   If $f(x) = x$, $D_f = R$, and $g(x) = |x|$, $D_g = R$,

then the sum of these two functions is the function whose domain is $D_f \cap D_g = R \cap R = R$ and the rule of correspondence is $x \rightarrow x + |x|$.

A useful graphical device is that the graph of the sum of two functions is found by adding the ordinates of points with the same abscissas to obtain the points of the new graph (see Figures 9–3, 9–4, 9–5).

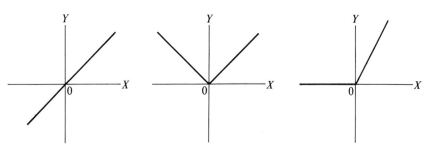

Figure 9–3 The function $f: y = x$, $D_f = R$

Figure 9–4 The function $g: y = |x|$, $D_g = R$

Figure 9–5 The function $x \rightarrow x + |x|$ defined on R

9.5 SUBTRACTION OF FUNCTIONS

Definition 1.   *Let f be a real-valued function with domain D_f and g be a real-valued function with domain D_g ($D_f \cap D_g \neq \emptyset$). Then $f - g$ is a real-valued function with domain $D_{f-g} = D_f \cap D_g$ and the rule of correspondence is $x \rightarrow f(x) - g(x)$ where $x \in D_{f-g}$.*

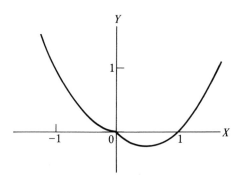

Figure 9–6   The function $x \rightarrow x^2 - x \, s(x)$

Definition 1 of 7.1, together with the result that the difference of two real numbers is a unique real number show that $f - g$ is a function.

Example 1. Let $f(x) = x^2,$ $D_f = R,$

$$g(x) = x\,s(x), \quad D_g = R.$$

By Definition 1, the domain of $f - g$ is R and the rule of correspondence is $x \rightarrow x^2 - x\,s(x)$. The graph of this function is shown in Figure 9-6. It is seen that the ordinate of the difference function at any x is equal to the difference of the ordinates of the separate functions.

9.6   MULTIPLICATION OF FUNCTIONS

Definition 1. *Let f be a real-valued function with domain D_f and g be a real-valued function with domain D_g ($D_f \cap D_g \neq \emptyset$). Then fg is a real-valued function with domain $D_{fg} = D_f \cap D_g$ and the rule of correspondence is $x \rightarrow f(x)\,g(x)$.*

The distinction between fg and $f \circ g$ should be kept clearly in mind. The mapping produced by fg is $x \rightarrow f(x)\,g(x)$, whereas the mapping produced by $f \circ g$ is $x \rightarrow f(g(x))$ (Definition 1 of 7.8). These two mappings will not, in general, be the same.

    Example 1. Let $f(x) = x^2$ $D_f = R,$

and $g(x) = x^{1/2}, \quad D_g = \{x \in R \mid x \geqslant 0\}.$

Then $f(x)g(x) = x^{5/2}, \quad D_{fg} = \{x \in R \mid x \geqslant 0\}.$

(On the other hand, $(f \circ g)(x) = x, \quad D_{f \circ g} = D_g.$)

The ordinate of the product function at any point is the product of the ordinates of the separate functions. (Figure 9-7)

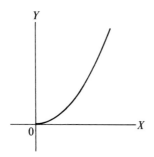

Figure 9–7   The function $x \rightarrow x^{5/2}$, $D = \{x \in R \mid x \geqslant 0\}$

9.7 DIVISION OF FUNCTIONS

Definition 1.   *Let f be a real-valued function with domain D_f and g be a real-valued function with domain D_g. Then f/g is a real valued function with domain*

$$D = D_{f/g} = D_f \cap D_g - \{x \in D_f \cap D_g | g(x) = 0\}$$

(provided $D_{f/g} \neq \varnothing$), and the rule of correspondence is $x \to \dfrac{f(x)}{g(x)}$.

In the definition it is clearly necessary to exclude from the domain those elements x for which $g(x) = 0$ since $f(x)/g(x)$ is not defined when $g(x) = 0$.

Example 1.   Let $f(x) = s(x)$, and

$$g(x) = x^2 - 3x + 2, \text{ both with domain } D = R.$$

Then

$$\frac{f}{g}(x) = \frac{s(x)}{x^2 - 3x + 2}, \qquad D = R - \{1, 2\}.$$

The domain excludes the points at which the function g is zero.

Example 2.   Let $f(x) = \sqrt{x}, \qquad D_f = \{x \in R \,|\, x \geqslant 0\}$

$$g(x) = \sqrt{(-x)}, \; D_g = \{x \in R \,|\, x \leqslant 0\}.$$

Then the quotient function f/g is not defined since

$$D_{f/g} = D_f \cap D_g - \{0\} = \{0\} - \{0\} = \varnothing.$$

9.8 RINGS AND FIELDS

From the definitions of 9.4 to 9.7, it follows that if f and g are real-valued functions, then, provided they are defined, $f + g, f - g, fg$, and f/g are real-valued functions. We shall show below that certain sets of real functions under these operations form a new type of mathematical system called a **commutative ring**.

Definition 1.   *A **ring** is a set that is closed under two operations, addition and multiplication, with the following properties. For all a, b, c that are elements of the ring:*

A1. $a + b = b + a$ *Commutative Property of Addition*

A2. $(a + b) + c = a + (b + c)$ *Associative Property of Addition*

A3. There exists an element 0 in the
     ring such that $0 + a = a$ *Property of Additive Identity*

A4. For each a, there is an element
     $-a$ in the ring such that
     $a + (-a) = 0$ *Property of Additive Inverse*

M2. $a(bc) = (ab)c$ *Associative Property of Multiplication*

D. $a(b + c) = ab + ac$,
     $(b + c)a = ba + ca$ *Distributive Property*

These properties should be compared with the basic properties of the natural numbers given in 2.2. It should be noted that M1, the commutative property of multiplication, is not assumed above, nor is M3, the property of multiplicative identity.

Recall the definition of a group from 5.3. An (additive) **commutative** group is a set which is closed under addition and which satisfies properties A1, A2, A3, A4. A ring, therefore, is an additive commutative group which additionally satisfies M2, and D for multiplication.

In 5.3 it was shown that C1, the cancellation property of addition, was a consequence of the group axioms. Thus, the cancellation property of addition is valid in a ring. Since the elements of a ring do not form a group under multiplication, C2, the cancellation property of multiplication is not necessarily valid (unless it is inserted as an extra assumption).

The concept of field was introduced in 6.1. By comparing the axioms for a ring and a field we see that a field is a ring, but a ring is not necessarily a field.

We will now discuss the relation between a ring and a field in more detail. We have already seen a number of examples of rings.

Example 1.   The set of integers forms a ring.

Example 2.   The set of rational numbers forms a ring.

Example 3.   The set of real numbers forms a ring.

In each of the above Examples, it is assumed that addition and multiplication have their usual meaning. The proofs of Examples 1, 2, 3 will be left as an exercise. It will be recalled that the set of rational numbers and the set of real numbers also form fields, while the set of integers does *not* form a field.

Example 4.   Let A, B be subsets of a universal set U in which addition is defined by $A + B = (A - B) \cup (B - A)$ (called the symmetric difference) and multiplication by $AB = A \cap B$. It will be left as an exercise for the reader to show that the subsets of U form a ring with respect to these operations. Note that each element in the ring is its own additive inverse, and $\emptyset$ is the additive identity.

This ring is called a *Boolean ring of sets*.

Example 5. Let I_m be the set of all nonnegative integers less than the positive integer m. For example $I_4 = \{0, 1, 2, 3\}$. If $a,b \in I_m$, we define $a + b$ and ab to be the remainders after the usual sum and product are divided by km for some k such that the remainder is between 0 and $m - 1$ inclusive. For example, considering again I_4, $3 \cdot 2 = 2$, $3 + 4 = 3$. The resulting system is a ring known as the **ring of integers modulo** m (in the sequel we shall abbreviate modulo as mod). The proof that this system is a ring will be left as an exercise.

Definition 2. *A ring is called a **commutative ring** if multiplication is commutative; that is, if*

M1. $ab = ba$ *for all elements a, b of the ring.*

All the rings of Examples 1–5 above are commutative rings. An example of a noncommutative ring will be given in 9.10.

Definition 3. *If, in a ring A, there is an element* 1 *such that* $a \cdot 1 = 1 \cdot a = a$ *for all* $a \in A$, *then the ring is called a **ring with unity** (that is, A satisfies* M3). *In this case* 1 *is called the **multiplicative identity** of A.*

In Examples 1, 2, 3, 5 above, the multiplicative identity is the number 1. In Example 4, U is the multiplicative identity.
Thus these are all examples of rings with unity.

Example 6. It is easily shown that the set of all even integers forms a ring. This ring contains no multiplicative identity.

Definition 4. *Let A be a ring with unity.*
M4.   *For each* $a \in A$, $a \neq 0$, *there exists an element* $a^{-1} \in A$ *such that* $a^{-1}a = aa^{-1} = 1$.

If A satisfies M4, *it is called a **division ring**.*

The element a^{-1} *is called the **multiplicative inverse** of a.*

A ring with unity may contain *some* elements a for which there exists a^{-1} such that $a^{-1}a = aa^{-1} = 1$. It is only if *all* the elements a ($\neq 0$) have this property that the ring is called a division ring.

In a division ring, all the nonzero elements form a group with respect to multiplication. (M2, M3, M4.)

Example 7. In the set I, the only elements for which an inverse exists are $+1$ and -1. Thus I is not a division ring.

Example 8. The sets Q and R form division rings since all nonzero elements have inverses.

In a division ring A, division can be defined as follows: Let $a, b \in A$. Then $a/b = ab^{-1}$. It is a simple matter for the reader to show that **a commutative division ring is a field.**

Example 9. The following is an example of a division ring which is not a field (that is, it is noncommutative).

We define a **quaternion**, A, as a "number" of the form $A = a_0 + a_1 i + a_2 j + a_3 k$ where $a_n \in R$, $n = 0, 1, 2, 3$, and i, j, k are three unit elements which have the following multiplication table:

	i	j	k
i	-1	k	$-j$
j	$-k$	-1	i
k	j	$-i$	-1

It is assumed that in using this table the elements in the vertical column at the left are taken to be the first factor.

Thus $jk = i$, but $kj = -i$; and $j^2 = -1$. Quaternions are a generalization of complex numbers. Two quaternions A and B are equal if and only if $a_n = b_n$, $n = 0, 1, 2, 3$ (where $B = b_0 + b_1 i + b_2 j + b_3 k$).

Addition is defined by

$$A + B = (a_0 + b_0) + (a_1 + b_1)i + (a_2 + b_2)j + (a_3 + b_3)k.$$

Multiplication is defined by multiplying out the expressions for A and B using the multiplication table above and the distributive law. As indicated above, the multiplication is noncommutative (for example, $ij \neq ji$).

There is a unit element $I = 1 + 0i + 0j + 0k$ and it can be shown that each nonzero element has a multiplicative inverse. It will be left to the reader to show that the resulting system is a noncommutative division ring (see Exercise 9 of 9.11).

We shall now give two basic properties of a ring which will be used in this section and 9.9.

Theorem 1. *If a is any element of a ring, then $0a = a0 = 0$.*

Proof. The proof is identical to the proof of Theorem 2 of 2.2. It will be left to the reader to check that only the axioms for a ring (that is, Definition 1 of this section) are used in the proof of Theorem 2 of 2.2.

Theorem 2. *In a ring with unity, both the multiplicative identity and the multiplicative inverse of an element if it exists are unique.*

Proof. The proof is identical to the proof of Exercise 5 of 2.7 and will be left to the reader.

It should be noted that essentially the same proof holds for the uniqueness of the additive identity and additive inverses in a ring.

It is important to note that the converse of Theorem 1 is not true. In other words, examples of rings can be found in which the product of two nonzero elements is zero without either of the elements being zero.

Example 10. (Cf. Example 5.) In the ring of integers mod 6, $2 \cdot 3 = 0$ and $2 \neq 0$, $3 \neq 0$.

Examples 1, 2, 3 show that in some rings the only way for the product of two elements to be zero is for one or other of the two elements to be zero.

Definition 5. *An element a ($\neq 0$) in a ring is called a **zero divisor** if there exists in the ring an element b ($\neq 0$) such that $ab = 0$ or $ba = 0$.*

Thus in Example 10, 2 and 3 are zero divisors.

It was shown in Theorem 2 of 6.1 that, in contrast with many rings, a field contains no zero divisors.

Theorem 3. *In a ring with unity, if an element a is a zero divisor, then a^{-1} does not exist.*

Proof. Let a be a zero divisor. Then there exists an element $b \neq 0$ such that $ab = 0$. To prove that a^{-1} does not exist we shall use the method of proof called *reductio ad absurdum* (see 2.5).

Suppose a^{-1} exists.

Then $ab = 0 \Rightarrow a^{-1}(ab) = a^{-1}0 = 0$ (Theorem 1 of 9.8).

But $a^{-1}(ab) = (a^{-1}a)b$ (M2)

$\qquad\qquad\quad = 1 \cdot b$

$\qquad\qquad\quad = b$ (Definition 3 of 9.8)

Thus $b = 0$ contradicting our assumption, and therefore a^{-1} does not exist.

It should be noted that the converse of Theorem 3 is not true, that is, if an element of a ring with unity is not a zero divisor this does not imply that its multiplicative inverse exists (cf. Example 7).

Theorem 4. *The set K of real valued functions defined on a common domain D forms a commutative ring with unity with respect to the operations of addition and multiplication defined above. The additive identity is the function $x \to 0$ for all $x \in D$; the multiplicative identity is the function $x \to 1$ for all $x \in D$.*

Proof. The proof of this theorem follows from the Definitions of 9.3–9.7 and Definitions 1, 2, and 3 of this section. The proof will be left to the reader.

It is easily seen that if D contains more than one element, then K does not form a field since, as in the following example, one can construct zero divisors in K.

Example 11. Let $a \in D$ and $f \in K$ be a function such that

$$f(a) = 1 \text{ and } f(x) = 0 \text{ for } x \in D, x \neq a.$$

Let $g \in K$ be any function such that $g(a) = 0$
Then $f(x)g(x) = 0$ for all $x \in D$.
Hence $fg = 0$. (We use 0 to stand for the number zero and the zero function; the distinction is always clear from context.) However, $f \neq 0$.

9.9 POLYNOMIAL AND RATIONAL FUNCTIONS

Definition 1. *Let $a_i \in R$, $i = 0, 1, 2, ..., n$, $a_n \neq 0$, and $x \in R$. Then $f(x) = a_0 + a_1 x + a_2 x^2 + \cdots + a_n x^n$ is a **polynomial function** of degree n. If all of the coefficients of $f(x)$ are 0, so that $f = 0$, then it is a polynomial function but has no degree.*

In order to justify this definition it is necessary to ensure that f is actually a function. The domain of f is clearly R, for, by the properties of the real numbers, $x \in R \Rightarrow x^i \in R$ and $x^i \in R \Rightarrow a_i x^i \in R$. Hence $f(x) \in R$, and since the mapping $x \to f(x)$ associates a unique real number $f(x)$ with any real number x, the result follows. Therefore it is clear that a polynomial function is a real function. From the definition we see that a constant function (including the zero function) and a linear function (that is, functions of the form $x \to c$ or $x \to ax + b$) are also polynomial functions (of degree 0 and 1 respectively when $a, c \neq 0$). However, not all real functions are polynomial functions.

For example, the function $s(x)$ (see Definition 2 of 9.3) is not a polynomial function. (It cannot be a polynomial function by Theorem 2 below because it vanishes on the entire set $\{x \in R \,|\, x < 0\}$.)

Lemma 1.   *If* f, g *are polynomial functions, then* $f + g$, $f - g$, $f \cdot g$ *are polynomial functions.*

Proof.   Let $f(x) = a_0 + a_1 x + \cdots + a_n x^n$

$$g(x) = b_0 + b_1 x + \cdots + b_m x^m$$

be two polynomial functions of degree n, m respectively.

Assume $n \geq m$ (this results in no loss of generality).

Now, $f + g$ is the mapping

$$x \to a_0 + b_0 + (a_1 + b_1)x + \cdots + (a_m + b_m)x^m + a_{m+1}x^{m+1} + \cdots + a_n x^n$$

by Definition 1 of 9.4, and this defines a polynomial function.

Similarly, using the definitions in 9.5 and 9.6, $f - g$ and $f \cdot g$ can be shown to be polynomial functions. See Exercise 13 of 9.11.

Definition 2.   *An **integral domain** is a commutative ring with unity in which the cancellation property of multiplication is valid.*

In other words an integral domain is a set closed under the operations of addition and multiplication, in which the properties A1, A2, A3, A4, M1, M2, M3, C2, and D are valid. These precisely are the properties discussed in 2.1 and 5.1 in connection with the properties of the integers. It is because the integers themselves have these properties that the system in Definition 2 is called an "integral" domain.

Example 1.   As we have remarked above, the integers form an integral domain.

Example 2.   The even integers do not form an integral domain since they contain no multiplicative identity.

Example 3. The set of numbers of the form $a + b\sqrt{5}$ where $a, b \in I$ form an integral domain. The proof will be left to the reader.

Theorem 1. *A commutative ring with unity is an integral domain if and only if it contains no zero divisors.*

Proof. Suppose that a, b are any two elements of an integral domain such that a is a zero divisor. Thus there is a b such that $ab = 0b = 0$ and $b \neq 0$. Apply cancellation to $ab = 0b$.

Hence, $a = 0$ and a is not a zero divisor (Definition 5 of 9.8 and C2), which is a contradiction.

Conversely, suppose that a, b, c $(c \neq 0)$ are three elements of a commutative ring with unity and without zero divisors, such that $ac = bc$.

Then $\qquad\qquad ac - bc = (a - b)c = 0 \qquad$ (D),

$$a - b = 0 \qquad \text{(Definition 5 of 9.8),}$$

and $\qquad\qquad\qquad a = b.$

Thus C2 holds, and the ring is an integral domain.

Example 4. In Examples 5 and 10 of 9.8 the ring of integers mod 6 was shown to be a commutative ring with unity. However, since it has zero divisors, by the preceding theorem it cannot be an integral domain.

For an integral domain to be a field, all nonzero elements must have multiplicative inverses. Under these circumstances the nonzero elements form a group under multiplication, and it follows from the Theorem of 5.3 that the cancellation law of multiplication is valid. Thus, C2 is redundant in the axioms for a field.

We shall show (Theorem 3) that the set of polynomials is an integral domain. To do this we need the following definition and some preliminary results.

Definition 3. *A zero of a real valued function f is an element x in the domain of f such that $f(x) = 0$.*

Theorem 2. *A polynomial function of degree $n \geqslant 0$ has at most n zeros.*

Proof. We shall prove this theorem by mathematical induction (see 2.5). First, a preliminary result is needed.

(a) Let $f_{n+1}(x) = a_0 + a_1 x + \cdots + a_{n+1} x^{n+1}$ define a polynomial function of degree $n + 1$. Then

$$f_{n+1}(x) - f_{n+1}(c) = a_1(x - c) + \cdots + a_{n+1}(x^{n+1} - c^{n+1}).$$

But $x - c$ is a factor of the right side. Thus we can write

$$f_{n+1}(x) - f_{n+1}(c) = (x - c) g_n(x)$$

where $g_n(x)$ is some polynomial of degree n.

Now if $x = c$ is a zero of f_{n+1}, then we have

$$f_{n+1}(x) = (x - c) g_n(x).$$

(b) BASIS. A polynomial function of degree 0 is a constant function whose value is different from 0, and has no zeros.

INDUCTION STEP. We assume that a polynomial function of degree n has at most n zeros. Thus $g_n(x)$ has at most n zeros. By (a) every polynomial function f_{n+1} of degree $n + 1$ can be written in the form

$$f_{n+1}(x) = (x - c) g_n(x)$$

where $g_n(x)$ has degree n. But $x - c$ and $g_n(x)$ are real numbers and $(x - c) g_n(x)$ can vanish only if $x - c = 0$ or $g_n(x) = 0$ (because of the absence of zero divisors in the field of reals). Thus $f_{n+1}(x)$ has at most $n + 1$ zeros.

This completes the induction step and proves the theorem.

It should be observed that Theorem 2 does not assert that a polynomial need have any zeros at all.

Lemma 2. *Let f be a polynomial function of degree $\leq n$. If f has more than n zeros, it vanishes for all x.*

Proof. This result is the contrapositive statement of Theorem 2.

Theorem 3. *The set of polynomial functions forms an integral domain.*

Proof. Using Lemma 1, it is a straightforward matter to show that A1, ..., M3, and D are satisfied. The proof rests ultimately on the fact that these properties are satisfied by the real numbers.

To complete the proof we must show that there are no zero divisors in the ring (Theorem 1). In other words, if f, g are two polynomial functions, then $fg = 0 \Rightarrow (f = 0 \lor g = 0)$. This follows from Theorem 2.

Definition 4. *A subset S of a ring A is called a **subring** of A if and only if S is itself a ring under the operations of addition and multiplication of the ring A.*

The integral domain formed by polynomial functions is a subring of the ring of real functions defined on R. That is to say, since the polynomial functions form a ring (as has been proved in Theorem 3) and are themselves real functions, they can be considered as a ring imbedded in the larger ring of real functions defined on R. The subring of polynomials contains the multiplicative and additive identities of the whole ring. The relation between the polynomial and real functions is shown diagrammatically as in Figure 9–8.

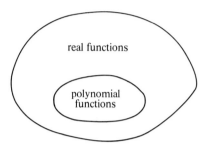

Figure 9–8   The polynomial functions form a subring of the ring of real functions

Definition 5.   *Let f, g be polynomial functions, $g \neq 0$.*

     *Then $p = f/g$ is a **rational function**.*

     The domain of p is $D_p = R - \{x \mid g(x) = 0\}$.

Clearly p is a real function,.

     Example 1.   Let $f(x) = x$, $g(x) = x^2 - 1$. Then $p(x) = x/(x^2 - 1)$ defines a rational function whose domain is the set $D_p = R - \{1, -1\}$.

Theorem 4.   *The rational functions form a field.*

     Proof.   The proof follows along the lines of the proof of Theorem 3. The only additional part of the proof is the observation of the obvious fact that each nonzero rational function has a multiplicative inverse.

9.10   INVERSE OF A REAL FUNCTION

It will be recalled from Definition 3 of 7.8 that if f is a one-to-one function, there exists an inverse function f^{-1} such that $y = f(x) \Leftrightarrow x = f^{-1}(y)$. In the particular case of real functions, f gives a one-to-one mapping of some subset B of R onto some subset C of R, and f^{-1} also gives a one-to-one mapping, but of C onto B.

     At this point it is worth emphasizing to the reader the necessity of distinguishing in his mind between f^{-1} and $1/f$.

The function $1/f$ is defined only if f is a real-valued function, and it is the mapping $x \to 1/f(x)$ where the domain is the set $\{x \in D_f | f(x) \neq 0\}$ (Definition 1 of 9.7). On the other hand, f^{-1} is defined if and only if f is one-to-one and onto, but f need not be a real-valued (or even a real) function. In this section, however, f will be considered to be a real function which is one-to-one (and, necessarily, onto its range). The function f^{-1} is then the mapping $y \to f^{-1}(y)$ where $f^{-1}(y) = x \Leftrightarrow y = f(x)$ (Definition 3 of 7.8). The domain of f^{-1} is the range of f and the range of f^{-1} is the domain of f. Thus $1/f$ and f^{-1} need not be the same. This is shown more clearly in Examples 1, 3, and 4 below.

In terms of the alternative definition of a function given in Theorem 2 of 7.6, a real function is a set of ordered pairs of real numbers (x, y) such that for each x there is a unique y. In the case of a real one-to-one function we have, in addition, for each y a unique x. Thus, if a real one-to-one function f is represented by the set of ordered pairs $\{(a, b) \in R \times R \,|\, a \in D_f \wedge b = f(a)\}$, the inverse function f^{-1} is represented by the set of ordered pairs

$$\{(b, a) \in R \times R \,|\, a \in D_f \wedge b = f(a)\}.$$

In 7.4 the graph of a relation was defined. Since a function is a relation, a graph of a real function f is the set of all ordered pairs of real numbers (x, y) such that $y = f(x)$. Hence a picture can be made in the usual way (as has been done throughout the book) by plotting the ordered pairs as points in the plane. It is usual to let the first element in the ordered pair represent the distance from a vertical axis and the second element represent the distance from a horizontal axis. We shall call this picture the graph of the function for short, although it is really (as explained in 7.4) a picture of the graph. In the case of a real one-to-one function, it follows from the above remarks that any horizontal or vertical line meets the graph in at most one point.

Example 1. Let a function f be given by the set of ordered pairs $f = \{(1, 2), (2, 3), (3, 4)\}$. This is a real one-to-one function with domain $\{1, 2, 3\}$ and range $\{2, 3, 4\}$. The inverse function is $f^{-1} = \{(2, 1), (3, 2),$

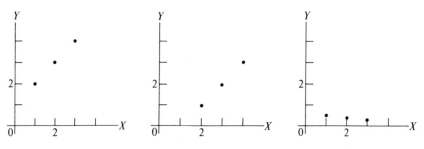

Figure 9–9 Graph of f **Figure 9–10** Graph of f^{-1} **Figure 9–11** Graph of $1/f$

(4, 3)}. The function $1/f$, on the other hand, is the set of ordered pairs $\{(1, \frac{1}{2}), (2, \frac{1}{3}), (3, \frac{1}{4})\}$. The graphs of these functions are shown in Figures 9–9, 9–10, and 9–11.

Example 2. Let f be defined by

$$y = \sqrt{(1 - x^2)}, \qquad D_f = \{x \in R| -1 \leqslant x \leqslant 1\}.$$

This is a real function with range $\{y \in R| 0 \leqslant y \leqslant 1\}$ but it is not a real one-to-one function and hence has no inverse. The graph of f is given in Figure 9–12.

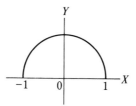

Figure 9–12   Graph of f: $y = \sqrt{(1 - x^2)}$, $D_f = \{x \in R| -1 \leqslant x \leqslant 1\}$

Example 3. Let f be defined by

$$y = \sqrt{(1 - x^2)}, \qquad D_f = \{x \in R| 0 \leqslant x \leqslant 1\}.$$

This is a real one-to-one function with range $\{y \in R| 0 \leqslant y \leqslant 1\}$. Since it is a one-to-one function (onto its range), it has an inverse f^{-1}, and it is easy to see that $f^{-1} = f$. The graph of this function is given in Figure 9–13. The function $1/f$ is defined by $y = 1/\sqrt{(1 - x^2)}$, $D = \{x \in R| 0 \leqslant x < 1\}$ and the graph is shown in Figure 9–14.

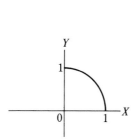

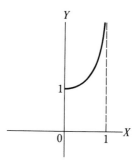

Figure 9–13   Graph of
f: $y = \sqrt{(1-x^2)}$,
$D_f = \{x \in R| 0 \leqslant x \leqslant 1\}$

Figure 9–14   Graph of
$1/f$: $y = 1/\sqrt{(1-x^2)}$,
$D = \{x \in R| 0 \leqslant x < 1\}$

Example 4. Let g be defined by $y = x^2$, $D_g = R$. This is a real function, but not one-to-one, and consequently has no inverse. If h is defined as $y = x^2$, $D_h = \{x \in R \mid x \geqslant 0\}$ the function is real and one-to-one and the inverse is $h^{-1}: x \to \sqrt{x}$, $D_{h^{-1}} = \{x \in R \mid x \geqslant 0\}$. The function $1/h$ is defined by $y = 1/x^2$, $D_{1/h} = \{x \in R \mid x > 0\}$. These functions are illustrated in Figures 9–15, 9–16, and 9–17.

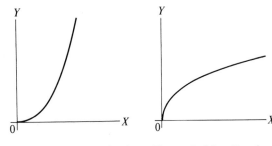

Figure 9–15 Graph of $h: x \to x^2$

Figure 9–16 Graph of $h^{-1}: x \to \sqrt{x}$

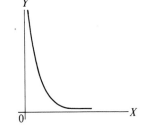

Figure 9–17 Graph of $1/h: x \to 1/x^2$

Example 5. The sequence $\{a_n\}$ where $a_n = 1 - 1/n$, $D = N - \{0\}$, is an example of a real one-to-one function. The inverse is given by the set of ordered pairs $\{(0, 1), (\frac{1}{2}, 2), (\frac{2}{3}, 3), ...\}$. What is the domain of this inverse? The graphs of these two functions are given in Figures 9–18 and 9–19.

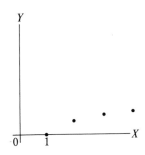

Figure 9–18 Graph of $\{a_n\}$, $D = N - \{0\}$

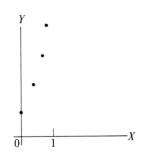

Figure 9–19 Graph of the inverse of the function in Figure 9–18

There is a simple geometric construction for the inverse of a real one-to-one function which is described in the following Lemma.

Lemma 1. *Suppose that f is a real one-to-one function defined by $y = f(x)$, $x \in D \subseteq R$. The graph of f^{-1} can be obtained by a reflection of f in the line $y = x$.*

Proof. The operation of reflection in $y = x$ associates with each point of the graph of $y = f(x)$ another point at the same distance from the line $y = x$ as the original point and on the perpendicular to the line $y = x$. If (a, b) are the coordinates of the original point, it follows by elementary geometry that the coordinates of the image point are (b, a) (see Figure 9–20). But by the alternative definition of a function given in Theorem 2 of 7.6, this interchange of the elements of the ordered pairs defining the original function gives the inverse function, and this completes the proof.

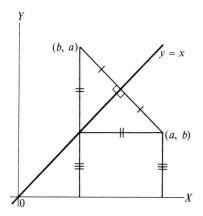

Figure 9–20   Reflection in the line $y = x$

The reader can easily verify that the inverse function in each of the preceding examples is obtained by reflection in the line $y = x$.

We close this section with an example of a noncommutative ring.

Example 6.   Consider the set of real functions that map the interval $\{x \in R \,|\, -1 \leqslant x \leqslant 1\}$ into R. We shall define equality and addition in the way already described in Definition 1 of 9.3 and 1 of 9.4. Consequently the properties A1, A2, A3, A4 are satisfied. We shall define "multiplication" as the composition operation defined in 7.8. That is, the "product" of two real functions f and g will be written $f \circ g$ and this is the function that maps $\{x \in R \,|\, -1 \leqslant x \leqslant 1\}$ into R defined by $(f \circ g)(x) = f(g(x))$.

Now it has been shown in Theorem 1 of 7.8 that, in general, $f \circ g \neq g \circ f$, thus "multiplication" is not commutative. However, $f \circ (g \circ h) = (f \circ g) \circ h$, that is, "multiplication" is associative. Since "multiplication" is distributive over addition, the resulting system is a (noncommutative) ring (Definition 1 of 9.8).

By Theorem 2 of 7.8 an identity function i exists such that $f \circ i = i \circ f = f$. Consequently the ring is a *noncommutative ring with unity*.

9.11   EXERCISES

1. Let $\mathscr{A}$ be an algebraic structure defined as follows. The elements of $\mathscr{A}$ are the subsets of a set U. Multiplication is defined by $AB = A \cap B$, and addition is defined by $A + B = A \cup B$, where $A, B \subseteq U$. Show that $\mathscr{A}$ is *not* a ring.

2. If a, b are elements of a ring, prove that $ab = (-a)(-b)$.

3. If a, b are elements of a ring, prove that $(-a)b = -(ab) = a(-b)$.

4. Let A, B be two rings. Consider the set of ordered pairs (a, b) where $a \in A$, $b \in B$. Addition and multiplication of the ordered pairs are defined by
$$(a_1, b_1) + (a_2, b_2) = (a_1 + a_2, b_1 + b_2),$$
$$(a_1, b_1)(a_2, b_2) = (a_1 a_2, b_1 b_2).$$
Prove that the set of ordered pairs forms a ring. Are there zero divisors in this ring?

5. Consider the set of ordered pairs (a, b) where $a \in A$, $b \in B$ and A and B are integral domains. Addition and multiplication of ordered pairs are defined as in 4. Show that the set of ordered pairs is not an integral domain.

6. Which of the following functions have inverses? What are the inverses, if they exist?
    (a) $f(x) = \log_{10} x$, $D = \{x \in R \,|\, x > 0\}$.
    (b) $f(x) = \sin x$, $D = \{x \in R \,|\, 0 \leqslant x \leqslant \frac{1}{2}\pi\}$.
    (c) $f(x) = \sin x$, $D = \{x \in R \,|\, -\pi \leqslant x \leqslant \pi\}$.
    (d) $f(x) = x$, $D = R$.

7. Prove that a subset S of a ring is a subring if and only if
    (1) $a, b \in S \Rightarrow a - b \in S$,
    (2) $a, b \in S \Rightarrow ab \in S$.

8. Complete the proof of all theorems and exercises which were indicated as left to the reader in this Chapter.

Exercises 9, 10, 11 refer to *quaternions*, which were introduced in connection with Example 9 of Section 9.8.

9. (a) If $A = a_0 + a_1 i + a_2 j + a_3 k$ is a quaternion, define $\bar{A} = a_0 - a_1 i - a_2 j - a_3 k$ as the conjugate quaternion and define $n = A\bar{A}$ as the norm of the quaternion.
    (b) Prove that the norm of a quaternion is a real non-negative number.
    (c) Prove that $\frac{1}{n}\bar{A}$ is the multiplicative inverse of A.

10. If $A = 1 + j + k$, $B = 1 + i + j + k$ calculate
$$A + B, \quad AB, \quad BA, \quad \bar{A}, \quad \bar{B}, \quad A^{-1}, \quad B^{-1}, \quad A/B.$$

11. Show that the multiplication table for quaternions can be derived from $i^2 = j^2 = k^2 = ijk = -1$.

12. (a) Obtain the multiplication and addition tables for I_3, I_4, I_5 (cf. Example 5 of 9.8).
    (b) Find the zero divisors (if any) in the rings of integers mod 4, mod 5, and mod 6. Find the elements which have multiplicative inverses.

13. Let f, g be polynomial functions of degree m, n respectively.
    (a) Show $f - g$ is a polynomial. What is its degree?
    (b) Illustrate with $m = 3$, $n = 2$, that $f \cdot g$ is a polynomial of degree $m + n$. (Observe that the method is general, but the details are tedious; we may assume Lemma 1 of 9.9 is now proved.)

14. (a) Prove the equality of the two functions
$$x \to s(x), \quad D = \{x \in R| \quad |x| < 1\},$$
$$x \to [x] + 1, \quad D = \{x \in R| \quad |x| < 1\}.$$
    (Cf. Sections 9.3 and 8.3.)
    (b) Draw a graph of $x \to [x]$, $D = \{x \in R| \quad |x| \leqslant 4\}$.
    (c) Draw graphs of $x \to x\,s(x)$, $x \to x\,s(-x)$ for $D = R$.
    (d) Using the functions in Part (c), prove that the ring of real functions has zero divisors.
    (e) Let $D = R$. Show that $x \to |x|$ and $x \to x\,s(x) - x\,s(-x)$ are equal.

15. Let $\mathscr{A}$ be an algebraic structure defined as follows: The elements of $\mathscr{A}$ are subsets of a set U. Multiplication is defined by $A \cdot B = A \cap B$ and addition by
$$A + B = (A \cup B) \cap (A \cap B)'.$$
    Show that $\mathscr{A}, +, \cdot$ forms a ring with the empty set as zero and U as the multiplicative identity, and that $A + A = \emptyset$ for all $A \in \mathscr{A}$.

10

Logic

Excepting notation introduced in 10.6, this chapter can be omitted or postponed. Its purpose is to provide a comprehensive treatment of the underlying logical principles of proof. Anyone wishing to rely on his intuition in these matters may pass on to Chapter 11 after digesting 10.6.

10.1 TRUTH FUNCTIONS

In Chapter 3 we began a discussion of certain principles of logic. Our discussion there was restricted to *sentences* and the *sentential* connectives: $\neg$, $\wedge$, $\vee$, $\Rightarrow$, $\Leftrightarrow$. It was further restricted to *truth valuations* of sentences, with very little reference to methods of proof. It is our intention in this chapter to extend these ideas in two directions.

1. We shall analyze sentence structure and truth valuations further to include *open sentences* (or *predicates*) such as "$x < y$"; and thence sentences such as "for all x, there exists a y, such that $x < y$".

2. We shall examine *methods of proof* which derive from truth valuations.

Sentences were defined in 3.1. What we called simple sentences in Chapter 3, we shall now call **sentence variables**. By this we mean that they are variables which take their values in either (a) a set of sentences or (b) the set $\{\mathcal{T}, \mathcal{F}\}$. Both points of view (a) and (b) are useful, and are not unrelated.

For example, let $\mathcal{P}$, $\mathcal{Q}$ be sentence variables, and consider the sentence

(1) $\mathcal{P} \Rightarrow \mathcal{Q}.$

(a) From Figure 10–1, let $\mathcal{P}$ take the value "$l_1 \parallel l_2$" and $\mathcal{Q}$ take the value "$\alpha_1 = \alpha_2$"; then (1) takes the value

(2) $l_1 \parallel l_2 \Rightarrow \alpha_1 = \alpha_2.$

(b) If $\mathcal{P}$, $\mathcal{Q}$ take the values $\mathcal{T}$, $\mathcal{F}$ respectively, then (1) takes the value $\mathcal{F}$ (by the definition in 3.6).

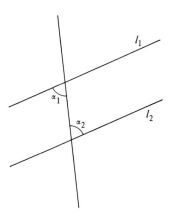

Figure 10–1

Sentence (2) might take the value $\mathcal{T}$ or $\mathcal{F}$ as well, but now its truth value depends on a particular context. If, for example, the particular sentences $l_1 \parallel l_2$ and $\alpha_1 = \alpha_2$ could be true and false respectively, then sentence (2) would be false. However, we know from geometry that (2) is always true (whence we can conclude: if $l_1 \parallel l_2$ is true, then $\alpha_1 = \alpha_2$ is true; if $\alpha_1 = \alpha_2$ is false, then $l_1 \parallel l_2$ is false).

From what we have been saying, it should be clear that a compound sentence as defined in 3.1 is a *function*. It is often called a **truth function**, but this name refers more to (b) than to (a). As a truth function, sentence (1) is completely described by its truth table (Figure 3–4 of 3.7); for each pair of values of the variables in the first two columns, the value of the function is given in the third column.

The truth function which a compound sentence defines, is always uniquely determined by its truth table.

A tautology is an identically true function.

Two compound sentences are logically equivalent if and only if they are equal regarded as truth functions (see 9.3 Definition 1 and following paragraph).

For example, $\neg(\mathscr{P} \wedge \neg \mathscr{Q})$ is logically equivalent to, and hence is the same truth function as, sentence (1). Compare Figures 3–3 and 3–4 of 3.7.

When one adopts the point of view in (a), it is really an intermediate step toward (b). The actual process is that of the composition of functions (7.8). For example, if we substitute in (1) the sentences "$2 < 1$", "$2 = 1$" for $\mathscr{P}$ and $\mathscr{Q}$ respectively, we get the sentence

$$(3) \qquad\qquad\qquad 2 < 1 \Rightarrow 2 = 1.$$

Now $2 < 1$ is false. Hence, by the definition of $\Rightarrow$, (3) is true.

The fact that (3) (and other sentences like it) is true is sometimes a point of confusion. The point is that it is true *by definition*, namely Definition 1 of 3.6, and experience shows that this definition is (at least for our purposes) the only sensible one. Some of the reasons were given in 3.6. Consider further

$$(4) \qquad\qquad\qquad 2 < 1 \Rightarrow 2 = 2.$$

Then (4) is true also. Now it is precisely because both (3) and (4) are true that the falsity of $2 < 1$ tells nothing about the consequent; it may be either true or false. Discuss what happens if one (or both) of the last two rows of Figure 3–4 of 3.7 is altered (see Exercise 1 of 10.11).

10.2   METHODS OF PROOF

Now we shall change our point of view from truth to proof.

In any mathematical proof, one begins with certain "true" sentences, called the **premises**, performs some sort of logical argument, and arrives at a final "true" sentence, the **conclusion**. The premises with which one begins fall into three categories:

(i) those which are taken as axioms in the given context,

(ii) those which have been previously proved in that context, and

(iii) those which are assumed to be true (whether or not they are).

If the premises belong to (i) and (ii) only, then the conclusion can be asserted to be true, or valid, in the given context. If the premises include some from (iii), then only in the case that these are valid can the conclusion be asserted to be valid.

Consider Theorem 1 of 2.2. If we let $\mathscr{P}$ stand for "$n = p$" and $\mathscr{Q}$ stand for "$m + n = m + p$", then (ignoring for now the "for all $m, n, p, \ldots$") Theorem 1 asserts $\mathscr{P} \Rightarrow \mathscr{Q}$.

Let $\mathscr{R}$ be "$m + n = m + n$"; $\mathscr{R}$ is an axiom (since it is an instance of an axiomatic property of equality).

Further, $(\mathscr{R} \wedge \mathscr{P}) \Rightarrow \mathscr{Q}$ is an axiom (since it is an instance of the replacement property).

The argument in the proof of Theorem 1 could now be summarized by saying: from the truth of the *premises* $\mathscr{R}, (\mathscr{R} \wedge \mathscr{P}) \Rightarrow \mathscr{Q}$, we infer the truth of the *conclusion* $\mathscr{P} \Rightarrow \mathscr{Q}$. Let us abbreviate this by $\mathscr{R}, (\mathscr{R} \wedge \mathscr{P}) \Rightarrow \mathscr{Q} \vDash \mathscr{P} \Rightarrow \mathscr{Q}$. (Since the premises are axioms, $\mathscr{P} \Rightarrow \mathscr{Q}$ is true in this context.) We shall give a precise definition of the symbol $\vDash$ below in 10.3. It can be read "yields".

The form of the proof of Theorem 1 of 2.2 might look like this.

  (i) $\mathscr{P}$ hypothesis (that is, temporary premise)

  (ii) $\mathscr{R}$ premise (axiom)

  (iii) $\mathscr{R} \wedge \mathscr{P}$ from (i) and (ii)

  (iv) $(\mathscr{R} \wedge \mathscr{P}) \Rightarrow \mathscr{Q}$ premise (axiom)

  (v) $\mathscr{Q}$ from (iii), (iv)

  (vi) $\mathscr{P} \Rightarrow \mathscr{Q}$ from (i), (v), discharging hypothesis (i)

The steps (i) to (iii) alone assert

(1) $$\mathscr{P}, \mathscr{R} \vDash \mathscr{R} \wedge \mathscr{P}.$$

Formula (1) appears to assert a correct method of proof. We shall see below (10.4) that it follows from the definition of $\vDash$ and $\wedge$.

Steps (iii) to (v) assert

(2) $$\mathscr{R} \wedge \mathscr{P}, (\mathscr{R} \wedge \mathscr{P}) \Rightarrow \mathscr{Q} \vDash \mathscr{Q}.$$

This is *modus ponens*, already discussed in 3.6.

Steps (i) to (v) assert

(3) $$\mathscr{P}, \mathscr{R}, (\mathscr{R} \wedge \mathscr{P}) \Rightarrow \mathscr{Q} \vDash \mathscr{Q}.$$

It must be the case then, that (3) follows from (1) and (2). Note that this is a transitive property of $\vDash$: the conclusion of (1) is a premise of (2), and is replaced by the premises of (1), yielding (3). We shall see below (10.3), that this property follows from the definition of $\vDash$.

Finally, steps (i) to (vi) assert

(4) $$\mathscr{R}, (\mathscr{R} \wedge \mathscr{P}) \Rightarrow \mathscr{Q} \vDash \mathscr{P} \Rightarrow \mathscr{Q}.$$

Hence, (4) follows from (3). Note that $\mathscr{P}$ does not appear as a premise in (4), whereas it did in (3). This is the meaning of the remark "discharging the hypothesis (i)" in (vi); that is, (i) to (vi) constitute a proof of (4) wherein only (ii) and (iv) are premises.

All methods of proof (involving the sentential connectives only) fall into three categories. These are named and illustrated below. They are discussed in detail in the following sections.

  I. *Basic Properties of* $\vDash$. For example, (3) follows from (1) and (2) by a transitive property of $\vDash$.

II. *Methods of Introducing a Sentential Connective.* For example, (1) is the method of *proving* a sentence *containing* the symbol $\wedge$. The way to *prove* a sentence *containing* the symbol $\Rightarrow$ is to proceed as we did above from (3) to (4).

III. *Methods of Eliminating a Sentential Connective.* For example, (2) is a method of proof *using* a sentence *containing* the symbol $\Rightarrow$ to prove some other sentence. (If the implication were in turn provable, then by the transitivity of $\vDash$, the symbol $\Rightarrow$ would be eliminated.)

The methods in I are called *structural* rules (or methods) of proof; those in II and III are called *logical*.

10.3   LOGICAL CONSEQUENCE

Example 1.   Let us examine (4) of 10.2

$$\mathcal{R}, (\mathcal{R} \wedge \mathcal{P}) \Rightarrow \mathcal{Q} \vDash \mathcal{P} \Rightarrow \mathcal{Q}$$

with premises $\mathcal{R}, (\mathcal{R} \wedge \mathcal{P}) \Rightarrow \mathcal{Q}$ and conclusion $\mathcal{P} \Rightarrow \mathcal{Q}$. We shall assume that the sentences $\mathcal{P}, \mathcal{Q}, \mathcal{R}$ are simple sentences in this context, or variables ranging over sentences. Table 10–1 lists all the truth values for the premises and

Table 10–1

$\mathcal{P}$	$\mathcal{Q}$	$\mathcal{R}$	Premises $(\mathcal{R} \wedge \mathcal{P}) \Rightarrow \mathcal{Q}$	Conclusion $\mathcal{P} \Rightarrow \mathcal{Q}$	
$\mathcal{T}$	$\mathcal{T}$	$\mathcal{T}$	$\mathcal{T}$	$\mathcal{T}$	*
$\mathcal{T}$	$\mathcal{T}$	$\mathcal{F}$	$\mathcal{T}$	$\mathcal{T}$	
$\mathcal{T}$	$\mathcal{F}$	$\mathcal{T}$	$\mathcal{F}$	$\mathcal{F}$	
$\mathcal{T}$	$\mathcal{F}$	$\mathcal{F}$	$\mathcal{T}$	$\mathcal{F}$	
$\mathcal{F}$	$\mathcal{T}$	$\mathcal{T}$	$\mathcal{T}$	$\mathcal{T}$	*
$\mathcal{F}$	$\mathcal{T}$	$\mathcal{F}$	$\mathcal{T}$	$\mathcal{T}$	
$\mathcal{F}$	$\mathcal{F}$	$\mathcal{T}$	$\mathcal{T}$	$\mathcal{T}$	*
$\mathcal{F}$	$\mathcal{F}$	$\mathcal{F}$	$\mathcal{T}$	$\mathcal{T}$	

conclusion. (Note, by the way, that $(\mathscr{R} \wedge \mathscr{P}) \Rightarrow \mathscr{Q}$ and $\mathscr{P} \Rightarrow \mathscr{Q}$ are *functions* of three and two variables respectively, where the domain (of the variables) is $\{\mathscr{T}, \mathscr{F}\}$ and the counter domain is $\{\mathscr{T}, \mathscr{F}\}$.) Note that every assignment of truth values which makes *both* premises true also makes the conclusion true (the starred rows). We say in this case that $\mathscr{P} \Rightarrow \mathscr{Q}$ is a **logical consequence** of $\mathscr{R}, (\mathscr{R} \wedge \mathscr{P}) \Rightarrow \mathscr{Q}$, and write $\mathscr{R}, (\mathscr{R} \wedge \mathscr{P}) \Rightarrow \mathscr{Q} \models \mathscr{P} \Rightarrow \mathscr{Q}$.

Definition. *Let $\mathscr{P}_1, \mathscr{P}_2, ..., \mathscr{P}_n, \mathscr{P}$ be (possibly) compound sentences each constructed from (some or all of) the simple sentences, or sentence variables, $\mathscr{R}_1, \mathscr{R}_2, ..., \mathscr{R}_m$. We say $\mathscr{P}$ is a **logical consequence** of $\mathscr{P}_1, \mathscr{P}_2, ..., \mathscr{P}_n$ and write*

$$\mathscr{P}_1, \mathscr{P}_2, ..., \mathscr{P}_n \models \mathscr{P}$$

if every assignment of truth values to the $\mathscr{R}_1, \mathscr{R}_2, ..., \mathscr{R}_m$ which makes all of $\mathscr{P}_1, \mathscr{P}_2, ..., \mathscr{P}_n$ true, also makes $\mathscr{P}$ true.

We shall refer to a statement of the form $\mathscr{P}_1, \mathscr{P}_2, ..., \mathscr{P}_n \models \mathscr{P}$ as a **logical derivation**. (More precisely, it asserts the existence of a logical derivation.)

Example 2. In 10.2 we observed that $\mathscr{P}, \mathscr{R} \models \mathscr{P} \wedge \mathscr{R}$. This obviously follows from the definition of $\wedge$ and $\models$.

Example 3. *Modus ponens* can be expressed by $\mathscr{P}, \mathscr{P} \Rightarrow \mathscr{Q} \models \mathscr{Q}$. This follows from the definition of $\Rightarrow$ and $\models$. Indeed, we defined $\Rightarrow$ in Chapter 3 precisely so that this would be the case.

There are five basic properties of $\models$, listed below, which constitute the fundamental structural methods of proof (I of 10.2). The first four, at least, are obvious, and the reader can supply the proof at a glance. It may seem that they are not worth mentioning, but if we are trying to make our methods of proof explicit, we need to record them at least once. When we use them in the sequel, we shall simply refer to "properties of $\models$" without being explicit. Subsequently, we may use them without reference at all. (See Exercise 1 of 10.11.)

(1) $\mathscr{P}, \mathscr{P}_1, \mathscr{P}_2, ..., \mathscr{P}_n \models \mathscr{P}$.

(2) If $\mathscr{P}_1, \mathscr{P}_2, ..., \mathscr{P}_n \models \mathscr{P}$, then
    $\mathscr{R}, \mathscr{P}_1, \mathscr{P}_2, ..., \mathscr{P}_n \models \mathscr{P}$ where $\mathscr{R}$ is any sentence.

(3) If $\mathscr{P}_1, \mathscr{P}_1, \mathscr{P}_2, ..., \mathscr{P}_n \models \mathscr{P}$, then
    $\mathscr{P}_1, \mathscr{P}_2, ..., \mathscr{P}_n \models \mathscr{P}$.

(4) $\mathscr{P}_1, \mathscr{P}_2, ..., \mathscr{P}_n \models \mathscr{P}$, then
    $\mathscr{P}_{i_1}, \mathscr{P}_{i_2}, ..., \mathscr{P}_{i_n} \models \mathscr{P}$, where $i_1, i_2, ..., i_n$ is any permutation of 1, 2, ..., n.

(5) If $\mathscr{P}_1, \mathscr{P}_2, ..., \mathscr{P}_n \vDash \mathscr{P}$, and $\mathscr{P}, \mathscr{Q}_1, \mathscr{Q}_2, ..., \mathscr{Q}_m \vDash \mathscr{Q}$, then $\mathscr{P}_1, \mathscr{P}_2, ..., \mathscr{P}_n, \mathscr{Q}_1, \mathscr{Q}_2, ..., \mathscr{Q}_m \vDash \mathscr{Q}$.

Proof of (5). Pick any assignment of truth values which makes all of $\mathscr{P}_1, ..., \mathscr{P}_n, \mathscr{Q}_1, ..., \mathscr{Q}_m$ true. Then, in particular, all of $\mathscr{P}_1, ..., \mathscr{P}_n$ are true; so $\mathscr{P}_1, ..., \mathscr{P}_n \vDash \mathscr{P}$ gives $\mathscr{P}$ true as well. Hence, all of $\mathscr{P}, \mathscr{Q}_1, ..., \mathscr{Q}_m$ are true, and so $\mathscr{P}, \mathscr{Q}_1, ..., \mathscr{Q}_m \vDash \mathscr{Q}$ gives $\mathscr{Q}$ true. Hence, $\mathscr{P}_1, ..., \mathscr{P}_n, \mathscr{Q}_1, ..., \mathscr{Q}_m \vDash \mathscr{Q}$.

10.4 INTRODUCTION AND ELIMINATION OF THE CONNECTIVES

The methods of proof which do not belong to the structural type discussed in I of 10.2 are called "logical" and fall into two main groups (II and III of 10.2).

II. There are four **introduction** rules which are used to *prove* sentences of the form $\neg \mathscr{P}$, $\mathscr{P} \wedge \mathscr{Q}$, $\mathscr{P} \vee \mathscr{Q}$, and $\mathscr{P} \Rightarrow \mathscr{Q}$; that is, they introduce the connectives. These are called respectively $\neg$-introduction, $\wedge$-introduction, $\vee$-introduction, and $\Rightarrow$-introduction.

III. There are four **elimination** rules which are used to prove certain sentences by *making use* of premises of the form $\neg \mathscr{P}$, $\mathscr{P} \wedge \mathscr{Q}$, $\mathscr{P} \vee \mathscr{Q}$, and $\mathscr{P} \Rightarrow \mathscr{Q}$. These are called $\neg$-elimination, etc. That these rules do actually eliminate the connectives is not so apparent, but depends, to a large extent, on the way they are used. This will become clearer as we study some examples. (See Table 10–2.)

Table 10–2

	Introduction	Elimination
$\neg$	If $\mathscr{P}, \mathscr{P}_1, ..., \mathscr{P}_n \vDash \mathscr{Q}$ and $\mathscr{P}, \mathscr{P}_1, ..., \mathscr{P}_n \vDash \neg \mathscr{Q}$, then $\mathscr{P}_1, ..., \mathscr{P}_n \vDash \neg \mathscr{P}$	$\neg \neg \mathscr{P} \vDash \mathscr{P}$
$\wedge$	$\mathscr{P}, \mathscr{Q} \vDash \mathscr{P} \wedge \mathscr{Q}$	$\mathscr{P} \wedge \mathscr{Q} \vDash \mathscr{P}$
$\vee$	$\mathscr{P} \vDash \mathscr{P} \vee \mathscr{Q}$	If $\mathscr{P}, \mathscr{P}_1, ..., \mathscr{P}_n \vDash \mathscr{R}$ and $\mathscr{Q}, \mathscr{P}_1, ..., \mathscr{P}_n \vDash \mathscr{R}$, then $\mathscr{P} \vee \mathscr{Q}, \mathscr{P}_1, ..., \mathscr{P}_n \vDash \mathscr{R}$
$\Rightarrow$	If $\mathscr{P}, \mathscr{P}_1, ..., \mathscr{P}_n \vDash \mathscr{Q}$ then $\mathscr{P}_1, ..., \mathscr{P}_n \vDash \mathscr{P} \Rightarrow \mathscr{Q}$	$\mathscr{P}, \mathscr{P} \Rightarrow \mathscr{Q} \vDash \mathscr{Q}$

These rules are of two types. One type ($\wedge$-intro, $\vee$-intro, $\neg$-elim, $\wedge$-elim, $\Rightarrow$-elim) asserts that a certain logical derivation is correct. The other type ($\neg$-intro, $\Rightarrow$-intro, $\vee$-elim) asserts that a certain logical derivation is correct given that certain other logical derivations, called the **subsidiary** derivations, are correct. The sentence $\mathcal{P}$ of $\neg$-intro and $\Rightarrow$-intro is a temporary premise, called a **hypothesis,** which is **discharged,** that is, ceases to be a premise. (Review the example in 10.2.) Similarly, the hypotheses $\mathcal{P}$, $\mathcal{Q}$ of $\vee$-elim are also discharged.

$\neg$-intro is a method of proof often referred to as *reductio ad absurdum.* (See, for example, the Proof in 2.5.) It is often used in conjunction with $\neg$-elim.

Example 1. Let A be a set of axioms for Euclidean Geometry. Then, referring to Figure 10–1,

(1) $$A, \neg(\alpha_1 = \alpha_2) \vDash \neg(l_1 \parallel l_2)$$

is one of the elementary theorems proved, but it is not usually stated this way. From (1), by properties of $\vDash$,

(2) $$A, \neg(\alpha_1 = \alpha_2), l_1 \parallel l_2 \vDash \neg(l_1 \parallel l_2).$$

By properties of $\vDash$,

(3) $$A, \neg(\alpha_1 = \alpha_2), l_1 \parallel l_2 \vDash l_1 \parallel l_2.$$

From (2), (3) by $\neg$-intro,

(4) $$A, l_1 \parallel l_2 \vDash \neg\neg(\alpha_1 = \alpha_2).$$

From (4), $\neg$-elim and properties of $\vDash$,

(5) $$A, l_1 \parallel l_2 \vDash \alpha_1 = \alpha_2.$$

This is the way the theorem is usually stated: from the assumption that $l_1 \parallel l_2$ (and the axioms) we derive the result that $\alpha_1 = \alpha_2$.

Proof of $\neg$-intro. We have

(6) $$\mathcal{P}, \mathcal{P}_1, \ldots, \mathcal{P}_n \vDash \mathcal{Q},$$

(7) $$\mathcal{P}, \mathcal{P}_1, \ldots, \mathcal{P}_n \vDash \neg\mathcal{Q}.$$

Pick an assignment of truth values which makes all of $\mathcal{P}_1, \ldots, \mathcal{P}_n$ true. We wish to show that $\neg\mathcal{P}$ is also true, and hence that

(8) $$\mathcal{P}_1, \ldots, \mathcal{P}_n \vDash \neg\mathcal{P}.$$

Suppose, on the contrary, that $\neg\mathscr{P}$ is false. Then $\mathscr{P}$ is true by the definition of $\neg$ in Chapter 3. Hence, $\mathscr{P}, \mathscr{P}_1, ..., \mathscr{P}_n$ are all true. By (6), (7) then, both $\mathscr{Q}$ and $\neg\mathscr{Q}$ are true, contradicting the definition of $\neg$ and the definition of sentence in Chapter 3. So $\neg\mathscr{P}$ is true and (8) is established. (See the remarks at the end of this section.)

The proof of $\neg$-elim follows trivially from the definition of $\neg$. (See Exercise 1 of 10.11.)

Example 1 (Continued). From (5), by $\Rightarrow$-intro,

(9) $$A \vDash l_1 \parallel l_2 \Rightarrow \alpha_1 = \alpha_2.$$

This is yet another way of stating the theorem: from the axioms we deduce, "If $l_1 \parallel l_2$, then $\alpha_1 = \alpha_2$."

Proof of $\Rightarrow$-intro. We have

(10) $$\mathscr{P}, \mathscr{P}_1, ..., \mathscr{P}_n \vDash \mathscr{Q}.$$

We wish to show that

(11) $$\mathscr{P}_1, ..., \mathscr{P}_n \vDash \mathscr{P} \Rightarrow \mathscr{Q}.$$

To this end, pick an assignment of truth values which makes all of $\mathscr{P}_1, ..., \mathscr{P}_n$ true. *Case* 1. $\mathscr{P}$ is false. Then $\mathscr{P} \Rightarrow \mathscr{Q}$ is true by the definition of $\Rightarrow$ in Chapter 3, and (11) is established. *Case* 2. $\mathscr{P}$ is true. Then, by (10), $\mathscr{Q}$ is true, and by the definition of $\Rightarrow$, $\mathscr{P} \Rightarrow \mathscr{Q}$ is true, proving (11).

We leave the proof of $\Rightarrow$-elim (*modus ponens*) to the reader, reminding him that $\Rightarrow$ was defined in Chapter 3 for this purpose.

The proofs of $\wedge$-intro, $\wedge$-elim, and $\vee$-intro follow trivially from the definitions of $\vDash$ and the connectives. (See Exercise 1 of 10.11.) We shall prove $\vee$-elim presently.

In the definition of $\vDash$ in 10.3 we said nothing about the possibility that the list of premises be empty ($n = 0$), but we did not explicitly state $n \neq 0$. It is a moot logical point, which many beginners find difficult to comprehend, that "every assignment of truth values which makes the empty set of premises true" is "*every* assignment of truth values." Consequently,

$$\vDash \mathscr{P} \quad means \ \mathscr{P} \ is \ a \ tautology$$

by the definition of tautology in Chapter 3 and the definition of $\vDash$.

Example 2. In 3.6 it was shown that

(12) $$\vDash (\mathscr{P} \Rightarrow \mathscr{Q}) \Leftrightarrow (\neg\mathscr{Q} \Rightarrow \neg\mathscr{P}).$$

However, $\Leftrightarrow$ is an abbreviation. Unabbreviating (12),

(12') $\vDash [(\mathcal{P} \Rightarrow \mathcal{Q}) \Rightarrow (\neg \mathcal{Q} \Rightarrow \neg \mathcal{P})] \wedge [(\neg \mathcal{Q} \Rightarrow \neg \mathcal{P}) \Rightarrow (\mathcal{P} \Rightarrow \mathcal{Q})]$.

By $\wedge$-elim,

(13) $\vDash (\neg \mathcal{Q} \Rightarrow \neg \mathcal{P}) \Rightarrow (\mathcal{P} \Rightarrow \mathcal{Q})$.

Example 1 (Concluded).
There is always more than one way to prove the same result. From (1) and
$\Rightarrow$-intro,

(14) $A \vDash \neg(\alpha_1 = \alpha_2) \Rightarrow \neg(l_1 \| l_2)$.

Since (13) holds for any sentences $\mathcal{P}$, $\mathcal{Q}$,

(15) $\vDash [\neg(\alpha_1 = \alpha_2) \Rightarrow \neg(l_1 \| l_2)] \Rightarrow (l_1 \| l_2 \Rightarrow \alpha_1 = \alpha_2)$.

By (14), (15), $\Rightarrow$-elim, and properties of $\vDash$ we get again

(9) $A \vDash l_1 \| l_2 \Rightarrow \alpha_1 = \alpha_2$.

Another application of $\Rightarrow$-elim yields (5).

Proof of $\vee$-elim. We have

(16) $\mathcal{P}, \mathcal{P}_1, ..., \mathcal{P}_n \vDash \mathcal{R}$,

(17) $\mathcal{Q}, \mathcal{P}_1, ..., \mathcal{P}_n \vDash \mathcal{R}$.

We wish to show

(18) $\mathcal{P} \vee \mathcal{Q}, \mathcal{P}_1, ..., \mathcal{P}_n \vDash \mathcal{R}$.

To this end, pick an assignment of truth values which makes all of $\mathcal{P} \vee \mathcal{Q}$,
$\mathcal{P}_1, ..., \mathcal{P}_n$ true. In particular, $\mathcal{P} \vee \mathcal{Q}$ is true. By the definition of $\vee$ in
Chapter 3 then, either $\mathcal{P}$ is true or $\mathcal{Q}$ is true. *Case 1.* $\mathcal{P}$ is true. Then all of
$\mathcal{P}, \mathcal{P}_1, ..., \mathcal{P}_n$ are true. By (16), $\mathcal{R}$ is true, proving (18). *Case 2.* $\mathcal{Q}$ is true.
This case is similar, completing the proof.

In the case that (16), (17) hold as well as

(19) $\mathcal{P}_1, \mathcal{P}_2, ..., \mathcal{P}_n \vDash \mathcal{P} \vee \mathcal{Q}$

we get

(20) $\mathcal{P}_1, \mathcal{P}_2, ..., \mathcal{P}_n \vDash \mathcal{R}$

by transitivity of ⊨. It is in this case that the symbol ∨ is eliminated. ∨-elim is nearly always used this way, and as such it is called *proof by cases* ((16) and (17) being the cases).

Example 3. Let B be a set of axioms for the arithmetic of the natural numbers. Let $a,b,p \in N$, $a > 0$, $b > 0$, p prime. Theorem 2 of 5.5 asserts

(21) $B \vDash p|ab \Rightarrow [(p|a) \lor (p|b)]$.

The proof actually demonstrated that

(22) $B, p|ab, \neg(p|a) \vDash p|b$.

Let us assume that (22) is proved (as indeed it is in 5.5), and see how (21) follows from it by our methods of proof.
From (22) and ∨-intro

(23) $B, p|ab, \neg(p|a) \vDash (p|a) \lor (p|b)$.

By properties of ⊨,

(24) $B, p|ab, p|a \vDash p|a$.

From (24) and ∨-intro

(25) $B, p|ab, p|a \vDash (p|a) \lor (p|b)$.

From (23), (25) by ∨-elim

(26) $B, p|ab, (p|a) \lor \neg(p|a) \vDash (p|a) \lor (p|b)$.

But

(27) $\vDash \mathscr{P} \lor \neg \mathscr{P}$

for every sentence $\mathscr{P}$. Hence

(28) $\vDash (p|a) \lor \neg(p|a)$.

Hence by (26), (28), and properties of ⊨

(29) $B, p|ab \vDash (p|a) \lor (p|b)$.

Finally, applying ⇒-intro to (29), we get (21).

Before we leave methods of proof temporarily, a few comments are in order.

1. The reader, if he has tried, has probably found it extremely difficult or impossible to formulate proofs of theorems in terms of the methods

of proof listed above. This is quite understandable, and the reason for it is twofold.

(a) It is almost impossible to find a theorem and its proof which do not involve quantification ("there exists . . .", "for all . . ."), and we have not yet discussed quantification.

(b) Nearly every proof involves *extralogical* axioms or methods of proof which are valid in the particular context, but not universally valid. For example, (22) was proved in 5.5 using the well-ordered property of N.

2. When we listed our basic methods of proof, and then proved them, the reader may well have asked himself, "Which is the chicken, and which is the egg?" This is a good point; when we proved $\vee$-elimination (which is proof by cases), we proved it by cases. The point is this: by the introduction of a special notation and a special language, we have codified and made explicit those methods of proof (for sentential connectives) which we do accept as valid.

When we "proved" them, we were simply demonstrating that these really are the methods of proof we normally use. It is not expected that the reader, or anyone else, would make extensive use of the rather formal treatment of proof methods given here, but it is hoped that a careful study of this exposition will give the reader a clearer insight into what really does constitute a proof.

10.5 THE RUSSELL PARADOX

In this section we shall examine one of the most fascinating problems of set theory, and in doing so, provide further justification for some of the conventions regarding sets which we have adopted earlier.

To begin with, it is important to note that one set can be a member (as opposed to subset) of another set. That is to say, a particular set is given the status of a single mathematical object and becomes, therefore, a candidate for membership in other sets. Example 2 of 7.3 illustrates this point. More than this, in building a foundation for mathematics, it unavoidably happens that we obtain sets of sets. An ordered pair of integers is a set (7.2). A rational number is a set (equivalence class) of ordered pairs of integers (Example 8 of 7.5). A real number is a Dedekind Cut (8.2), that is, an ordered pair of sets of rationals. A real function or relation is a set of ordered pairs of real numbers (7.4, 7.6).

If sets can (and must) be members of sets, one can ask: Can a set be a member of itself? It may appear bizarre to have a set A such that $A \in A$, but this situation cannot be ruled out without examination. It seems difficult to think of a genuinely mathematical example, but at least one nonmathematical example exists: the set of abstract ideas. The "set of abstract ideas" is itself an abstract idea, and is, therefore, a member of itself.

In 1.4, we adopted the convention of naming sets

$$\{x \in A \mid \mathscr{P}(x)\}$$

where A is a *previously defined* set.

Let us violate this convention temporarily and "define" the "set"

(1) $V = \{x \mid x \notin x\}.$

Hence, for any a,

(2) $\vdash (a \in V) \Leftrightarrow \neg (a \in a).$

V is the "set" of all sets which do not contain themselves as elements. If there are no sets which contain themselves as elements, or if one insists that we adopt this point of view as a convention, then V is still "defined"; it is then the "set" of all sets.

Replacing a by V in (2),

(3) $\vdash (V \in V) \Leftrightarrow \neg (V \in V).$

This looks suspicious. By truth valuations, we have a sentence (viz., $V \in V$) which is true if and only if it is false. Let us continue and see what we can derive by our proof methods.

Unabbreviating $\Leftrightarrow$ in (3),

(3′) $\vdash [(V \in V) \Rightarrow \neg (V \in V)] \land [\neg (V \in V) \Rightarrow (V \in V)].$

By $\land$-elim,

(4a) $\vdash (V \in V) \Rightarrow \neg (V \in V),$
(4b) $\vdash \neg (V \in V) \Rightarrow (V \in V).$

By properties of $\vdash$,

(5) $V \in V \vdash V \in V.$

By (4a) and $\Rightarrow$-elim,

(6) $V \in V \vdash \neg (V \in V).$

By (5), (6), and $\neg$-intro,

(7) $\vdash \neg (V \in V).$

So we have proved $V \notin V$; this is asserted by (7). So far so good; the "set" of all sets which do not contain themselves as elements, does not contain itself as an element.

However, by (4b) and ⇒-elim,

(8) $\neg (V \in V) \vdash V \in V.$

By (7), (8), and properties of ⊢,

(9) $\vdash V \in V.$

Now things are not so good; (7) and (9) give us a sentence and its negation *both true!*

This contradiction is called the Russell Paradox (named after its discoverer, Bertrand Russell). It is only one of several paradoxes of set theory, but it is the only one we can introduce at this stage.

The Russell Paradox, and the other paradoxes, demonstrate that set theory, approached naively, is *inconsistent.* On the other hand, much set theory appears to be quite safe from contradiction, and this fragment of it (whatever it is) is much too valuable a part of mathematics to be discarded. The problem then is to try to locate the source of the difficulty and to reformulate set theory accordingly.

During the past fifty years or so, a great deal of research in mathematics and logic has been devoted to the paradoxes. No universally accepted solution has been found.

If we attempt to locate the source of the difficulty, we obtain at least the following possibilities.

1. There is an error in reasoning in the paradoxes. All attempts to find such an error have been fruitless.

2. A set should not be a member of itself. But then (1) is the "definition" of the "set" of all sets. So (9) holds by definition, and (7) holds by assumption.

3. Russell pointed out that the "set" V, and each of the other "sets" used to obtain the other paradoxes, is defined in a way which he called **impredicative.** A set is defined impredicatively if its definition makes reference to another set to which the set being defined belongs. The "set" W, of all sets is impredicatively defined. The "set" V is impredicatively defined because, in defining it, we are dividing W into those members which contain themselves as elements, and those which do not. All of the paradoxes could be avoided by disallowing impredicative definitions. Unfortunately, a great deal of mathematics which is not suspect would then be discarded also; *the definition of the least upper bound of a set of real numbers is impredicative*, and no one has been able to find a predicative (that is, not impredicative)

definition of it. Much research has been devoted to predicative set theory, but most of the important questions remain unanswered.

4. The "sets" V and W have a very obvious feature: they are extremely "large sets" in some sense. Each of the other "sets" used to obtain the other paradoxes has this feature as well. The paradoxes can be avoided by not allowing certain "large sets" to constitute objects which may then be members of other sets. These "large sets" are sometimes called **classes.** In this approach V and W are classes, but not sets. Set theory has been reformulated on an *axiomatic* basis by a number of mathematicians (beginning with A. A. Fraenkel) to incorporate this feature. This has been the most successful solution of the problem to date. (A few mathematicians do not agree.) It has been successful in the sense that all the known paradoxes have been avoided, and no known mathematics has been discarded. However, no one has succeeded in *proving* the consistency of any existing axiomatic theory of sets, and it seems unlikely that anyone will prove it.

Taking our cue from this last (partial) solution to the problem, we have introduced certain conventions into our informal (that is, nonaxiomatic) treatment of set theory.

We adopted the convention in 1.4 that we would name or define new sets by the device

$$B = \{x \in A \mid \mathscr{P}(x)\}$$

where A is any previously defined set and $\mathscr{P}$ is any open sentence which is well defined for the elements of A. B is a subset of the previously defined set A, and cannot, therefore be "too large".

Similarly, we introduced the notion of a universal set in 4.5. As long as we work within a previously defined universal set, U, about which we have no doubts, then the difficulties of the paradoxes cannot arise.

In closing, it is important to note that there are many other "safe" ways to define new sets from previously defined ones. For example:

(i) The Cartesian product of two previously defined sets;

(ii) The set of all subsets of a previously defined set;

(iii) The set of all functions from one previously defined set into another.

(For a complete discussion of these and related matters, see *The Foundations of Set Theory*, by Fraenkel and Bar-Hillel, North-Holland Publishing Co., 1958.)

10.6 PREDICATES AND QUANTIFIERS

In 10.2 we examined the proof of Theorem 1 of 2.2, but we deliberately ignored certain essential features. The statement of the theorem was not

$$"(n = p) \Rightarrow (m + n = m + p)",$$

but rather,

"for all natural numbers m, n, p, $(n = p) \Rightarrow (m + n = m + p)$".

Strictly speaking, we cannot (as we did there) let "$n = p$" and "$m + n = m + p$" be replaced by sentences $\mathscr{P}$ and $\mathscr{Q}$ respectively. If we do this the effect of the assertion "for all ..." is lost.

The expression "$n = p$" is *not* a sentence, for it is neither true nor false. On the other hand "$1 = 2$" and "$2 = 2$" are sentences, false and true respectively. The expression "$n = p$" is a *function* (of two variables) whose domain is $N \times N$. Sometimes we say the domain of each variable (n and p) is N. The counter domain of the function is either (a) a set of sentences, or (b) $\{\mathscr{T}, \mathscr{F}\}$, depending on how we choose to look at it. The situation is similar to, but not exactly the same as, that discussed in 10.1.

If we substitute 1 and 2 respectively for n and p we obtain either (a) the sentence "$1 = 2$", or (b) the value $\mathscr{F}$. If we substitute 2 for both n and p we obtain either (a) the sentence "$2 = 2$", or (b) the value $\mathscr{T}$.

An expression like "$n = p$" is called an **open sentence** or **predicate**.

Definition 1. *A **predicate** is a function (of one or more variables with some specified non-empty domain) whose values are sentences, and hence ultimately whose values are in $\{\mathscr{T}, \mathscr{F}\}$.*

A relation (Chapter 7) is a predicate of two variables.

Let $\mathscr{P}(n, p)$ stand for "$n = p$", and let $\mathscr{Q}(m, n, p, q)$ stand for "$m + n = q + p$", where it is understood that the domain of the variables is N. Then $\mathscr{P}(n, p)$ is a predicate of two variables, and $\mathscr{Q}(m, n, p, q)$ is a predicate of four variables. On the other hand $\mathscr{Q}(m, n, p, m)$ (which stands for "$m + n = m + p$") and

$$\mathscr{P}(n, p) \Rightarrow \mathscr{Q}(m, n, p, m)$$

are predicates of three variables.

Theorem 1 of 2.2 asserts

For all m, n, p, $\mathscr{P}(n, p) \Rightarrow \mathscr{Q}(m, n, p, m)$.

This is often abbreviated as

$$\forall m \, \forall n \, \forall p [\, \mathscr{P}(n, p) \Rightarrow \mathscr{Q}(m, n, p, m)].$$

Note that this is a sentence, not a predicate; it is either true or false. (In fact, it is true.) It is a sentence because none of the variables m, n, p are **free** (for substitution); they have been **bound** by the **universal quantifiers** $\forall m, \forall n, \forall p$. $(\forall 1 \, \forall 2 \, \forall 3[\,\mathscr{P}(2, 3) \Rightarrow \mathscr{Q}(1, 2, 3, 1)]$ is meaningless; see the discussion on bound variables in 1.4.) On the other hand, $\forall n \, \forall p[\,\mathscr{P}(n, p) \Rightarrow \mathscr{Q}(m, n, p, m)]$ is a predicate of one variable, m.

Definition 2. *Let $\mathscr{P}(..., x, ...)$ be a predicate in the variable x, and possibly other variables. $\forall x$ is called the **universal quantifier** in x. $\forall x \, \mathscr{P}(..., x, ...)$ is a predicate in the remaining variables only; the variable x is now called a **bound** variable. If x is the only variable in $\mathscr{P}(x)$, then $\forall x \, \mathscr{P}(x)$ is a sentence. The universal quantifier $\forall x$ has the meaning "for all x".*

"For all", "for each", "for every", and "for any" are synonymous. The truth valuation of the universal quantifier as a logical operation is clear. For example:

If $\mathscr{P}(x, y)$ is a predicate with domain $D \times D$, then $\forall x \, \mathscr{P}(x, y)$ is a predicate with domain D such that: for each $y \in D$, $\forall x \, \mathscr{P}(x, y)$ takes the value $\mathscr{T}$ if and only if $\mathscr{P}(x, y)$ takes the value $\mathscr{T}$ for every $x \in D$. (That is, we have defined what function of y $\forall x \, \mathscr{P}(x, y)$ is.)

If $\mathscr{P}(x)$ is a predicate with domain D, then $\forall x \, \mathscr{P}(x)$ is a sentence which is $\mathscr{T}$ if and only if $\mathscr{P}(x)$ has the value $\mathscr{T}$ for every $x \in D$.

Consider $x|y$ defined on N. $\forall x(x|y)$ is a predicate in y; it takes the value $\mathscr{F}$ for every y except 0. $\forall y(x|y)$ is a predicate in x; $\forall y(1|y)$ is true and $\forall y(x|y)$ is false for every $x \neq 1$. $\forall x \, \forall y(x|y)$ is a false sentence.

Now consider the predicate of two variables "$m < n$". Let $\mathscr{L}_N(m, n)$ be "$m < n$" defined on N, and let $\mathscr{L}_Q(m, n)$ be "$m < n$" defined on Q. So, for example, $\mathscr{L}_N(1, 2)$ is true, $\mathscr{L}_N(3, 1)$ is false, $\mathscr{L}_Q(\frac{1}{2}, \frac{4}{3})$ is true, $\mathscr{L}_Q(10, -\frac{1}{2})$ is false, and $\mathscr{L}_N(10, \frac{1}{2})$ is undefined.

Consider the sentences:

There exists an x, such that $\mathscr{L}_N(x, 0)$.

There exists an x, such that $\mathscr{L}_Q(x, 0)$.

These are often abbreviated as:

(1a) $\qquad\qquad\qquad\qquad \exists x \; \mathscr{L}_N(x, 0).$
(1b) $\qquad\qquad\qquad\qquad \exists x \; \mathscr{L}_Q(x, 0).$

(1a) is false; there is no natural number x, such that $x < 0$. (1b) is true; for example, $-\frac{1}{2} < 0$.

Definition 3. *Let $\mathscr{P}(..., x, ...)$ be a predicate in the variable x, and possibly other variables. $\exists x$ is called the **existential quantifier** in x. $\exists x \, \mathscr{P}(..., x, ...)$ is a* *predicate in the remaining variables only; the variable x is now called a **bound** variable. If x is the only variable in $\mathscr{P}(x)$, then $\exists x\, \mathscr{P}(x)$ is a sentence. The existential quantifier $\exists x$ has the meaning "there exists an x such that".*

Any variables appearing in a predicate which are not bound (by $\forall$ or $\exists$) are called **free** variables. A predicate is a function of its free variables only.

The truth valuation of the existential quantifier is based on its meaning, as is the case with the universal quantifier. For example:

If $\mathscr{P}(x)$ is a predicate with domain D, then $\exists x\, \mathscr{P}(x)$ is a sentence which is $\mathscr{T}$ if and only if there is at least one $x \in D$ such that $\mathscr{P}(x)$ has the value $\mathscr{T}$.

Consider the following examples.

(2a) $\forall x\, \exists y\, \mathscr{L}_N(x, y)$.

(2b) $\forall x\, \exists y\, \mathscr{L}_Q(x, y)$.

Both (2a) and (2b) are true: for any natural number (or for any rational number) x which we choose, we can always find one which is larger. On the other hand, consider

(3a) $\forall x\, \exists y\, \mathscr{L}_N(y, x)$,

(3b) $\forall x\, \exists y\, \mathscr{L}_Q(y, x)$.

(Notice carefully how (3a,b) differ from (2a,b).) (3a) is false, whereas (3b) is true.

With certain precuations, the name of a bound variable is immaterial: $\forall y\, \exists x\, \mathscr{L}_N(x, y)$ means the same as (3a); so does $\forall u\, \exists v\, \mathscr{L}_N(v, u)$. On the other hand $\forall x\, \exists x\, \mathscr{L}_N(x, x)$ is quite different. All the variables in $\exists x\, \mathscr{L}_N(x, x)$ are already bound; $\exists x\, \mathscr{L}_N(x, x)$ is a false sentence. Adding the quantifier "$\forall x$" to get $\forall x\, \exists x\, \mathscr{L}_N(x, x)$ adds nothing; we can say either that it is meaningless, or that it means the same as $\exists x\, \mathscr{L}_N(x, x)$. It is best to avoid it.

As one final example, consider the predicate

$$\exists z[\mathscr{L}_N(x, z) \wedge \mathscr{L}_N(z, y)].$$

From this we can construct the new predicate

$$\mathscr{L}_N(x, y) \Rightarrow \exists z[\mathscr{L}_N(x, z) \wedge \mathscr{L}_N(z, y)],$$

and thence the sentence

(4a) $\forall x\, \forall y(\mathscr{L}_N(x, y) \Rightarrow \exists z[\mathscr{L}_N(x, z) \wedge \mathscr{L}_N(z, y)])$.

Sentence (4a) asserts that N is densely ordered; it is false. On the other hand

(4b) $\forall x\, \forall y(\mathscr{L}_Q(x, y) \Rightarrow \exists z[\mathscr{L}_Q(x, z) \wedge \mathscr{L}_Q(z, y)])$

asserts that Q is densely ordered, and it is true (6.3).

10.7   TRUTH VALUATIONS

In this section we shall deal with sentences and predicates generally. In addition to a list of sentence variables, $\mathscr{P}, \mathscr{Q}, \ldots$ we shall also have a list of **predicate variables** $\mathscr{P}(x), \mathscr{Q}(x), \ldots, \mathscr{P}(x,y), \mathscr{Q}(x,y,), \ldots, \mathscr{P}(x,y,z), \ldots$. The name "predicate variable" is not to be confused with the variables x, y. z, etc. in the predicates. We shall assume that the predicates are defined on some fixed nonempty domain D.

From the sentence and predicate variables, we can construct more complicated predicates and sentences; for example,

(1) $\mathscr{P}(x) \Rightarrow [\exists y \, \mathscr{Q}(x,y) \wedge \mathscr{R}]$,

(2) $\forall x (\mathscr{P}(x) \Rightarrow [\exists y \, \mathscr{Q}(x,y) \wedge \mathscr{R}])$,

(3) $\forall x \, \mathscr{P}(x) \Rightarrow [\exists y \, \mathscr{Q}(x,y) \wedge \mathscr{R}]$.

Now (1) is a predicate of one free variable, x. This x has two *occurrences* in the definition of (1). The variable y in (1) is bound (by $\exists y$).

All variables in (2) are bound, so (2) is a sentence. Both occurrences of x are bound by the leading $\forall x$. The *scope* of the quantifier $\forall x$ in (2) is

$$\mathscr{P}(x) \Rightarrow \exists y [\mathscr{Q}(x,y) \wedge \mathscr{R}].$$

Notice that (2) and (3) are *not* the same. In fact, (3) is a predicate of one free variable, x, but the first occurrence of x (in $\forall x \, \mathscr{P}(x)$) is bound. It is the second occurrence only (in $\exists y \, \mathscr{Q}(x,y)$) that is free, and (3) is a predicate in this occurrence of x only. The scope of $\forall x$ in this case is $\mathscr{P}(x)$. (The scope of $\exists y$ in all three is $\mathscr{Q}(x,y)$.)

Predicate (3) means the same as

$$\forall w \, \mathscr{P}(w) \Rightarrow [\exists y \, \mathscr{Q}(x,y) \wedge \mathscr{R}].$$

(And in practice, this is a better way to write it. See (*) and (9) in 10.8.)

In making truth valuations of (1), (2), or (3) we assign $\mathscr{T}$ or $\mathscr{F}$ to $\mathscr{R}$. The situation is more complicated for $\mathscr{P}$ and $\mathscr{Q}$. We assign to $\mathscr{P}$ one of the functions from D into $\{\mathscr{T}, \mathscr{F}\}$ and to $\mathscr{Q}$ one of the functions from $D \times D$ into $\{\mathscr{T}, \mathscr{F}\}$. Then we apply the rules of Chapter 3 and 10.6.

To illustrate this, let $D = \{a, b\}$, and make the following assignments. (This is only one of 128 different assignments. Prove this. See Exercise 1 of 10.11.)

(i) $\mathscr{R}$ is assigned the value $\mathscr{T}$.

(ii) $\mathscr{P}$ is assigned the value which is the function given in Table 10–3.

Table 10–3

x	$\mathcal{P}(x)$
a	$\mathcal{T}$
b	$\mathcal{T}$

(iii) $\mathcal{Q}$ is assigned the value which is the function given in Table 10–4.

Table 10–4

x	y	$\mathcal{Q}(x, y)$
a	a	$\mathcal{F}$
a	b	$\mathcal{F}$
b	a	$\mathcal{T}$
b	b	$\mathcal{F}$

The predicate $\exists y\, \mathcal{Q}(x, y)$ is the following function based on the definition of $\exists y$ in 10.6.

x	$\exists y\, \mathcal{Q}(x, y)$
a	$\mathcal{F}$ (there is no y, such that $\mathcal{Q}(a, y)$ is true)
b	$\mathcal{T}$ (there is a y, viz., a, such that $\mathcal{Q}(b, y)$ is true)

By the definition of $\wedge$,

x	$\exists y\, \mathcal{Q}(x, y) \wedge \mathcal{R}$
a	$\mathcal{F}$
b	$\mathcal{T}$

Hence, by the definition of $\Rightarrow$, the predicate (1) has the following function as its value under the assignments (i), (ii), and (iii):

x	$\mathscr{P}(x) \Rightarrow [\exists y\, \mathscr{Q}(x,y) \wedge \mathscr{R}]$
a	$\mathscr{F}$
b	$\mathscr{T}$

It follows from this, by 10.6, that the value of sentence (2) is $\mathscr{F}$ under the assignments (i), (ii), and (iii).

Similarly, the value of $\forall x\ \mathscr{P}(x)$ is $\mathscr{T}$, and hence, the predicate (3) is $\mathscr{F}$ at a and $\mathscr{T}$ at b.

10.8 TAUTOLOGIES

A sentence is a **tautology** if, for every assignment of truth values as described in 10.7 for any nonempty domain D, it always has the value $\mathscr{T}$. Tautologies asserting logical equivalences tell us, among other things, those manipulations of the logical symbols that are correct regardless of the particular context. Tautologies also tell us that certain sentences are true because of their logical form alone. We shall find it convenient occasionally to add predicates to our definition of tautology: a predicate, $\mathscr{S}(x)$, is a tautology if for every assignment of truth values it is the identically true function of x. In other words $\mathscr{S}(x)$ is a tautology if it has the value $\mathscr{T}$ regardless of the truth assignments we make or the value x takes. The same is true for several variables. For example, $\mathscr{P}(x) \Rightarrow \mathscr{P}(x)$ is a tautology. We shall use the symbol $\vDash$:

$$\vDash \mathscr{P}(x) \Rightarrow \mathscr{P}(x).$$

In the following list of tautologies, the letters $\mathscr{P}$, $\mathscr{Q}$, $\mathscr{R}$ stand for sentence and predicate variables. Only certain bound variables in the predicates are shown. All the formulas are correct if additional free and bound variables are present, provided the following condition is satisfied.

(*) All other variables are distinct from all of the bound variables which are exhibited.

(1) $\qquad\qquad\qquad \vDash \neg\neg\,\mathscr{P} \Leftrightarrow \mathscr{P}.$

(2) $\qquad\qquad \vDash \neg(\mathscr{P} \vee \mathscr{Q}) \Leftrightarrow (\neg\mathscr{P} \wedge \neg\mathscr{Q}).$

(3) $$\vDash \neg(\mathscr{P} \wedge \mathscr{Q}) \Leftrightarrow (\neg \mathscr{P} \vee \neg \mathscr{Q}).$$

(4) $$\vDash (\mathscr{P} \Rightarrow \mathscr{Q}) \Leftrightarrow \neg(\mathscr{P} \wedge \neg \mathscr{Q}).$$

(5) $$\vDash (\mathscr{P} \Rightarrow \mathscr{Q}) \Leftrightarrow (\neg \mathscr{P} \vee \mathscr{Q}).$$

(6) $$\vDash [(\mathscr{P} \wedge \mathscr{Q}) \Rightarrow \mathscr{R}] \Leftrightarrow [\mathscr{P} \Rightarrow (\mathscr{Q} \Rightarrow \mathscr{R})].$$

(7) $$\vDash [\mathscr{P} \wedge (\mathscr{Q} \vee \mathscr{R})] \Leftrightarrow [(\mathscr{P} \wedge \mathscr{Q}) \vee (\mathscr{P} \wedge \mathscr{R})].$$

(8) $$\vDash [\mathscr{P} \vee (\mathscr{Q} \wedge \mathscr{R})] \Leftrightarrow [(\mathscr{P} \vee \mathscr{Q}) \wedge (\mathscr{P} \vee \mathscr{R})].$$

The following is a list of most of the basic results involving the quanti-fiers.

(9) $$\vDash \forall x\ \mathscr{P}(x) \Leftrightarrow \forall y\ \mathscr{P}(y).$$

(10) $$\vDash \exists x\ \mathscr{P}(x) \Leftrightarrow \exists y\ \mathscr{P}(y).$$

(11) $$\vDash \neg \forall x\ \mathscr{P}(x) \Leftrightarrow \exists x \neg \mathscr{P}(x).$$

(12) $$\vDash \neg \exists x\ \mathscr{P}(x) \Leftrightarrow \forall x \neg \mathscr{P}(x).$$

(13) $$\vDash \forall x[\mathscr{P}(x) \wedge \mathscr{Q}(x)] \Leftrightarrow [\forall x\ \mathscr{P}(x) \wedge \forall x\ \mathscr{Q}(x)].$$

(14) $$\vDash \forall x[\mathscr{P}(x) \wedge \mathscr{Q}] \Leftrightarrow [\forall x\ \mathscr{P}(x) \wedge \mathscr{Q}].$$

(15) $$\vDash \forall x[\mathscr{P}(x) \vee \mathscr{Q}] \Leftrightarrow [\forall x\ \mathscr{P}(x) \vee \mathscr{Q}].$$

(16) $$\vDash [\forall x\ \mathscr{P}(x) \vee \forall x\ \mathscr{Q}(x)] \Rightarrow \forall x[\mathscr{P}(x) \vee \mathscr{Q}(x)].$$

(17) $$\vDash \exists x[\mathscr{P}(x) \vee \mathscr{Q}(x)] \Leftrightarrow [\exists x\ \mathscr{P}(x) \vee \exists x\ \mathscr{Q}(x)].$$

(18) $$\vDash \exists x[\mathscr{P}(x) \vee \mathscr{Q}] \Leftrightarrow [\exists x\ \mathscr{P}(x) \vee \mathscr{Q}].$$

(19) $$\vDash \exists x[\mathscr{P}(x) \wedge \mathscr{Q}] \Leftrightarrow [\exists x\ \mathscr{P}(x) \wedge \mathscr{Q}].$$

(20) $$\vDash \exists x[\mathscr{P}(x) \wedge \mathscr{Q}(x)] \Rightarrow [\exists x\ \mathscr{P}(x) \wedge \exists x\ \mathscr{Q}(x)].$$

(21) $$\vDash \forall x[\mathscr{P} \Rightarrow \mathscr{Q}(x)] \Leftrightarrow [\mathscr{P} \Rightarrow \forall x\ \mathscr{Q}(x)].$$

(22) $$\vDash \exists x[\mathscr{P} \Rightarrow \mathscr{Q}(x)] \Leftrightarrow [\mathscr{P} \Rightarrow \exists x\ \mathscr{Q}(x)].$$

(23) $$\vDash \forall x[\mathscr{P}(x) \Rightarrow \mathscr{Q}] \Leftrightarrow [\exists x\ \mathscr{P}(x) \Rightarrow \mathscr{Q}].$$

(24) $$\vDash \exists x[\mathscr{P}(x) \Rightarrow \mathscr{Q}] \Leftrightarrow [\forall x\ \mathscr{P}(x) \Rightarrow \mathscr{Q}].$$

(25) $$\vDash \forall x \forall y\ \mathscr{P}(x, y) \Leftrightarrow \forall y \forall x\ \mathscr{P}(x, y).$$

(26) $$\vDash \exists x \exists y\ \mathscr{P}(x, y) \Leftrightarrow \exists y \exists x\ \mathscr{P}(x, y).$$

(27) $$\vDash \exists x \forall y\ \mathscr{P}(x, y) \Rightarrow \forall y \exists x\ \mathscr{P}(x, y).$$

These results are of the utmost importance. Every nontrivial mathe-matical statement involves quantifiers, and its proof or disproof involves manipulations of quantifiers according to these rules.

We shall prove only a few of these, leaving the rest as exercises (Exercise 1 of 10.11). They can all be proved on the basis of their meaning, without going into a detailed analysis of truth valuations based on 10.6, 10.7, and Chapter 3. The truth definitions given there were to make explicit the meanings of the sentential connectives and the quantifiers. The reader should try to reason flexibly with their meanings.

Tautologies (9) and (10) assert, as we have observed before, that the name of a bound variable may be changed (provided that the new name is distinct from all other free and bound variables in the predicate).

In (11) and (12) we have results of considerable importance. Almost every theorem or conjecture of mathematics is of the form $\forall x \ \mathscr{P}(x)$ or $\exists x \ \mathscr{P}(x)$, or combinations thereof such as (4a), (4b) of 10.6. Hence (11) and (12) (with (1)) provide us with more flexibility in trying to prove a conjecture. They also tell us how to disprove a conjecture: to disprove $\forall x \ \mathscr{P}(x)$, we must prove $\neg \forall x \ \mathscr{P}(x)$; hence we may instead prove $\exists x \ \neg \ \mathscr{P}(x)$; that is, find an x such that $\mathscr{P}(x)$ is false. This is called disproving $\forall x \ \mathscr{P}(x)$ by finding a **counter example.**

To prove (11), assume $\neg \forall x \ \mathscr{P}(x)$ is $\mathscr{T}$, that is, $\forall x \ \mathscr{P}(x)$ is $\mathscr{F}$. By the definition of $\forall x$ then, $\mathscr{P}(x)$ is $\mathscr{F}$ for at least one $x \in D$; that is, $\neg \ \mathscr{P}(x)$ is $\mathscr{T}$ for at least one $x \in D$. By the definition of $\exists x$ then, $\exists x \ \neg \ \mathscr{P}(x)$ is $\mathscr{T}$. The proof of the converse implication is just as easy, as is the proof of (12).

In (13) note that the scope of the first quantifier is $\mathscr{P}(x) \wedge \mathscr{Q}(x)$, whereas the scopes of the second and third quantifiers respectively are $\mathscr{P}(x)$ and $\mathscr{Q}(x)$. In other words $\forall x$ distributes over $\wedge$. Similarly, (17) asserts that $\exists x$ distributes over $\vee$. However, $\forall x$ does not distribute over $\vee$ in general; (16) is the best we can do. The converse of (16) is false, as we shall see presently. Similarly the converse of (20) is false.

The important feature of (14) and (15) is that $\mathscr{Q}$ does not contain the variable x. In the weaker sense of (15), $\forall x$ does distribute over $\vee$.

Similar remarks apply to (18) and (19).

The reader should have no difficulty proving (13) to (20). As an example, we shall prove (16).

Suppose $\forall x \ \mathscr{P}(x) \vee \forall x \ \mathscr{Q}(x)$ is $\mathscr{T}$. Then either $\forall x \ \mathscr{P}(x)$ is $\mathscr{T}$ or $\forall x \ \mathscr{Q}(x)$ is $\mathscr{T}$. *Case* 1. $\forall x \ \mathscr{P}(x)$ is $\mathscr{T}$. Then, by the definition of $\forall x$, $\mathscr{P}(x)$ is $\mathscr{T}$ for every $x \in D$. Hence, by the definition of $\vee$, $\mathscr{P}(x) \vee \mathscr{Q}(x)$ is $\mathscr{T}$ for every $x \in D$. Hence $\forall x[\mathscr{P}(x) \vee \mathscr{Q}(x)]$ is $\mathscr{T}$ by the definition of $\forall x$. *Case* 2. Similar treatment.

In order to demonstrate that the converse of (16) is not a tautology, we need to exhibit a particular domain D and particular predicates $\mathscr{P}(x)$, $\mathscr{Q}(x)$ for which the converse of (16) is false. Let $D = N$, let $\mathscr{P}(x)$ be "x is even", and let $\mathscr{Q}(x)$ be "x is odd". Then (i) $\forall x[\mathscr{P}(x) \vee \mathscr{Q}(x)]$ is true, but (ii) $\forall x \ \mathscr{P}(x) \vee \forall x \ \mathscr{Q}(x)$ is false. (Note, incidentally, how our symbolism renders language precise. The sentence "Every natural number is either even or odd" is ambiguous, except that one usually supplies the intended meaning, viz., (i), which is true. It could mean (ii), which is false.)

Tautologies (21) to (24) relate the quantifiers to $\Rightarrow$; (21) and (22) are what one would naively guess, whereas (23) and (24) are often somewhat of a surprise. We shall leave it to the reader to try to prove these on the basis of their meaning (it can be done!). We shall prove (23) a different way, thereby demonstrating a new approach.

We shall assume (1) to (20) have already been proved. We shall also assume that the following **replacement principle** is intuitively obvious:

In a compound sentence or predicate, any one of its component sentences or predicates can be replaced by another one with which it is logically equivalent (without changing the truth value of the original sentence or predicate).

Hence, by (5)

$$\models \forall x[\, \mathscr{P}(x) \Rightarrow \mathscr{Q}] \Leftrightarrow \forall x[\neg \, \mathscr{P}(x) \vee \mathscr{Q}].$$

Whence, by (15),

$$\models \forall x[\, \mathscr{P}(x) \Rightarrow \mathscr{Q}] \Leftrightarrow [\forall x \neg \, \mathscr{P}(x) \vee \mathscr{Q}].$$

Whence, by (12),

$$\models \forall x[\, \mathscr{P}(x) \Rightarrow \mathscr{Q}] \Leftrightarrow [\neg \, \exists x \, \mathscr{P}(x) \vee \mathscr{Q}].$$

Whence, by (5) again,

(23) $$\models \forall x[\, \mathscr{P}(x) \Rightarrow \mathscr{Q}] \Leftrightarrow [\exists x \, \mathscr{P}(x) \Rightarrow \mathscr{Q}].$$

The meaning and importance of (25) and (26) are obvious. Quantifiers of like kind commute. The proofs are obvious from the meanings of the quantifiers.

Quantifiers not of like kind do not commute; (27) holds, but the converse is not universally true. To prove (27), suppose $\exists x \, \forall y \, \mathscr{P}(x, y)$ is $\mathscr{T}$. Then for some $x \in D$, call it x', $\forall y \, \mathscr{P}(x', y)$ is $\mathscr{T}$. Let $y \in D$ be arbitrary. Then $\mathscr{P}(x', y)$ is $\mathscr{T}$. By the meaning of $\exists x$ then, $\exists x \, \mathscr{P}(x, y)$ is $\mathscr{T}$. But $y \in D$ was arbitrary, so, by the meaning of $\forall y$, $\forall y \, \exists x \, \mathscr{P}(x, y)$ is $\mathscr{T}$.

To disprove the converse of (27), let $D = I$ and $\mathscr{P}(x, y)$ be "$x \leqslant y$". Then $\forall y \, \exists x \, (x \leqslant y)$ is true (for every integer there is a smaller one). However, $\exists x \, \forall y \, (x \leqslant y)$ is false (there does not exist a smallest integer).

The failure of the converse of (27) is worth a second look. The reason for this failure is easily discovered when we examine the meanings of the antecedent and consequent of (27). The consequent means: no matter what $y \in D$ we choose, there is always an $x \in D$, *depending in general on the choice of y*, such that $\mathscr{P}(x, y)$ is $\mathscr{T}$; choose a different y and you may have to choose a different x. On the other hand, the antecedent means: there is an $x \in D$ (*independent of y*) such that, no matter what $y \in D$ we choose, $\mathscr{P}(x, y)$ is $\mathscr{T}$. It is conceivable, therefore, that the consequent be true and the antecedent false. The example of the preceding paragraph bears this out.

Before we leave this section, we shall consider a few examples of how combinations of (1) to (24) easily yield additional useful and important tautologies. (We have already seen how (21) to (24) follow from (1) to (20); even (1) to (20) is an unnecessarily long list in this regard.) By (21)

$$\vDash [\forall x\ \mathscr{P}(x) \Rightarrow \forall y\ \mathscr{Q}(y)] \Leftrightarrow \forall y[\forall x\ \mathscr{P}(x) \Rightarrow \mathscr{Q}(y)].$$

Hence, by (24),

(28) $\vDash [\forall x\ \mathscr{P}(x) \Rightarrow \forall y\ \mathscr{Q}(y)] \Leftrightarrow \forall y\ \exists x[\mathscr{P}(x) \Rightarrow \mathscr{Q}(y)].$

(Note that we implicitly used the replacement property and transitivity of $\Leftrightarrow$.)
In a similar way (reversing the order of (21), (24)),

(29) $\vDash [\forall x\ \mathscr{P}(x) \Rightarrow \forall y\ \mathscr{Q}(y)] \Leftrightarrow \exists x\ \forall y[\mathscr{P}(x) \Rightarrow \mathscr{Q}(y)].$

Tautologies (28) and (29) are additional useful ones, derived from the original list.
From (28), (29) (and the transitivity of $\Leftrightarrow$),

$$\vDash \forall y\ \exists x[\mathscr{P}(x) \Rightarrow \mathscr{Q}(y)] \Leftrightarrow \exists x\ \forall y[\mathscr{P}(x) \Rightarrow \mathscr{Q}(y)].$$

Here we have an instance of the converse of (27) which does hold.
The following illustrates an important point in the use of bound variables.
By (10),

(30) $\vDash [\forall x\ \mathscr{P}(x) \Rightarrow \forall x\ \mathscr{Q}(x)] \Leftrightarrow [\forall x\ \mathscr{P}(x) \Rightarrow \forall y\ \mathscr{Q}(y)].$

By (29), (30),

(31) $\vDash [\forall x\ \mathscr{P}(x) \Rightarrow \forall x\ \mathscr{Q}(x)] \Leftrightarrow \exists x\ \forall y[\mathscr{P}(x) \Rightarrow \mathscr{Q}(y)].$

This illustrates an important point. In the predicate

$$\forall x\ \mathscr{P}(x) \Rightarrow \forall x\ \mathscr{Q}(x)$$

we can advance, for example, the first quantifier by (24) to get

$$\exists x[\mathscr{P}(x) \Rightarrow \forall x\ \mathscr{Q}(x)].$$

This is quite all right. Note the scope of $\exists x$ is

(32) $\mathscr{P}(x) \Rightarrow \forall x\ \mathscr{Q}(x),$

but only the first (occurrence of) x is free in this predicate, and it is the only one bound by $\exists x$. $\forall x\ \mathscr{Q}(x)$ is just another sentence (unless it contains free variables *other than* x), and it could just as well be $\forall y\ \mathscr{Q}(y)$ (provided $\mathscr{Q}(x)$ does not also contain the free variable y, for in this case y would become bound by $\forall y$, and $\forall y\, \mathcal{Q}(y)$ would then have a different meaning). Now, if we attempt to advance the $\forall x$ to the front in (32), we are stopped by a violation of condition (*). This, of course, is as it should be, and illustrates the reason for condition (*). If we did use (21) incorrectly, we would obtain

(33) $$\forall x[\, \mathcal{P}(x) \Rightarrow \mathcal{Q}(x)]$$

which is certainly not logically equivalent to (32).

To summarize then: not only *may* we change the names of bound variables according to (9) and (10), but we *must* do so if the change is needed to avoid violating condition (*).

In general, the reader should try to reason flexibly in these and similar matters according to the meanings of the logical symbols. For example,

$$\forall x\, \forall y\, [\, \mathcal{P}(x) \Rightarrow \mathcal{Q}(y)]$$

has quite a different meaning than does (33), and could not, therefore, be logically equivalent to it.

10.9   RESTRICTED QUANTIFICATION

Consider Theorem 2 of 8.3. It would certainly be incorrect to express this theorem in logical symbolism as $\forall x\, \forall y\, \exists n(n \cdot x > y)$, and yet in some way this expresses the essence of the theorem. If we let the domain be R, a little thought will convince the reader that the theorem is properly expressed by

$$\forall x\{x \in R^+ \Rightarrow \forall y\, \exists n[(n \in I^+) \land (n \cdot x > y)]\},$$

where I^+ is the set of positive integers, and R^+ is the set of positive reals.

This is frequently abbreviated in the form:

$$\forall x_{x \in R+} \forall y\, \exists n_{n \in I+}(n \cdot x > y).$$

More generally, let D by any nonempty domain on which all predicates under discussion are defined. Let D_1 be any subdomain of D (that is, $D_1 \subset D$), and let $\mathcal{P}$ be a particular predicate defined on D.

1. "For all $x \in D_1$, $\mathcal{P}(x)$ is true" is expressed in logical symbolism by

$$\forall x[x \in D_1 \Rightarrow \mathcal{P}(x)],$$

and may be abbreviated as

$$\forall x_{x \in D_1} \mathcal{P}(x).$$

2. "There exists an $x \in D_1$ such that $\mathscr{P}(x)$ is true" is expressed by

$$\exists x[x \in D_1 \wedge \mathscr{P}(x)],$$

and may be abbreviated as

$$\exists x_{x \in D_1} \mathscr{P}(x).$$

The **restricted quantifiers**, $\forall x_{x \in D_1}$, $\exists x_{x \in D_1}$, have the advantage that they often make the meaning of a complicated predicate or sentence more easily discernible. Furthermore, *they satisfy all the tautologies* (9) *to* (24) *of* 10.8, with restricted quantifiers replacing the unrestricted ones. For example,

$$\vDash \neg \forall x_{x \in D_1} \mathscr{P}(x) \Leftrightarrow \exists x_{x \in D_1} \neg \mathscr{P}(x).$$

We shall prove this, and leave the rest to be proved by the reader as an exercise (Exercise 1 of 10.11).

$$\begin{aligned}
\neg \forall x_{x \in D_1} \mathscr{P}(x) &\Leftrightarrow \neg \forall x[x \in D_1 \Rightarrow \mathscr{P}(x)] && \text{Definition,} \\
&\Leftrightarrow \neg \forall x(x \notin D_1 \vee \mathscr{P}(x)) && \text{(5) of 10.8,} \\
&\Leftrightarrow \exists x \neg (x \notin D_1 \vee \mathscr{P}(x)) && \text{(11) of 10.8,} \\
&\Leftrightarrow \exists x(x \in D_1 \wedge \neg \mathscr{P}(x)) && \text{(2), (1) of 10.8,} \\
&\Leftrightarrow \exists x_{x \in D_1} \neg \mathscr{P}(x) && \text{Definition}
\end{aligned}$$

In practice, we shall write $\exists_{x \in D_1}$ instead of $\exists x_{x \in D_1}$ etc. in the sequel.

Additional important properties of the restricted quantifiers appear in Exercise 8 of 10.11.

10.10   METHODS OF PROOF INVOLVING QUANTIFIERS

As was the case in 10.7 (which the reader is urged to reread), we shall have an arbitrary nonempty domain D, lists of sentence and predicate variables (defined on D), and variables x, y, z whose domain is D. In addition, we shall let the letters a, b, c stand for *particular* elements of D.

Given a predicate variable $\mathscr{P}(x)$, recall that a truth assignment means assigning to $\mathscr{P}$ a function from D into $\{\mathscr{T}, \mathscr{F}\}$. Once we have made such an assignment, $\mathscr{P}(a)$ will be either $\mathscr{T}$ or $\mathscr{F}$, depending on the function we assigned to $\mathscr{P}$ and the element "a" chosen. It is important to note that $\mathscr{P}(a)$ *is a sentence*; it is either $\mathscr{T}$ or $\mathscr{F}$. $\mathscr{P}(a)$ is called the result of substituting a for x in $\mathscr{P}(x)$.

Compound predicates and sentences are constructed from the sentence and predicate variables, the logical symbols, and substitutions of a, b, c.

Consider for example the predicate (1) of 10.7. Substituting c for x, we get the sentence

$$\mathscr{P}(c) \Rightarrow [\exists y \mathscr{Q}(c, y) \wedge \mathscr{R}],$$

where c denotes, not a variable, but a particular element of D.

When we say that a **predicate has the value** $\mathscr{T}$ under a particular truth assignment, we shall mean that it is the identically true function of its free variables.[1] With this convention, the definitions of **logical consequence** and ⊨ in 10.3 are extended simply by allowing the $\mathscr{P}_1, \mathscr{P}_2, ..., \mathscr{P}_n, \mathscr{P}$ to be *predicates or sentences*, and the $\mathscr{R}_1, ..., \mathscr{R}_m$ to be *predicate variables or sentence variables*. We shall use the name **logical derivation** as before. If $n = 0$, $\mathscr{P}$ is a tautology. (See 10.8.)

The five basic properties of ⊨ (10.3) still hold.

An additional property of ⊨ which now holds is

$$\mathscr{P}(x) \vDash \mathscr{P}(a) \text{ for any } a \in D.[2]$$

This is called the **rule of substitution.** The proof is obvious: any assignment of truth values which makes $\mathscr{P}(x)$ the identically true function of x certainly makes $\mathscr{P}(a)$ true for any $a \in D$.

Example 1. We shall show that (cf. (21) of 10.8)

(1) $\qquad\qquad \forall x[\mathscr{P}(y) \Rightarrow \mathscr{Q}(x)] \vDash \mathscr{P}(y) \Rightarrow \forall x \, \mathscr{Q}(x),$

where $\mathscr{P}$ and $\mathscr{Q}$ are predicate variables of one free variable each. In the premise and conclusion of (1), y is free and x is bound. To make an assignment of truth values in (1) is to assign to each of $\mathscr{P}$ and $\mathscr{Q}$ a function from D into $\{\mathscr{T}, \mathscr{F}\}$. Accordingly, let $\mathscr{P}$ and $\mathscr{Q}$ be functions from D into $\{\mathscr{T}, \mathscr{F}\}$ such that

(2) $\qquad\qquad \forall x[\mathscr{P}(y) \Rightarrow \mathscr{Q}(x)]$

has the value $\mathscr{T}$ (for all $y \in D$); we wish to show that this assignment makes

(3) $\qquad\qquad \mathscr{P}(y) \Rightarrow \forall x \, \mathscr{Q}(x)$

have the value $\mathscr{T}$ (for all $y \in D$). Since (2) has the value $\mathscr{T}$,

$$\mathscr{P}(y) \Rightarrow \mathscr{Q}(x)$$

---

[1] From this point on, the symbol ⊨, its meaning and all related notions are defined relative to an unspecified but fixed domain D. This was done for pedagogical reasons, but is at variance with the customary, *correct* approach (which would add at this point "over every nonempty set D") in which D is arbitrary and variable. For example, the rule of substitution (below) should say "... for any a in any D". Similar remarks apply to ∀-intro and ∃-elim (below), and elsewhere.

[2] See footnote 1.

has the value $\mathcal{T}$ for all $x \in D$ by the definition of $\forall x$. Hence, by the definition of $\Rightarrow$, either (a) $\mathscr{P}(y)$ is $\mathscr{F}$, or (b) $\mathscr{Q}(x)$ is $\mathscr{T}$ for every $x \in D$.

(a) If $\mathscr{P}(y)$ is $\mathscr{F}$, then (3) is $\mathscr{T}$ by the definition of $\Rightarrow$.

(b) In this case $\forall x \mathscr{Q}(x)$ is $\mathscr{T}$ by the definition of $\forall x$, and hence, (3) is $\mathscr{T}$ by the definition of $\Rightarrow$.

Example 2.   By almost the same argument,

(1′) $\forall x [\mathscr{P}(b) \Rightarrow \mathscr{Q}(x)] \vDash \mathscr{P}(b) \Rightarrow \forall x \mathscr{Q}(x),$

where b is a particular element of D. In (1′), the premise and conclusion are sentences.

Rather than discuss logical derivations in terms of truth valuations, we shall use the extended definition of $\vDash$ to get four new rules of proof similar to those of 10.4. We shall also examine how the eight rules of 10.4 need to be modified so that they still hold for the extended definition of $\vDash$. When we have all twelve rules of proof (plus the six basic properties of $\vDash$), we shall be able to talk in terms of "proof" rather than "truth," thus bringing us more in line with the methods of mathematics.[3]

At this point, the reader should reread 10.4.

The four rules of proof in Table 10–5 below introduce and eliminate the universal and existential quantifiers. They are correct provided that the following conditions are satisfied.

(i) $\mathscr{P}_1, \mathscr{P}_2, \ldots, \mathscr{P}_n, \mathscr{Q}$ are arbitrary sentences or predicates (but sometimes subject to certain other conditions which are discussed below).

(ii) $\mathscr{P}(x)$ is a predicate containing x as a free variable (and possibly other free variables).

(iii) The letter a denotes any element of D; it is *not a free variable*.

(iv) $\mathscr{P}(a)$ is the result of substituting a for x in $\mathscr{P}(x)$.

Table 10–5[4]

	Introduction	Elimination
$\forall$	If $\mathscr{P}_1, \mathscr{P}_2, \ldots, \mathscr{P}_n \vDash \mathscr{P}(a)$ for every $a \in D$, then $\mathscr{P}_1, \mathscr{P}_2, \ldots, \mathscr{P}_n \vDash \forall x \, \mathscr{P}(x).$	$\forall x \, \mathscr{P}(x) \vDash \mathscr{P}(a)$
$\exists$	$\mathscr{P}(a) \vDash \exists x \, \mathscr{P}(x)$	If $\mathscr{P}_1, \mathscr{P}_2, \ldots, \mathscr{P}_n, \mathscr{P}(a) \vDash \mathscr{Q}$ for every $a \in D$, then $\mathscr{P}_1, \mathscr{P}_2, \ldots, \mathscr{P}_n, \exists x \, \mathscr{P}(x) \vDash \mathscr{Q}.$

[3] See footnote 1.

[4] See footnote 1.

$\forall x\ \mathscr{P}(x)$ is sometimes viewed as the "conjunction" of the sentences $\mathscr{P}(a)$ for $a \in D$. In fact, if $D = \{1, 2, 3\}$, then $\forall x\ \mathscr{P}(x)$ means the same as $\mathscr{P}(1) \wedge \mathscr{P}(2) \wedge \mathscr{P}(3)$. However, if D is infinite (and in most interesting cases it is), then $\forall x\ \mathscr{P}(x)$ is an "infinite conjunction". From this point of view, $\forall$-intro is a generalization of $\wedge$-intro. We could have stated $\wedge$-intro as: If $\mathscr{P}_1, ..., \mathscr{P}_n \vDash \mathscr{P}$ and $\mathscr{P}_1, ..., \mathscr{P}_n \vDash \mathscr{Q}$, then $\mathscr{P}_1, ..., \mathscr{P}_n \vDash \mathscr{P} \wedge \mathscr{Q}$. In this form it resembles $\forall$-intro. By $\forall$-intro and the rule of substitution:

$$\mathscr{P}(x) \vDash \forall x\ \mathscr{P}(x).$$

We shall call this $\forall$-intro as well. This resembles $\wedge$-intro: just as the premises $\mathscr{P}$, $\mathscr{Q}$ yield the conclusion $\mathscr{P} \wedge \mathscr{Q}$, so the premises $\mathscr{P}(x)$ as x varies over D yield the conclusion $\forall x\ \mathscr{P}(x)$. We stated $\forall$-intro as we did to avoid certain technical difficulties.

$\forall$-intro is the rule whereby one proves a sentence or predicate of the form $\forall x\ \mathscr{P}(x)$. It emphasizes the fact that, to prove $\forall x\ \mathscr{P}(x)$, one must prove $\mathscr{P}(x)$ independently of the element x chosen from D.

This rule has subsidiary derivations, one for each $a \in D$, of the form

$$\mathscr{P}_1, \mathscr{P}_2, ..., \mathscr{P}_n \vDash \mathscr{P}(a).$$

In general, one could not write down a proof of each one. (Certainly not if D is infinite.) So one must show that these derivations hold by the form of $\mathscr{P}(a)$. (For example, $\vDash a = a$ for all $a \in D$ is axiomatic.)

There is a tacit assumption in $\forall$-intro which should be made explicit. Since the subsidiary deductions hold for each $a \in D$, and since the premises are the same in each case, the premises could not depend on, or contain, the letter a. For example, $a = 0 \vDash a = 0$ for all $a \in N$, but if we misapply $\forall$-intro to get $a = 0 \vDash \forall x\ (x = 0)$ for all $a \in N$, then in particular $0 = 0 \vDash \forall x\ (x = 0)$. But certainly $\vDash 0 = 0$, and hence $\vDash \forall x\ (x = 0)$, which is nonsense.

The rule of $\forall$-elim is the analog of $\wedge$-elim. It is the rule whereby one *uses* a sentence or predicate of the form $\forall x\ \mathscr{P}(x)$. If $\forall x\ \mathscr{P}(x)$ is proved (or true) then so is $\mathscr{P}(a)$ for every $a \in D$.

It follows that $\forall$-elim can also be stated in the form

$$\forall x\ \mathscr{P}(x) \vDash \mathscr{P}(x),$$

or

$$\forall x\ \mathscr{P}(x) \vDash \mathscr{P}(y).$$

$\exists$ is to $\vee$ as $\forall$ is to $\wedge$. Indeed, if $D = \{1, 2, 3\}$, then $\exists x\ \mathscr{P}(x)$ means $\mathscr{P}(1) \vee \mathscr{P}(2) \vee \mathscr{P}(3)$. The "disjunction" is infinite if D is infinite, however.

The rule of $\exists$-intro then is the analog of $\vee$-intro. The proof is simple: any assignment of truth values which makes $\mathscr{P}(a)$ true certainly makes $\exists x\ \mathscr{P}(x)$ true by the meaning of the existential quantifier. Stated as a rule of proof, it emphasizes that to prove $\exists x\ \mathscr{P}(x)$, one must prove $\mathscr{P}(a)$ for some element $a \in D$.

Observe that ∃-intro and substitution give

$$\mathscr{P}(x) \vDash \exists x \ \mathscr{P}(x)$$

or even

$$\mathscr{P}(y) \vDash \exists x \ \mathscr{P}(x).$$

We shall call these ∃-intro as well.

It should be noted that ∃-intro can be used in a way illustrated as follows:

$$x = x \vDash \exists y (x = y).$$

We say that we have applied ∃-intro to the second occurrence of x.

The rule of ∃-elim is a rule of the subsidiary derivation type. There are several (perhaps infinitely many) subsidiary derivations of the form

(4) $\mathscr{P}(a) \vDash \mathscr{Q}$, for each $a \in D$.

(Assuming for simplicity that there are no other hypotheses.) The hypotheses $\mathscr{P}(a)$ are discharged.

If D is infinite, one could not write down a proof of (4). Instead one must show, by the form of $\mathscr{P}(a)$, that (4) holds for any $a \in D$.

There is an important distinction between (4) and

(5) $\mathscr{P}(x) \vDash \mathscr{Q}$.

Formula (5) means that every assignment of truth values which makes $\mathscr{P}(x)$ the *identically true function of x*, also makes $\mathscr{Q}$ true; whereas (4) means, *pick any $a \in D$ first*, and then any assignment which makes $\mathscr{P}(a)$ true, also makes $\mathscr{Q}$ true.

Example 3. If (4) holds then (5) holds. By the rule of substitution,

(6) $\mathscr{P}(x) \vDash \mathscr{P}(a)$, for each $a \in D$.

Now (6), (4), and transitivity of $\vDash$ give (5).

Example 4. *Formula* (5) *can hold when* (4) *fails for some* $a \in D$. For example, let $D = N$, and let $\mathscr{P}(x)$ be $x = 0$. Then

(5′) $x = 0 \vDash 1 = 0$

is an instance of (5), and it holds by the rule of substitution. However,

(4?) $0 = 0 \vDash 1 = 0$

is an instance of (4) for a particular $a \in N$ (viz., 0). Also,

(6′) $\vDash 0 = 0$

holds in N. Now (4?) and (6′) give $\vDash 1 = 0$, which certainly does not hold in N. So (4?) does not hold, and hence it is false that "$a = 0 \vDash 1 = 0$ for all $a \in D$". (It is false when a is zero, but it is true otherwise. Prove this; Exercise 1 of 10.11.)

The reader may be disturbed that (5′) holds, but remember that it means: If "$x = 0$" is identically true, then "$1 = 0$" is true; however, "$x = 0$" is not identically true in this case.

Now take a look at the similarity between $\vee$-elim and $\exists$-elim. Remember that we have subsidiary derivations of the form (4) for every $a \in D$.

Example 5. The following modification of $\exists$-elim is *not* a valid rule of proof.

($\exists$-elim?)   If $\mathscr{P}(x) \vDash \mathscr{Q}$, then $\exists x \, \mathscr{P}(x) \vDash \mathscr{Q}$.

For, returning to Example 4, we can apply ($\exists$-elim?) to (5′) to get

(5′?) $\exists x (x = 0) \vDash 1 = 0.$

However, (6′) and $\exists$-intro (applied to the first occurrence of 0) give

(7) $\vDash \exists x (x = 0).$

From (5′?) and (7), we again get the contradiction $\vDash 1 = 0$.

Proof of $\exists$-elim. We have

(4) $\mathscr{P}(a) \vDash \mathscr{Q}$, for each $a \in D$.

(For simplicity, we assume that there are no other hypotheses.)
We wish to show

(8) $\exists x \, \mathscr{P}(x) \vDash \mathscr{Q}.$

To this end, pick any assignment of truth values which makes the sentence $\exists x \, \mathscr{P}(x)$ have the value $\mathscr{T}$. By the definition of $\exists x$, $\mathscr{P}(a)$ has the value $\mathscr{T}$ for at least one $a \in D$. By (4) then, $\mathscr{Q}$ has the value $\mathscr{T}$. This proves (8).

Just as in the case of $\forall$-intro, there is a tacit assumption concerning $\exists$-elim which should be made explicit. In the subsidiary derivations (4), the conclusion $\mathscr{Q}$ is the same for each subsidiary derivation; that is, for each $a \in D$. $\mathscr{Q}$ could not therefore depend on, or contain, the letter a. For example, $a = 0 \vDash a = 0$ for every $a \in N$. But the conclusion varies with a. If we misapply $\exists$-elim to get $\exists x (x = 0) \vDash a = 0$ for every $a \in N$, then (7) gives $\vDash a = 0$ for every $a \in N$.

Example 6. The following expresses an axiomatic property (reflexiveness) of equality.

(9) $\models x = x.$

By (9), ∃-intro (applied to the second occurrence of x) and properties of $\models$,

(10) $\models \exists y\,(x = y).$

By (10), ∀-intro and properties of $\models$,

(11) $\models \forall x\,\exists y\,(x = y).$

On the other hand, $\exists x\,\forall y\,(x = y)$ is not a tautology. (It is false for every D that contains more than one element.) See the discussion of (27) in 10.8.

Example 7. Let $\mathscr{P}(x, y)$ be any predicate.

(12) $\forall y\,\mathscr{P}(a, y) \models \mathscr{P}(a, y)$ by ∀-elim.

(13) $\mathscr{P}(a, y) \models \exists x\,\mathscr{P}(x, y)$ by ∃-intro.

(14) $\forall y\,\mathscr{P}(a, y) \models \exists x\,\mathscr{P}(x, y)$ by (12), (13).

(15) $\exists x\,\forall y\,\mathscr{P}(x, y) \models \exists x\,\mathscr{P}(x, y)$ by (14), ∃-elim.

(16) $\exists x\,\mathscr{P}(x, y) \models \forall y\,\exists x\,\mathscr{P}(x, y)$ by ∀-intro.

(17) $\exists x\,\forall y\,\mathscr{P}(x, y) \models \forall y\,\exists x\,\mathscr{P}(x, y)$ by (15), (16).

Now, if it is permissible to apply ⇒-intro to (17), we get

$$\models \exists x\,\forall y\,\mathscr{P}(x, y) \Rightarrow \forall y\,\exists x\,\mathscr{P}(x, y),$$

which is (27) of 10.8.

It is now time to see if the methods of proof of 10.4 are still applicable with our extended definition of $\models$.

Example 8. Let $D = N$. By substitution,

(18) $x = 0 \models 1 = 0.$

If it is permissible to apply ⇒-intro,

(19?) $\models x = 0 \Rightarrow 1 = 0.$

But, by substitution

(20) $x = 0 \Rightarrow 1 = 0 \models 0 = 0 \Rightarrow 1 = 0.$

Whence

$$\models 0 = 0 \Rightarrow 1 = 0,$$

and, since $\vDash 0 = 0$ in N, we get $\vDash 1 = 0$ by $\Rightarrow$-elim.

So we have our answer: the rules of proof of 10.4 no longer apply as they stand.

Example 8 (revisited).   By $\forall$-elim,

(21) $\forall x(x = 0) \vDash x = 0.$

(22) $\forall x(x = 0) \vDash 1 = 0$ by (18), (21).

(23) $\vDash \forall x(x = 0) \Rightarrow 1 = 0$ if $\Rightarrow$-intro is permissible.

Notice that the rule of substitution cannot be applied to (23). Notice also that, by truth valuations, (23) does hold (the antecedent is false), whereas (19?) does not (prove this; Exercise 1 of 10.11).

Applying $\forall$-intro to (19?),

(24?) $\vDash \forall x(x = 0 \Rightarrow 1 = 0).$

By $\forall$-intro and $\forall$-elim, (24?) holds if and only if (19?) does. Notice the difference between (23) and (24?). We cannot apply $\forall$-elim to (23), but we can to (24?).

Applying $\Rightarrow$-intro to (18) gave a false result. Applying $\Rightarrow$-intro to (22) gave a correct result. The difference is that in the latter case the *hypothesis which was discharged was a sentence.*

All the rules of proof of 10.4 *which do not have subsidiary derivations* ($\wedge$-intro, $\vee$-intro, $\neg$-elim, $\wedge$-elim, $\Rightarrow$-elim) *still hold.*

The rules of proof of 10.4 *which do have subsidiary derivations* ($\neg$-intro, $\Rightarrow$-intro, $\vee$-elim) *still hold provided the hypotheses which are discharged are sentences.*

We shall leave it to the reader to check (Exercise 1 of 10.11) that the rules of 10.4 hold as modified. However, let us examine how the proof of $\neg$-intro in 10.4 breaks down if $\mathscr{P}$ is a predicate $\mathscr{P}(x)$. The statement "We wish to to show that $\neg \mathscr{P}$ is also true, ..." now means: We wish to show that $\neg \mathscr{P}(x)$ is the identically true function of x. To suppose, on the contrary, that this is false, is to suppose that $\neg \mathscr{P}(a)$ is false for some $a \in D$; that is, that $\mathscr{P}(a)$ is true for some $a \in D$. *It does not follow that* $\mathscr{P}(x)$ *is the identically true function of x.* The reader should examine the fact that what is involved here is the difference between $\forall x \neg \mathscr{P}(x)$ and $\neg \forall x \mathscr{P}(x)$. (See (11), (12) of 10.8.) Similarly the difference between (23) and (24?) above underlies the break-down of $\Rightarrow$-intro when $\mathscr{P}$ is a predicate.

Example 9.   See 6.1. Let F be any field and let $D = F$. Axiom M4 states

$$\vDash \forall x[\neg (x = 0) \Rightarrow \exists z(zx = 1)],$$

or by $\forall$-elim

(i) $\vDash \neg (a = 0) \Rightarrow \exists z(za = 1))$, for any $a \in F$.

Theorem 2 of 6.1 asserts

(ii) $\vDash \forall x \, \forall y[xy = 0 \Rightarrow (x = 0 \lor y = 0)].$

We shall now exhibit a proof of (ii), using the methods of this chapter.

(iii) $ab = 0, a = 0 \vDash a = 0.$

(iv) $ab = 0, a = 0 \vDash (a = 0) \lor (b = 0),$ (iii), $\lor$-intro.

(v) $\vDash da = 1 \Rightarrow (da) b = b,$ M3.

(vi) $da = 1, ab = 0 \vDash (da) b = b,$ (v), $\Rightarrow$-elim.

(vii) $\vDash ab = 0 \Rightarrow d(ab) = 0,$ Theorem 1 (of 6.1).

(viii) $da = 1, ab = 0 \vDash d(ab) = 0,$ (vii), $\Rightarrow$-elim.

(ix) $da = 1, ab = 0 \vDash (da) b = 0,$ (viii), M2, Replacement.

(x) $\vDash ((da) b = b \land (da) b = 0) \Rightarrow b = 0,$ transitivity of $=$.

(xi) $da = 1, ab = 0 \vDash ((da) b = b) \land ((da) b = 0),$ (vi), (ix), $\land$-intro.

(xii) $da = 1, ab = 0 \vDash b = 0,$ (x), (xi), $\Rightarrow$-elim.

(xiii) $\exists z(za = 1), ab = 0 \vDash b = 0,$ (xii), $\exists$-elim.

(xiv) $ab = 0, \neg (a = 0) \vDash b = 0,$ (i), (xiii), $\Rightarrow$-elim.

(xv) $ab = 0, \neg (a = 0) \vDash a = 0 \lor b = 0,$ (xiv), $\lor$-intro.

(xvi) $ab = 0, a = 0 \lor \neg (a = 0) \vDash a = 0 \lor b = 0,$ (iv), (xv), $\lor$-elim.

(xvii) $\vDash a = 0 \lor \neg (a = 0),$ tautology.

(xviii) $ab = 0 \vDash a = 0 \lor b = 0,$ (xvi), (xvii).

(xix) $\vDash ab = 0 \Rightarrow (a = 0 \lor b = 0),$ (xviii), $\Rightarrow$-intro.

(xx) $\vDash \forall x \, \forall y[xy = 0 \Rightarrow (x = 0 \lor y = 0)],$ $\forall$-intro twice.

(The approach to methods of proof in this Chapter is a truth-functional modification of a formal approach in *Introduction to Metamathematics*, by S. C. Kleene, Van Nostrand, 1952.)[5]

[5] See footnote 1.

10.11 EXERCISES

1. Many results in this chapter were left unproved. The reader should prove them.
2. Show that the tautologies of 10.8 follow from the rules of proof. (See Example 6 of 10.10.)
3. Select a theorem from elsewhere in this book and analyze its proof as in this chapter. (See Example 9 of 10.10.)
4. Consider the following rule of proof:
   If $\vDash \mathscr{P}(0)$, and $\mathscr{P}(x) \vDash \mathscr{P}(x+1)$, then $\vDash \forall x\ \mathscr{P}(x)$.
   Show that it holds for N, but not for I.
5. (a) Let $D = R$. Disprove $\forall x\ \forall y\ ((x+y)^2 = x^2 + y^2)$.
   (b) Let D be any field. Can you disprove $\forall x\ \forall y\ ((x+y)^2 = x^2 + y^2)$? Be careful; see Exercise 5 of 6.4.
6. Prove, by *reductio ad absurdum,* that there are infinitely many prime numbers (Euclid).
7. (a) The following is another amusing paradox.
       Let $\mathscr{A}$ be the sentence: Sentence $\mathscr{B}$ is true.
       Let $\mathscr{B}$ be the sentence: Sentence $\mathscr{A}$ is false.
       Show that $\mathscr{A}$ is both true and false.
   (b) Can you find a similar paradox in St. Paul's "Epistle to Titus", I, 12 where Paul wrote, "One of themselves, even a prophet of their own, said, 'The Cretans are always liars' "?
       This is called the "liar" or "Epimenides" paradox. (The Cretan prophet is believed to have been Epimenides.)
8. (a) Prove the following
       (i) $(A \subseteq B) \Rightarrow (\forall_{x \in B}\ \mathscr{P}(x) \Rightarrow \forall_{x \in A}\ \mathscr{P}(x))$.
       (ii) $(A \subseteq B) \Rightarrow (\exists_{x \in A}\ \mathscr{P}(x) \Rightarrow \exists_{x \in B}\ \mathscr{P}(x))$.
   (b) Is the converse of (i) true? Is the converse of (ii) true?
   (c) If (i) is modified by replacing the consequent by its converse, is the resulting sentence true?
       How about (ii)?
9. Let $\mathscr{P}, \mathscr{Q}$ be any predicates defined on A. Let
$$P = \{x \in A \mid \mathscr{P}(x)\},$$
$$Q = \{x \in A \mid \mathscr{Q}(x)\}.$$
Match each of the expressions in the left column below with exactly one of the expressions in the right column and explain exactly what the connection is.

$\forall x\ \mathscr{P}(x) \Rightarrow \forall x\ \mathscr{Q}(x)$	$P \cup \overline{Q}$
$\forall x\ \mathscr{Q}(x) \Rightarrow \forall x\ \mathscr{P}(x)$	$P \subseteq Q$
$\forall x\ [\mathscr{P}(x) \Rightarrow \mathscr{Q}(x)]$	$Q \subseteq P$
$\forall x\ [\mathscr{Q}(x) \Rightarrow \mathscr{P}(x)]$	$\overline{P \cap \overline{Q}}$
$\{x \in A \mid \mathscr{P}(x) \Rightarrow \mathscr{Q}(x)\}$	
$\{x \in A \mid \mathscr{Q}(x) \Rightarrow \mathscr{P}(x)\}$	

11

Sequences and Series

11.1 SEQUENCES OF REAL NUMBERS

Sequences have already been discussed in Chapters 7 and 8. In 7.7 a sequence was defined as a function whose domain is N or $N - \{0\}$ and whose counter domain is some set of real numbers. In Chapter 8 sequences with Q, the rational numbers, as counter domain were discussed in some detail, with special attention paid to the concept of the limit of such sequences. In this chapter we investigate intensively real-valued sequences, that is, sequences whose counter domains are subsets of the real number system.

Aside from their value in the development of mathematics, sequences are a useful tool in applied mathematics. An example of the type of problem to which sequences can be applied is the description of events which occur at stated instants in time. Thus the growth of a biological population is studied through annual censuses. The mean monthly air temperature at a given location is obtained by averaging the temperature measured at specified times throughout the month (say, every three hours). In general if the times of measurement are $t_1, t_2, \ldots, t_n, \ldots$, and if $f(t_n)$ is the measured quantity (size of population, or temperature, in the preceding examples) at time t_n, then $\{a_n\}$, where $a_n = f(t_n)$, $n \in N - \{0\}$, is a sequence. As mentioned in 7.7 we sometimes consider sequences from N into the reals, and at other times sequences from $N - \{0\}$ into the reals. Which domain is used is usually clear from the context.

Example 1. Suppose we put an amount of money, say, A dollars, in a bank at $r\%$ interest compounded annually. Then at the end of one year we have

$$A + A\frac{r}{100} = A\left(1 + \frac{r}{100}\right) \text{ dollars.}$$

At the end of two years we have

$$A\left(1 + \frac{r}{100}\right) + A\left(1 + \frac{r}{100}\right)\frac{r}{100} = A\left(1 + \frac{r}{100}\right)^2 \text{ dollars.}$$

It can be easily verified by induction that at the end of n years we have

$$A\left(1 + \frac{r}{100}\right)^n$$

dollars in the bank. The bank balance is easily exhibited as a sequence $\{b_n\}$ where

$$b_n = A\left(1 + \frac{r}{100}\right)^n.$$

Since the amount of money in the bank keeps on increasing indefinitely, this sequence has no upper bound (recall Definition 1 of 8.4), but it has a lower bound, namely, A.

The sequence obtained by measuring the air temperature at specified times will have an upper bound (say, 212°F, the boiling point of water) and a lower bound (say, $-459°F$, absolute zero). Such a sequence is called a **bounded sequence** and is defined as follows:

Definition 1.   *A sequence $\{a_n\}$ is **bounded** if*

$$\exists l \ \exists u \ \forall n \in N \quad (l \leqslant a_n \leqslant u).$$

*Otherwise a sequence is said to be **unbounded.***

More succinctly a sequence is bounded if it has an upper and a lower bound (Definition 1 of 8.4).

Example 2.   The following are bounded sequences,

(a) $\left\{1 - \dfrac{1}{n}\right\}$,

(b) $\{(-1)^n\}$,

(c) $\{\sin n\}$,

(d) $\left\{\dfrac{n}{n^2 + 3}\right\}$.

The sequence $\{b_n\}$ in Example 1 has no upper bound as remarked, and consequently by Definition 1 it is unbounded. The fact that no upper bound exists can be readily verified as follows. For any preassigned constant M we show that there is an integer n_0 such that for $n > n_0$

$$b_n = A\left(1 + \frac{r}{100}\right)^n > M.$$

By Exercise 10 of 11.9, $b_n \geqslant A + nAr/100$ for all $n \in N$. Now if $n > 100 \times (M - A)/Ar$, it is readily shown that $b_n > M$. Thus for the given M we can choose $n_0 = [100\,(M - A)/Ar]$ where $[x]$ indicates the greatest integer in x (see 8.3).

Sequences may be defined in at least three different ways.

(i) By a formula indicating how to obtain the nth term, $a_n = f(n)$. This is illustrated in Examples 1 and 2.

(ii) By giving the first few terms and then a **recursion formula** expressing the nth term as a function of terms of lower index. Examples of this are the sequence for $\sqrt{2}$ given in 8.5 and the sequence for $\sqrt{a}$ given in Exercise 18 of 8.7. Further examples are:

Example 3.   (a) $a_n = 2a_{n-1}$, $n \geqslant 1$, $a_0 = 3$.

(b) $a_n = 3a_{n-1} - 2a_{n-2}$, $n \geqslant 2$, $a_0 = 1$, $a_1 = 2$.

(c) $a_n = a_{n-1}^2 - a_{n-2}$, $n \geqslant 2$, $a_0 = 1$ $a_1 = 0$.

In each of these examples, we can show by induction that the formula determines uniquely each term of the sequence.

(iii) A sequence may be defined by giving the nth term as a sum. Thus

$$a_n = 1 + 2 + \cdots + n \quad \text{for } n \geqslant 1,$$

and

$$a_n = 1 + 4 + 9 + \cdots + n^2 \quad \text{for } n \geqslant 1.$$

Examples 7 and 8 of 8.4 are also sequences of this type.

11.2   DIFFERENCES AND ANTIDIFFERENCES

There are certain operations on sequences which are useful in describing their behavior. The local behavior, whether the sequence is increasing or decreasing or constant for a given n, can be discussed in terms of the **difference** defined below.

Definition 1.   *Let $\{a_n\}$ be a sequence. If we set $\Delta a_n = a_{n+1} - a_n$, then the differences $\{\Delta a_n\}$ form a new sequence, which is called the **difference** of the sequence $\{a_n\}$.*

Example 1.   Let $\{a_n\}$ be a sequence in which $a_n = 2^n$.

By Definition 1, $\Delta a_n = a_{n+1} - a_n = 2^{n+1} - 2^n = 2^n (2 - 1) = 2^n$. Thus the difference of the sequence $\{2^n\}$ is the same sequence $\{2^n\}$.

Example 2.   Let $\{a_n\}$ be a sequence in which $a_n = (-1)^n n$. By Definition 1, $\Delta a_n = (-1)^{n+1} (n + 1) - (-1)^n n = (-1)^{n+1} (2n + 1)$. Thus the difference of $\{(-1)^n n\}$ is $\{(-1)^{n+1} (2n + 1)\}$.

Whether a given sequence $\{a_n\}$ is increasing or decreasing or constant for a given n can be determined as follows:

(a) The sequence $\{a_n\}$ is *increasing* at n if $a_{n+1} > a_n$, that is, if $\Delta a_n > 0$.

(b) The sequence $\{a_n\}$ is *decreasing* at n if $a_{n+1} < a_n$, that is, if $\Delta a_n < 0$.

(c) The sequence $\{a_n\}$ is *constant* at n if $a_{n+1} = a_n$, that is, if $\Delta a_n = 0$.

The number Δa_n can be interpreted as the rate of increase of the sequence $\{a_n\}$ from n to $n + 1$. Thus in Example 1 the sequence $\{2^n\}$ is increasing for all n, and it is therefore a monotone increasing sequence. The sequence in Example 2 is increasing if n is odd and decreasing if n is even. This sequence is thus not monotonic.

The difference operation is used extensively in actuarial science. Some of the more important properties of the difference are given in Theorem 1 below.

Theorem 1.   *If $\{a_n\}$ and $\{b_n\}$ are any sequences and $\{c\}$ is a constant sequence, then*

(a) $\Delta c = 0$,

(b) $\Delta(c a_n) = c(\Delta a_n)$,

(c) $\Delta(a_n + b_n) = \Delta a_n + \Delta b_n$.

The proof follows immediately from the definition of Δa_n, and is left to the reader.

It is clear that (c) can be extended to the sum of any number of sequences. The following examples illustrate the use of Theorem 1.

Example 3.   Let $\{a_n\}$ be a sequence in which $a_n = n^2 - 4n + 1$. Then, by Theorem 1(c),

$$\begin{aligned}
\Delta a_n &= \Delta(n^2 - 4n + 1) = \Delta n^2 - \Delta(4n) + \Delta 1 \\
&= \Delta n^2 - 4\Delta n && \text{by Theorem 1 (a), (b),} \\
&= (n + 1)^2 - n^2 - 4(n + 1 - n) && \text{by Definition 1} \\
&= (n^2 + 2n + 1) - n^2 - 4 = 2n - 3.
\end{aligned}$$

The sequence $\{a_n\}$ thus decreases for $n = 0, 1$; it increases for $n \geqslant 2$.

Example 4. In Example 1 of 11.1 we found that if A is the amount deposited in a bank account, and if it earns interest at the rate of $r\%$ per period compounded periodically, at the end of n periods its amount will be

$$b_n = A\left(1 + \frac{r}{100}\right)^n.$$

Now

$$\Delta b_n = \Delta A\left(1 + \frac{r}{100}\right)^n = A\Delta\left(1 + \frac{r}{100}\right)^n \qquad \text{(Theorem 1 (b))}$$

$$= A\left\{\left(1 + \frac{r}{100}\right)^{n+1} - \left(1 + \frac{r}{100}\right)^n\right\}$$

$$= A\left(1 + \frac{r}{100}\right)^n\left(1 + \frac{r}{100} - 1\right).$$

Hence we find what we expected, that in one interest period the amount b_n grows by the amount

$$(1) \qquad\qquad\qquad \Delta b_n = \frac{r}{100}\,b_n.$$

Thus $\{b_n\}$ is a monotone increasing sequence and its rate of growth is proportional to b_n, the capital at time n. Any sequence $\{a_n\}$ in which $a_n = ca^n$ will have the same property (cf. Example 1).

Inversely related to the difference operator Δ is the **antidifference** operator Δ^{-1} defined as follows.

Definition 2. $\qquad\qquad \Delta^{-1}b_n = a_n \Leftrightarrow \Delta a_n = b_n;$

*the sequence $\{a_n\}$ is called an **antidifference** of $\{b_n\}$.*

Thus a sequence $\{a_n\}$ is called an antidifference of the sequence $\{b_n\}$ if $\Delta a_n = b_n$. We denote an antidifference of the sequence $\{b_n\}$ symbolically as $\{\Delta^{-1}b_n\}$, the n signifying the nth term of a particular antidifference sequence.

In Definition 2, $\{a_n\}$ is called *an* antidifference of $\{b_n\}$, because it is not unique. This point is brought out in the following examples.

Example 5. We know by Theorem 1 (a), that $\Delta c = 0$. Hence by Definition 2, $\Delta^{-1}0 = c$ where c is *any* constant.

Example 6. We wish to find $\Delta^{-1}n$. Suppose we try n^2. Now $\Delta n^2 = 2n + 1$. Since $\Delta n = n + 1 - n = 1$, we evaluate $\Delta(n^2/2 - n/2)$ in the hope that this will lead to the result. By Theorem 1 and Definition 1 we find that

$$\Delta\left(\frac{n^2}{2} - \frac{n}{2}\right) = \frac{1}{2}\Delta n^2 - \frac{1}{2}\Delta n = \left(n + \frac{1}{2}\right) - \frac{1}{2} = n.$$

Thus

$$\{\Delta^{-1}n\} = \left\{\frac{n^2}{2} - \frac{n}{2}\right\}$$

is an antidifference of $\{n\}$ and, by Theorem 1,

$$\left\{\frac{n^2}{2} - \frac{n}{2} + c\right\}$$

is also an antidifference of $\{n\}$.

The preceding examples indicate that finding an antidifference is partly guesswork. The connection between two antidifferences of the same sequence is shown in the following lemma and theorem.

Lemma 1.   *If $\Delta a_n = 0$, then $a_n = c$ for some fixed real number c.*

Proof.   If $\Delta a_n = 0$, then by Definition 1, $a_{n+1} = a_n$ for all $n \in N$. Thus, by induction, $a_n = a_0$, for all $n \in N$, and the result follows if we set $c = a_0$.

Theorem 2.   *If $\{a_n\}$ and $\{b_n\}$ are antidifferences of $\{c_n\}$, then there is a constant c such that*

$$a_n = b_n + c.$$

Proof.   Since $a_n = \Delta^{-1}c_n$ and $b_n = \Delta^{-1}c_n$, by Definition 2

$$\Delta a_n = \Delta b_n = c_n.$$

Thus $0 = c_n - c_n = \Delta a_n - \Delta b_n = \Delta(a_n - b_n)$ by Theorem 1. By Lemma 1, $a_n - b_n = c$ or $a_n = b_n + c$, where c is some constant.

Theorem 2 tells us that if we find one antidifference, any other anti-difference is determined by merely adding an arbitrary constant to this antidifference. Thus in Example 6, we found $\Delta^{-1}n = \frac{1}{2}(n^2 - n)$. By Theorem 2 all antidifferences of n are given by

(2) $$\{\Delta^{-1}n\} = \{\tfrac{1}{2}(n^2 - n) + c\}$$

where c is an arbitrary constant.

Antidifferences are often useful in evaluating sums.

Suppose a sequence is defined, as in 11.1 (iii), by giving the nth term as a sum. If $\{a_n\}$ is the given sequence and if

(3) $$a_n = b_0 + b_1 + \cdots + b_{n-1}.$$

then

(4) $\qquad \Delta a_n = (b_0 + b_1 + \cdots + b_n) - (b_0 + \cdots + b_{n-1})$

$\qquad\qquad\quad = b_n.$

In order to evaluate a sum of the form $b_1 + b_2 + \cdots + b_n$, which has no zeroth term, let $b_0 = 0$ so that (3) takes the form

$$a_n = 0 + b_1 + \cdots + b_{n-1}.$$

In this way various notational problems can be obviated. We have

(5) $\qquad\qquad\qquad\qquad a_n = \Delta^{-1} b_n,$

and the sum can be evaluated by finding an antidifference of b_n. If the given sum is of the form

$$b_m + b_{m+1} + \cdots + b_n,$$

for $n > m$, then

(6) $\qquad b_m + b_{m+1} + \cdots + b_n = (b_0 + \cdots + b_n) - (b_0 + \cdots + b_{m-1})$

$\qquad\qquad\qquad\qquad\qquad\quad = a_{n+1} - a_m,$

or

(7) $\qquad\qquad b_m + \cdots + b_n = \Delta^{-1} b_{n+1} - \Delta^{-1} b_m.$

Example 7. As an example of the use of equation (5), we evaluate $1 + 2 + \cdots + n$. (This has already been treated in 2.5 by mathematical induction.) Let $b_n = n$ and let

$$a_n = b_0 + b_1 + \cdots + b_{n-1}$$
$$= 0 + 1 + 2 + \cdots + (n - 1).$$

Then from (5), $a_n = \Delta^{-1} b_n = \Delta^{-1} n = \frac{1}{2}(n^2 - n) + c$ for all $n = 1, 2, \ldots$ (by Example 6). But $a_1 = 0$ and so c must be 0. Thus $a_n = \frac{1}{2} n(n - 1)$ and

$$a_{n+1} = 1 + 2 + \cdots + n = \tfrac{1}{2} n(n + 1), \text{ as before.}$$

Example 8. Let $b_n = ar^n$, and let

$$a_n = a + ar + ar^2 + \cdots + ar^{n-1};$$

then from (5)

$$a_n = \Delta^{-1}(ar^n).$$

Now (cf. Example 1) we have $\Delta r^n = r^n(r - 1)$, and by Theorem 1,

$$\Delta\left(\frac{ar^n}{r - 1}\right) = ar^n, \qquad r \neq 1.$$

Thus $a_n = ar^n/(r-1) + c$ where c is an arbitrary constant. But $a_1 = a$, and $c = a - ar/(r-1) = a/(1-r)$. Therefore we find that

$$a + ar + \cdots + ar^{n-1} = a\frac{(r^n - 1)}{r - 1}, \qquad r \neq 1,$$

the well-known result for the sum of n terms of a geometric sequence.

Example 9. As an example of the use of equation (6), consider the problem of finding

$$m + (m + 1) + (m + 2) + \cdots + 2m$$

for all $m > 1$. In Example 7 we observed that

$$\Delta^{-1}b_n = \Delta^{-1}n = \tfrac{1}{2}(n^2 - n) + c.$$

From equation (7) we have

$$m + (m + 1) + \cdots + 2m = \Delta^{-1}b_{2m+1} - \Delta^{-1}b_m$$
$$= \tfrac{1}{2}[(2m + 1)^2 - (2m + 1)] + c - \tfrac{1}{2}[m^2 - m] - c$$
$$= \tfrac{1}{2}(3m^2 + 3m) = \tfrac{3}{2}m(m + 1).$$

Example 10. Find

$$1 + 4 + 9 + \cdots + k^2.$$

To do this let

$$a_k = 0 + 1 + 4 + 9 + \cdots + (k - 1)^2.$$

Then we must find some $f(k)$ such that $\Delta f(k) = k^2$. A little searching yields the result that

$$\Delta k^3 = 3k^2 + 3k + 1$$
$$= 3k^2 + 3\Delta\left(\frac{k^2 - k}{2}\right) + \Delta k$$

by Theorem 1 and the fact that $\Delta k = 1$. Therefore

$$k^2 = \tfrac{1}{3}\Delta k^3 - \Delta\left(\frac{k^2 - k}{2}\right) - \tfrac{1}{3}\Delta k.$$

If we apply Theorem 1, we obtain

$$k^2 = \Delta\left(\frac{k^3}{3} - \frac{k^2}{2} + \frac{k}{2} - \frac{k}{3}\right)$$
$$= \Delta\left(\frac{k^3}{3} - \frac{k^2}{2} + \frac{k}{6}\right)$$

$$= \Delta \left(\frac{k}{6}(2k^2 - 3k + 1) \right)$$

$$= \Delta \left(\frac{k}{6}(k - 1)(2k - 1) \right).$$

Hence

$$\Delta^{-1}k^2 = \frac{k(k - 1)(2k - 1)}{6} + c,$$

and

$$a_{k+1} = 1 + 4 + 9 + \cdots + k^2 = \Delta^{-1}(k + 1)^2$$

$$= \frac{(k + 1)k(2k + 1)}{6}$$

$$= \frac{k(k + 1)(2k + 1)}{6}.$$

11.3   LIMITS OF SEQUENCES

In the preceding sections of this chapter we have made no use of the completeness property, OC, of the real number field R (see 8.2). In property OC lies the basic difference between R and other ordered fields; it plays a fundamental role in the remainder of this book. In particular, it makes possible the notions of *convergence* and *limit* of a sequence of real numbers.

Sequences of rationals have already been discussed in Chapter 8, but it was noted there that convergent sequences of rationals do not necessarily have rational limits. On the other hand, convergent sequences of real numbers do have real numbers as limits. This remark constitutes another kind of completeness enjoyed by the real numbers and is, in fact, equivalent to OC. (See Theorem 4 and Remark 3 below.)

We shall take Definition 3 of 8.4 (applied now to sequences of real numbers) as our basic definition of convergence. What it says is: $\{a_n\}$ is convergent with limit a, if we can make all the terms a_n, $n > n_0$ as close to a as we wish by making n_0 sufficiently large (see Fig. 11–1). Or, corresponding to any positive distance $d > 0$, we can find a term such that the distance between a_n and a is less than d for all terms later than a_{n_0}, or, more precisely such that

$$|a_n - a| < d \text{ whenever } n > n_0.$$

Definition.   *A sequence $\{a_n\}$ **converges** with **limit** a, written* $\lim_n a_n = a$, *if*

$$\forall \varepsilon > 0 \, \exists n_0 \in N \, \forall n > n_0 \, (|a_n - a| < \varepsilon).$$

If $\{a_n\}$ does not converge, it is said to *diverge*.

Recall from Chapter 10 that this definition means: for all ε, if $\varepsilon > 0$, then we can find a natural number n_0, depending on the choice of ε, with the following property; for every $n > n_0$, a_n is within a distance ε of a. This is a precise way of rendering the statements preceding the definition (see Fig. 11–1).

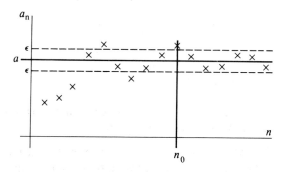

Figure 11–1   All terms of the sequence $\{a_n\}$ beyond a_{n_0}
are less than a distance ε from a

Example 1.   Consider the sequence $\{1/\sqrt{n}\}$, $n > 0$. We can show that $\lim_{n} (1/\sqrt{n}) = 0$. It is intuitively clear that $1/\sqrt{n}$ can be made as small as we please by making n sufficiently large. To apply the Definition precisely, pick any $\varepsilon > 0$. We want to find n_0 such that

$$n > n_0 \Rightarrow |1/\sqrt{n}| < \varepsilon.$$

But

$$|1/\sqrt{n}| < \varepsilon \text{ if and only if } 1/\varepsilon < \sqrt{n}, \text{ or } 1/\varepsilon^2 < n.$$

So

$$n > 1/\varepsilon^2 \Rightarrow |1/\sqrt{n}| < \varepsilon,$$

and we can let $n_0 = [1/\varepsilon^2]$. Thus by the Definition $\lim_{n} (1/\sqrt{n}) = 0$.

Example 2.   The sequence $\{\sqrt{n}\}$ is clearly divergent; $\sqrt{n}$ can be made as large as we wish by making n sufficiently large. (See Lemma below.)

Example 3.   Consider the sequence $\{r^n\}$ for a given $r \in R$. If $r = 1$, the sequence is constant (all terms are equal to 1) and hence converges with limit 1. We now discuss the two cases (a) $|r| < 1$ and (b) $|r| > 1$.

(a)   We show that if $|r| < 1$, $\lim_{n} r^n = 0$. Pick any $\varepsilon > 0$. We want $|r^n| < \varepsilon$ for all sufficiently large n. Now $|r^n| = |r|^n$. We choose a real number

q_0, such that

$$|r|^{q_0} = \varepsilon.$$

This is always possible, for q_0 is given explicitly by

$$q_0 = \frac{\log \varepsilon}{\log |r|},$$

where the logarithms are to any convenient base. Now, for any integer $n > [q_0] = n_0$,

$$|r|^n < r^{[q_0]} \leqslant r^{q_0} = \varepsilon.$$

Thus $|r|^n < \varepsilon$ if $n > n_0$ and consequently $\lim_n r^n = 0$.

(b) By similar arguments it can be shown that $\{r^n\}$ is unbounded (and hence divergent; see Lemma below) if $|r| > 1$.

It should be observed that the convergence or divergence of a sequence depends only on how the sequence behaves ultimately and is quite independent of any initial segment of the sequence; if it is convergent, the limit is also independent of any initial segment. These facts are an immediate consequence of the Definition. Thus, for example, the sequence $\{b_n\}$ defined by $b_n = \sqrt{n}$ if $n < 100^{100}$, $b_n = (\frac{1}{3})^n$ if $n > 100^{100}$ is convergent, with limit zero, by Example 3. (And in spite of Example 2.)

Example 4. The sequence $\{(-1)^n\}$ is divergent because no matter how large n is taken, the series alternates between $+1$ and -1, and hence does not stay close to any single number. By a similar argument, $\{\sin \frac{1}{2}n\pi\}$ is divergent, since, for integral values of n, $\sin \frac{1}{2}n\pi$ takes on periodically the values $\{0, 1, 0, -1\}$.

Lemma. *If $\{a_n\}$ is convergent, then it is bounded.*

Proof. Let $\lim_n a_n = a$. By the Definition (with $\varepsilon = 1$), there is an n_0 such that

$$n > n_0 \Rightarrow |a_n - a| < 1, \text{ or}$$
$$n > n_0 \Rightarrow -1 < a_n - a < 1, \text{ or}$$
$$n > n_0 \Rightarrow a - 1 < a_n < a + 1.$$

Let $u = \max(a_0, a_1, ..., a_{n_0}, a + 1)$, and

$$l = \min(a_0, a_1, ..., a_{n_0}, a - 1).$$

Then u is an upper bound of $\{a_n\}$ and l is a lower bound.

The sequence $\{(-1)^n\}$ of Example 4 is bounded and divergent. Hence, the converse of the Lemma is false.

The equivalence of Definitions 3 and 4 of 8.4 carries over to the case of real sequences.

Theorem 1. *Let $\{a_n\}$ be any sequence. Then $\lim\limits_n a_n = a$ if and only if there exist sequences $\{p_n\}$, $\{q_n\}$ such that*

(a) *$\{q_n\}$ is monotone nonincreasing and $\{p_n\}$ is monotone nondecreasing,*

(b) *$p_n \leqslant a_n \leqslant q_n$ for all $n \in N$, and*

(c) *l.u.b. $\{p_n\}$ = g.l.b. $\{q_n\}$ = a.*

Proof. If there exist sequences $\{p_n\}$, $\{q_n\}$ satisfying (a), (b), and (c), then $\lim\limits_n a_n = a$ is proved exactly as in Theorem 1 (a) of 8.4.

Conversely, suppose $\lim\limits_n a_n = a$.

Let the sequence $\{A_i^n\} = \{a_i \mid i \geqslant n\}$, $n \in N$.

$$= \{a_n, a_{n+1}, a_{n+2}, \ldots\}.$$

That is, $\{A_i^n\}$ is the sequence formed from $\{a_0, a_1, a_2, \ldots\}$ by deleting the first n members. We see, for example, that

$$\{A_i^0\} = \{a_i\}, \{A_i^1\} = \{a_1, a_2, \ldots\} \text{ and}$$
$$\{A_i^k\} = \{a_k, a_{k+1}, \ldots\}.$$

Now $\{A_i^n\} \subseteq \{A_i^0\}$; hence by the Lemma $\{A_i^n\}$ is bounded. By OC let

$$q_n = \text{l.u.b. } \{A_i^n\},$$
$$p_n = \text{g.l.b. } \{A_i^n\}.$$

(a) Since $\{A_i^{n+1}\} \subseteq \{A_i^n\}$, $q_{n+1} \leqslant q_n$ and $p_{n+1} \geqslant p_n$.

(b) Since $a_n \in \{A_i^n\}$, g.l.b. $\{A_i^n\} = p_n \leqslant a_n \leqslant q_n =$ l.u.b. $\{A_i^n\}$.

(c) Let $q =$ g.l.b. $\{q_n\}$ ($=\lim\limits_n q_n$). Suppose (for *reductio ad absurdum*) that $|q - a| > 0$. Let $\delta = |q - a|$. For all sufficiently large n (say $n > n_1$), $|a_n - a| < \delta/3$ by the Definition. Similarly, for all sufficiently large n (say $n > n_2$), $|q_n - q| < \delta/3$. (See Figure 11–2.) Let $n_0 = \max (n_1, n_2) + 1$. Then $|q_{n_0} - q| < \delta/3$ and $|a_n - a| < \delta/3$ for all $n > n_0$. But this implies that $|a_n - q_{n_0}| > \delta/3$ for all $n > n_0$ (otherwise a_n and q_{n_0} could not satisfy the preceding inequalities). But if q_{n_0} is more than a distance $\delta/3$ away from all the terms which belong to $\{A_i^{n_0}\}$, q_{n_0} could not be a l.u.b. of $\{A_i^{n_0}\}$. This is a contradiction, and so $|q - a| = 0$; that is, $q = a$. Similarly we can show that $p =$ l.u.b. $\{p_n\} = a$.

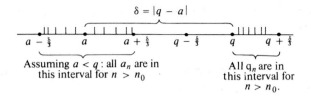

Figure 11-2

Corollary 1. *In the statement of Theorem* 1, *the sequences* $\{p_n\}$ *and* $\{q_n\}$ *can be chosen to be sequences of rationals.*

Proof. This follows from Theorem 1 above and Corollary 1 of 8.3.

The following Corollary also follows directly from Theorem 1.

Corollary 2. *If* $\{a_n\}$ *is a monotone nondecreasing (nonincreasing) sequence with an upper bound (lower bound), then it converges to its* l.u.b. (g.l.b.).

Example 5. For the sequence $\{(-1)^n/n\}$ we have

(a) $\quad -\dfrac{1}{n} < \dfrac{-1}{n+1}$ and $\dfrac{1}{n} > \dfrac{1}{n+1}$, for all $n \in N$;

(b) $\quad -\dfrac{1}{n} \leqslant \dfrac{(-1)^n}{n} \leqslant \dfrac{1}{n}$, for all $n \in N$.

(c) l.u.b. $\left\{-\dfrac{1}{n}\right\}$ = g.l.b. $\left\{\dfrac{1}{n}\right\}$ = 0.

Hence,

$$\lim_n \frac{(-1)^n}{n} = 0 \text{ by Theorem 1.}$$

Remark 1. The number q in the proof of Theorem 1 is called lim sup $\{a_n\}$ (superior limit); the number p is called lim inf $\{a_n\}$ (cf. the footnote in 2.6). By OC any bounded sequence has a unique lim inf and a unique lim sup. Clearly, when they exist, lim inf $\{a_n\} \leqslant$ lim sup $\{a_n\}$. Theorem 1 can be restated as: A sequence $\{a_n\}$ converges if and only if lim inf $\{a_n\}$ = lim sup $\{a_n\}$ = $\lim_n a_n$.

Example 6. (a) $\lim \inf \{1/\sqrt{n}\} = \lim \sup \{1/\sqrt{n}\} = \lim_{n} (1/\sqrt{n}) = 0$.

(b) $\lim \inf \{(-1)^n\} = -1 < 1 = \lim \sup \{(-1)^n\}$, and $\{(-1)^n\}$ has no limit.

(c) $\lim \inf \{\sqrt{n}\}$ and $\lim \sup \{\sqrt{n}\}$ do not exist.

(d) Let $\{a_n\}$ be the sequence (2) of 7.7. Then

$$\lim \inf \{a_n\} = 0 < 1 = \lim \sup \{a_n\},$$

and by Remark 1 this sequence does not converge.

Theorem 2. *Let $\{a_n\}$, $\{b_n\}$ be convergent sequences, and let $c \in R$. Then*

(1) $$\lim_{n} ca_n = c \lim_{n} a_n,$$

(2) $$\lim_{n} (a_n + b_n) = \lim_{n} a_n + \lim_{n} b_n,$$

(3) $$\lim_{n} (a_n b_n) = \lim_{n} a_n \cdot \lim_{n} b_n,$$

(4) $$\lim_{n} (a_n/b_n) = \lim_{n} a_n/\lim_{n} b_n \text{ (division by zero excepted)}.$$

Proof. Parts (1) and (2) were proved in 8.4 for rational sequences, and (3) and (4) were Exercises in 8.7. The same proofs apply here. We shall prove (3) by a different approach using the Lemma above. The reader can prove (4) similarly.

By the Lemma, there is a $c > 0$ such that

(5) $$|b_n| < c \text{ for all } n \in N.$$

Let $\lim_{n} b_n = b$, $\lim_{n} a_n = a \neq 0$ (see below). Then, for all n,

$$\begin{aligned} |a_n b_n - ab| &= |a_n b_n - ab_n + ab_n - ab| \\ &\leqslant |a_n b_n - ab_n| + |ab_n - ab| \\ &= |a_n - a||b_n| + |b_n - b||a|. \end{aligned}$$

Hence by (5),

(6) $$|a_n b_n - ab| < |a_n - a|c + |b_n - b||a| \text{ for all } n.$$

Pick $\varepsilon > 0$; there exist n_1, n_2 such that

(7a) $$n > n_1 \Rightarrow |a_n - a| < \varepsilon/2c,$$

(7b) $$n > n_2 \Rightarrow |b_n - b| < \varepsilon/2|a|.$$

Let $n_0 = \max (n_1, n_2)$. By (6), (7a), (7b),

(8) $n > n_0 \Rightarrow |a_n b_n - ab| < (\varepsilon/2c)\,c + (\varepsilon/2|a|)\,|a| = \varepsilon.$

But (8) asserts that $\lim_n (a_n b_n) = ab.$

If $\lim_n a_n = a = 0,$ then for any $\varepsilon > 0$ choose n_0 such that

$$|a_n| < \varepsilon/2 \text{ if } n > n_0.$$

Now $|a_n b_n| = |a_n|\,|b_n| < (\varepsilon/c)c = \varepsilon$ if $n > n_0,$ and again $\lim_n (a_n b_n) = ab.$

Example 7. Consider the sequence $\{3n^2/(2n^2+1)\}.$ Since $\lim_n (3n^2)$ and $\lim_n (2n^2+1)$ do not exist, we cannot apply (4). This, however does not imply that $\lim_n \left(3n^2/(2n^2+1)\right)$ does not exist. Let us write

$$\frac{3n^2}{2n^2+1} = \frac{3}{2} - \frac{3/2}{2n^2+1}.$$

Hence by (2),

$$\lim_n [3n^2/(2n^2+1)] = \lim_n (3/2) - \lim_n [(3/2)/(2n^2+1)]$$

provided the limits on the right exist. But

$$\lim_n (3/2) = 3/2, \text{ and}$$

$$0 \leqslant \frac{1}{2n^2+1} \leqslant \frac{1}{n} \text{ for all } n \in N.$$

Since $\lim_n (0) = 0$ and $\lim_n (1/n) = 0,$ by Theorem 1, $\lim_n \left(1/(2n^2+1)\right) = 0.$
Hence

$$\lim_n [3n^2/(2n^2+1)] = 3/2 - 0 = 3/2.$$

Alternatively,

$$\frac{3n^2}{2n^2+1} = \frac{3}{2+(1/n^2)}, \quad n \neq 0,$$

so, by (4), (2),

$$\lim_n [3n^2/(2n^2+1)] = \lim_n (3)/\left[\lim_n (2) + \lim_n (1/n^2)\right]$$

$$= 3/(2+0) = 3/2.$$

Example 8. On the other hand, for the sequence $\{3n^2/(2n+1)\}$

$$\frac{3n^2}{2n+1} = n\left(\frac{3}{2+1/n}\right).$$

The second factor has the limit $\frac{3}{2}$, but n increases indefinitely, therefore, by (3) the sequence has no limit. Alternatively,

$$\frac{3n^2}{2n + 1} = \frac{3}{2/n + 1/n^2}.$$

Since $\lim_n [(2/n) + (1/n^2)] = \lim_n (2/n) + \lim_n (1/n^2) = 0 + 0 = 0$, it is clear that the given sequence is unbounded.

Theorem 3. $\lim_n a_n = a \Rightarrow \lim_n \Delta a_n = 0.$

  Proof.   We have $|\Delta a_n - 0| = |\Delta a_n| = |a_{n+1} - a_n|$

$$= |a_{n+1} - a + a - a_n| \leqslant |a_{n+1} - a| + |a - a_n|.$$

We leave the rest of the proof to the reader. (See, for example, the proof of (3).)

The converse of Theorem 3 is false: The sequence $\{\sqrt{n}\}$, for example, is unbounded (and hence, divergent), whereas

$$\Delta\sqrt{n} = \sqrt{(n + 1)} - \sqrt{n} = \frac{1}{\sqrt{(n + 1)} + \sqrt{n}},$$

so that $\lim_n (\Delta\sqrt{n}) = 0.$

Theorem 3'.

$$\lim_n a_n = a \Rightarrow \forall_{p \in N} \left(\lim_n |a_n - a_{n+p}| = 0 \right).$$

  Proof.   This is similar to the proof of Theorem 3 (which is a special case).

The converse of Theorem 3' is also false: The same sequence, $\{\sqrt{n}\}$, is a counter example.

There is a stronger condition than $\forall_{p \in N} \left(\lim_n |a_n - a_{n+p}| = 0 \right)$ which *does* imply convergence. This condition, stated in Theorem 4, is due to Cauchy, and is another important consequence of OC.

Theorem 4.   (*Cauchy's Convergence Criterion*). *A sequence $\{a_n\}$ converges if corresponding to an arbitrary $\varepsilon > 0$, there is an $n_0 \in N$ such that $|a_n - a_m| < \varepsilon$ whenever both $m > n_0$ and $n > n_0$. Or more formally,*

$$[\forall \varepsilon > 0 \,\exists n_0 \in N \,\forall_{m > n_0} \,\forall_{n > n_0} (|a_n - a_m| < \varepsilon)] \Rightarrow \exists a \left(\lim_n a_n = a \right).$$

Discussion. The antecedent or hypothesis of Theorem 4 is Cauchy's convergence criterion. It says that, by taking m and n sufficiently large, we can make $|a_n - a_m|$ as small as we please. We could generalize the definition of convergence and write this as

$$\lim_{m,n} |a_n - a_m| = 0.$$

Before proving the theorem, we shall give an expository outline of the proof.

(i) We first show that the criterion implies that the sequence $\{a_n\}$ is bounded.

(ii) As in the proof of Theorem 1, we let $\{A_i^n\} = \{a_i | i \geqslant n\}$, $p_n = $ g.l.b. $\{A_i^n\}$, $q_n = $ l.u.b. $\{A_i^n\}$, and let $p = \lim \inf \{a_n\} = $ l.u.b. $\{p_n\}$, $q = \lim \sup \{a_n\} = $ g.l.b. $\{q_n\}$. By (i) and OC, p and q exist.

(iii) We next show that infinitely many a_n are arbitrarily close to p (and to q). That is, no matter what $\varepsilon > 0$ we choose, we can find terms a_n as far out in the sequence as we like such that $|a_n - p| < \varepsilon$ (similarly for q).

(iv) By Remark 1, $[\exists a (\lim a_n = a)] \Leftrightarrow p = q$. So suppose $\lim a_n$ does not exist and that $p < q$. We show that the convergence criterion is then false because of (iii). This will prove the Theorem by *reductio ad absurdum*.

Proof of Theorem 4.

(i) Let $\varepsilon = 1$. There is a natural number n_0 such that

$$m, n > n_0 \Rightarrow |a_n - a_m| < 1, \text{ or}$$
$$m, n > n_0 \Rightarrow -1 < a_n - a_m < 1,$$

whence

$$n > n_0 \Rightarrow -1 < a_n - a_{n_0+1} < 1, \text{ or}$$
$$n > n_0 \Rightarrow a_{n_0+1} - 1 < a_n < a_{n_0+1} + 1.$$

Hence, let

$$u = \max (a_0, a_1, ..., a_{n_0}, a_{n_0+1} + 1),$$
$$l = \min (a_0, a_1, ..., a_{n_0+1} - 1).$$

Then u is an upper bound of $\{a_n\}$, and l is a lower bound.

(iii) Pick any $\varepsilon > 0$. For sufficiently large n (say $n > n_0$), $|p_n - p| < \varepsilon/2$. Consider any $n > n_0$. Since $p_n = $ g.l.b. $\{A_i^n\}$ and $a_k \in \{A_i^n\}$ if $k > n$, there is an $m > n$ such that $|a_m - p_n| < \varepsilon/2$. Whence

$$|a_m - p| = |a_m - p_n + p_n - p| \leqslant |a_m - p_n| + |p_n - p| < \varepsilon/2 + \varepsilon/2 = \varepsilon.$$

That is, for any n we have found an $m > n$ such that $|a_m - p| < \varepsilon$. Similarly, for any n we can find another integer $m > n$ such that $|a_m - q| < \varepsilon$.

(iv) Suppose $\{a_n\}$ has no limit. By Remark 1, $p < q$. Let $|p - q| = \delta$. By (iii), for any $n_0 \in N$, we can find $m > n_0$, $n > n_0$ such that

$|a_m - p| < \delta/3$, $|a_n - q| < \delta/3$. It follows that $|a_m - a_n| \geqslant \delta/3$. We have found an ε $(=\delta/3)$ such that for any $n_0 \in N$,

$$(m > n_0 \wedge n > n_0) \Rightarrow \neg \, (|a_m - a_n| < \varepsilon).$$

In other words (see 10.9)

$$\exists \varepsilon > 0 \; \forall n_0 \in N \; \exists m > n_0 \; \exists n > n_0 \, [\neg \, (|a_m - a_n| < \varepsilon)].$$

But by (11), (12) of 10.8 (using also 10.9), this is the negation of the convergence criterion. This contradiction completes the proof.

The converse of Theorem 4 is also true. Its proof is similar to the proofs of Theorems 3, 3′.

Remark 2. The primary importance of Cauchy's convergence criterion is that it enables one to verify that $\{a_n\}$ converges without knowing $\lim_n a_n$.

This is essential in the Cantor–Cauchy development of the real numbers from the rationals discussed in 8.5. For, in this case, real numbers are defined in terms of convergent sequences of rationals, but the limit of a convergent sequence of rationals may possibly not exist (for example, if it is irrational) until after the real numbers have been defined.

Remark 3. In addition to providing a "limit free" criterion for convergence, Theorem 4 also asserts that a convergent sequence of real numbers has a real number as its limit. (In contrast to the situation whereby convergent sequences of rationals may possibly not have a rational number as a limit.)

In applying the convergence criterion

$$\lim_{m,n} |a_m - a_n| = 0,$$

there is no loss of generality in assuming that $m > n$. Hence the criterion is equivalent to

$$\lim_{n,p} |a_n - a_{n+p}| = 0.$$

Note that this is not equivalent to

$$\forall p \left(\lim_n |a_n - a_{n+p}| = 0 \right),$$

which, as we saw above, does not in general imply convergence. However there is a version of Theorem 4 somewhat like this which *does* imply convergence. It is the following:

Theorem 4′.

$$\forall \varepsilon > 0 \, \exists n_0(\varepsilon) \in N \, \forall n > n_0 \, \forall p \in N \, (|a_{n+p} - a_n| < \varepsilon) \Rightarrow \exists a \left(\lim_n a_n = a \right).$$

The proof is an immediate deduction from Theorem 4. To emphasize that n_0 depends only on ε (and not, for instance, on p) we have written $n_0(\varepsilon)$.

Let us apply Theorem 4′ to the sequence $\{\sqrt{n}\}$. We find

$$|a_{n+p} - a_n| = |\sqrt{(n+p)} - \sqrt{n}| = \sqrt{(n+p)} - \sqrt{n}$$

$$= \frac{p}{\sqrt{(n+p)} + \sqrt{n}}.$$

Hence,

$$|a_{n+p} - a_n| = \frac{p}{\sqrt{(n+p)} + \sqrt{n}} < \frac{p}{\sqrt{n}} < \varepsilon \quad \text{if} \quad n > n_0,$$

provided

$$n_0 = \left[\frac{p^2}{\varepsilon^2} \right] + 1.$$

This follows since, if $n > n_0$,

$$\frac{p}{\sqrt{n}} < \frac{p}{\sqrt{n_0}} = \frac{p}{\sqrt{\left\{ \left[\frac{p^2}{\varepsilon^2} \right] + 1 \right\}}} < \frac{p}{\sqrt{\frac{p^2}{\varepsilon^2}}} = \varepsilon.$$

At first sight the result that $|a_{n+p} - a_n| < \varepsilon$ if $n > n_0$ would appear to imply convergence; however this is not the case for we see that n_0 depends on p as well as ε. Given ε we must be able to choose an n_0 depending *only* on ε such that $|a_{n+p} - a_n| < \varepsilon$ for *all* $p \in N$. The sequence actually diverges as we have previously remarked, since it is unbounded.

Example 9. Consider the sequence $\{\log n\}$. The logarithm is to any convenient base b where $2 < b$. In this case

$$|a_{n+p} - a_n| = \log \left(1 + \frac{p}{n} \right) < \frac{p}{n} < \varepsilon$$

if $n > n_0$ where $n_0 = \left[\frac{p}{\varepsilon} \right] + 1$

(we have used the inequality $\log (1 + x) < x$ if $x > 0$; see Example 1 of 13.6).

As in the above discussion, Theorem 4′ does not imply convergence in this example. It is clear that the sequence diverges since it is unbounded.

Example 10. Let

$$a_n = -1 + \tfrac{1}{2} - \tfrac{1}{3} + \cdots + (-1)^n/n.$$

We shall show, by Theorem 4', that $\{a_n\}$ converges.

$$|a_n - a_{n+p}| = \left| \frac{(-1)^{n+1}}{n+1} + \frac{(-1)^{n+2}}{n+2} + \cdots + \frac{(-1)^{n+p}}{n+p} \right|$$

$$= \frac{1}{n+1} - \frac{1}{n+2} + \frac{1}{n+3} - \cdots + \frac{(-1)^{p+1}}{n+p}$$

$$< \frac{1}{n+1} - \frac{1}{n+2} + \frac{1}{n+2} - \frac{1}{n+4} + \frac{1}{n+4} - \cdots + q$$

where $q = 1/(n + p - 1)$ if p is odd, and $q = (-1)/(n + p)$ if p is even. In either case,

$$|a_n - a_{n+p}| < \frac{1}{n+1}.$$

Thus, given ε we choose $n_0 = [1/\varepsilon]$; then $|a_{n+p} - a_n| < \varepsilon$ if $n > n_0$ for all p, and by Theorem 4' the sequence converges.

11.4   SUMMATION NOTATION

In the sequel we will find that the following abbreviated notation for sums is extremely useful.

Suppose we have a sequence $\{a_n\} = \{a_0, a_1, a_2, \dots \}$;

we define $\sum\limits_{i=p}^{q} a_i$ as

$$\sum_{i=p}^{q} a_i = a_p + a_{p+1} + \dots + a_q, \quad p \leqslant q.$$

We adopt the following nomenclature: for each i, a_i is called a **summand**; i is the **variable (or index) of summation**; p, q are the **limits of summation**. We read $\sum_{i=p}^{q} a_i$ as "the summation of a_i as i ranges from p to q". If $p = q$ we adopt the convention

$$\sum_{i=p}^{p} a_i = a_p.$$

Using this notation we write:

(a) the sum of the first n natural numbers is

$$\sum_{i=1}^{n} i,$$

(b) the sum of n terms of a geometric sequence (Example 8 of 11.2) is

$$\sum_{i=0}^{n-1} ar^i,$$

(c) the expansion of $(1 + x)^n$, $n \in N$, $x \in R$ is

$$(1 + x)^n = \sum_{i=0}^{n} \binom{n}{i} x^i,$$

where $\displaystyle \binom{n}{i} = \frac{n!}{i!\,(n-i)!} = \frac{n(n-1)\,\dots\,(n-i+1)}{i!}$ (The Binomial Theorem).

Note: $\binom{n}{i}$ is sometimes written as $_nC_i$ and is called a **binomial coefficient**.

(d) $\displaystyle \frac{1}{1} + \frac{1}{2^2} + \frac{1}{3^3} + \cdots + \frac{1}{i^i} = \sum_{n=1}^{i} \frac{1}{n^n},$

(e) $\displaystyle 1 \cdot 2 + 2 \cdot 3 + 3 \cdot 4 + 4 \cdot 5 + \cdots + (p-1)p = \sum_{i=2}^{p} (i-1)\,i.$

From the Definition it follows that

$$\sum_{i=p}^{q} a_i = \sum_{j=p}^{q} a_j;$$

that is, the name of the variable of summation does not affect the sum. The variable of summation is a bound variable (cf. 1.4, 10.6). Not only *may* we change the name of the variable of summation but sometimes we *must* change it as in (d) above where i is already used for another purpose.

From the Definition and the properties of real numbers, the following formulas can be deduced:

(1)   If $p \leqslant r < q$, then

$$\sum_{i=p}^{q} a_i = (a_p + a_{p+1} + \cdots + a_r) + (a_{r+1} + a_{r+2} + \cdots + a_q)$$

$$= \sum_{i=p}^{r} a_i + \sum_{i=r+1}^{q} a_i.$$

(2) $\displaystyle \sum_{i=p}^{q} (a_i + b_i) = \sum_{i=p}^{q} a_i + \sum_{i=p}^{q} b_i,$ $q \geqslant p.$

(3) $\displaystyle\sum_{i=p}^{q} 1 = \sum_{i=1}^{q} 1 - \sum_{i=1}^{p-1} 1 = q - (p-1) = q - p + 1, \quad q \geqslant p.$

(4) $\displaystyle\sum_{i=p}^{q} (a_{i+1} - a_i) = \sum_{i=p}^{q} \Delta a_i = a_{q+1} - a_p, \qquad\qquad q \geqslant p.$

(5) $\displaystyle\sum_{i=p}^{q} ca_i = c\left(\sum_{i=p}^{q} a_i\right), \qquad\qquad\qquad\qquad q \geqslant p.$

This result can be proved by successive application of D, the distributive property of the reals.

(6) $\displaystyle\left(\sum_{i=p}^{q} a_i\right)\left(\sum_{j=m}^{n} b_j\right) = \sum_{i=p}^{q}\left(\sum_{j=m}^{n} a_i b_j\right) = \sum_{j=m}^{n}\left(\sum_{i=p}^{q} a_i b_j\right), \quad q \geqslant p, n \geqslant m.$

To prove (6) we use (5) $q - p + 1$ times; the left side becomes

$$a_p \sum_{j=m}^{n} b_j + a_{p+1} \sum_{j=m}^{n} b_j + \cdots + a_q \sum_{j=m}^{n} b_j = \sum_{i=p}^{q}\left[a_i\left(\sum_{j=m}^{n} b_j\right)\right].$$

Applying (5) again $q - p + 1$ times gives

$$\sum_{j=m}^{n} a_p b_j + \sum_{j=m}^{n} a_{p+1} b_j + \cdots + \sum_{j=m}^{n} a_q b_j,$$

and we write this as

$$\sum_{i=p}^{q}\left(\sum_{j=m}^{n} a_i b_j\right)$$

that is, the sum of all possible products of the a's and the b's. It is readily shown that this can also be written as

$$\sum_{j=m}^{n}\left(\sum_{i=p}^{q} a_i b_j\right).$$

It should be noted again that we were forced to use another index of summation in stating this result, for

$$\sum_{i=p}^{q} \sum_{i=m}^{n} a_i b_i$$

is meaningless.

(7) $\displaystyle\left|\sum_{i=p}^{q} a_i\right| \leqslant \sum_{i=p}^{q} |a_i|.$

This result is known as the (generalized) **triangle inequality**. The special case of $q = p + 1$ was proved in Theorem 6 of 8.3; and (7) can be proved by successive applications of this Theorem.

An example which illustrates the use of (6) is the following:

Example. By the Binomial Theorem (see Exercise 11 of 11.7) we have:

$$(1 + x)^m = \sum_{i=0}^{m} \binom{m}{i} x^i, \quad (1 + x)^n = \sum_{i=0}^{n} \binom{n}{i} x^i, \quad \text{and}$$

$$(8) \qquad (1 + x)^{m+n} = \sum_{i=0}^{m+n} \binom{m+n}{i} x^i, \qquad m, n \in N.$$

We shall apply (6) to the first two sums. In order to do this, we change the index of summation in the second sum to j.

We then have

$$(9) \qquad (1 + x)^m (1 + x)^n = (1 + x)^{m+n} = \sum_{i=0}^{m} \sum_{j=0}^{n} \binom{m}{i} \binom{n}{j} x^{i+j}.$$

We now simplify (9) by putting $i + j = k$, replacing the old variable of summation i by a new variable of summation k.

We find

$$(10) \qquad (1 + x)^{m+n} = \sum_{k=0}^{m+n} \sum_{j=0}^{k} \binom{m}{k-j} \binom{n}{j} x^k.$$

In verifying (10) the only non-obvious aspect is that we use new limits of summation. We should note, to begin, that we are implicitly making use of the convention $\binom{n}{j} = 0$ if $j < 0$ or $j > n$ (and similarly $\binom{m}{k-j} = 0$ if $k - j < 0$ or $k - j > m$).

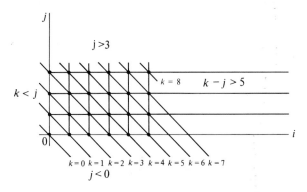

Figure 11–3 The set of points (i, j), $0 \leqslant i \leqslant 5$, $0 \leqslant j \leqslant 3$

From Figure 11–3 (drawn for the special case $m = 5$, $n = 3$), we see that the double summation (9) is over all points (i, j) where i ranges from 0 to m and j ranges from 0 to n. In the summation (9) a particular k picks out the points on the line $i + j = k$. However, points outside the rectangle $0 \leqslant i \leqslant m$, $0 \leqslant j \leqslant n$ are not present in the sum by the conventions above.

Thus (9) and (10) are equivalent. Now changing k back to i (this is possible since i is no longer in use) and comparing (10) with (8) we find finally

$$\binom{m + n}{i} = \sum_{j=0}^{i} \binom{m}{i - j} \binom{n}{j}.$$

This result is known as the addition theorem for the binomial coefficients.

11.5 SERIES

An ancient paradox attributed to the Greek philosopher Zeno asserts that a runner cannot reach the end of a race course (in a finite length of time) if one accepts the usual intuition concerning the infinite divisibility of time and space. For the runner must go half the distance before he can go the whole way. Of the remaining half, he must go half that distance (viz. a quarter of the whole distance) before he can go the whole way. Continuing in this fashion, we see that the runner always has half the remaining distance to go *ad infinitum*, and consequently, never finishes the race! Simple experience, of course, tells us that most runners do finish their races.

Zeno has actually expressed the *total* distance covered by the runner after certain intervals of time as a sequence $\{s_n\}$, where

$$s_j = \frac{1}{2} + \frac{1}{4} + \frac{1}{8} + \cdots + \frac{1}{2^j} = \sum_{k=1}^{j} \frac{1}{2^k}$$

We see that the terms of the sequence $\{s_n\}$ are formed by the successive addition of the terms of the sequence $\{1/2^n\}$; more precisely the jth term of $\{s_n\}$ is the sum of the first j terms of the sequence $\{1/2^n\}$

$$s_1 = \tfrac{1}{2}$$
$$s_2 = \tfrac{1}{2} + \tfrac{1}{4}$$
$$s_3 = \tfrac{1}{2} + \tfrac{1}{4} + \tfrac{1}{8},$$
$$\vdots \quad \vdots \quad \vdots \quad \vdots$$
$$s_j = \sum_{k=1}^{j} \frac{1}{2^k}$$

Now if it should happen that $\lim_{n} s_n = 1$, both the total distance to be covered and the total time required to cover it are well defined, and at least some of Zeno's sophistry disappears. This is, in fact the case.

Definition 1.   *An **infinite series** (or simply a **series**) is an expression of the form*

$$(1) \qquad\qquad a_1 + a_2 + a_3 + \dots .$$

where $\{a_n\}$ is a given sequence of numbers.

Unless the contrary is stated explicitly, we shall assume that $a_n \in R$. *The number a_n is the **nth term** of the series.* The series can be denoted by

$$a_1 + a_2 + a_3 + \cdots \quad \text{or simply} \sum_n a_n.$$

Since it is not possible to find an ordinary sum of infinitely many terms, as it stands the expression (1) has no meaning. In order to give it meaning we proceed as follows. We form the sum of the first n terms of the series. We call this sum the **nth partial sum** of the series, and denote it by s_n, so that

$$(2) \qquad\qquad s_n = \sum_{k=1}^{n} a_k = a_1 + a_2 + \cdots + a_n.$$

We have thus formed the sequence $\{s_n\}$ of partial sums of the series.

Definition 2.   *The series $\sum_n a_n = a_1 + a_2 + a_3 + \dots$ is said to **converge** if the sequence of its partial sums $\{s_n\} = \{\sum_{k=1}^{n} a_k\}$ converges; if $\lim\limits_{n} s_n = s$, then s is also called the **limit** or **sum** of the series and we write $s = \sum_{n=1}^{\infty} a_n$; if $\{s_n\}$ diverges, the series is said to **diverge,** and it has no sum.*

As a matter of notation, it is often convenient to denote the first term of a series by a_0 instead of by a_1, such a change in notation clearly has no effect in our study of the convergence or divergence of a series.

Definition 3.   *A **geometric series** is a series of the form*

$$(3) \quad \sum ar^{n-1} = a + ar + ar^2 + \dots \; (\text{or} \sum_n ar^n = a + ar + ar^2 + ar^3 + \dots),$$

*where $a \in R$, $r \in R$; r is called the **ratio** of the geometric series.*

By division, if $r \neq 1$, we obtain

$$(4) \qquad a\frac{1 - r^n}{1 - r} = a(1 + r + r^2 + \cdots + r^{n-1}) = a + ar + ar^2 + \cdots + ar^{n-1},$$

a result that is also easily established by induction for all $n \in N$. We shall assume that $a \neq 0$. We denote by s_n the nth partial sum of the geometric series (3). By (4) we have

$$s_n = a\frac{1 - r^n}{1 - r} = \frac{a}{1 - r}(1 - r^n), \qquad r \neq 1, \quad a \neq 0.$$

If $|r| > 1$, $|r^n|$ increases without limit as n increases, (Example 3 of 11.3) and $\{s_n\}$ diverges. If $|r| < 1$, $\lim\limits_n r^n = 0$, and $\lim\limits_n s_n = a/(1 - r)$. If $|r| = 1$, either $r = 1$ or $r = -1$. If $r = 1$, $s_n = a + a + \cdots + a = na$ and $|s_n|$ increases without limit. If $r = -1$, $s_n = a - a + a - \cdots + (-1)^{n-1} a$, which equals a when n is odd and 0 when n is even; consequently s_n oscillates between the values a and 0, and $\{s_n\}$ diverges. We have established the following theorem.

Theorem 1.   *If $a \neq 0$, the geometric series*

$$a + ar + ar^2 + \cdots$$

diverges if $|r| \geq 1$ and converges to the value $s = \dfrac{a}{1 - r}$ if $|r| < 1$.

As an example, the series

$$1 + \tfrac{1}{2} + \tfrac{1}{4} + \tfrac{1}{8} + \cdots + \frac{1}{2^{n-1}} + \cdots$$

is a geometric series in which $a = 1$ and $r = \tfrac{1}{2}$. Since $|r| = \tfrac{1}{2} < 1$, the series converges to the value $1/(1 - \tfrac{1}{2}) = 2$. The series

$$1 - \tfrac{1}{2} + \tfrac{1}{4} - \tfrac{1}{8} + \cdots$$

is a geometric series with $a = 1$, $r = -\tfrac{1}{2}$, which converges to the value

$$\frac{1}{1 - (-\tfrac{1}{2})} = \frac{1}{\tfrac{3}{2}} = \frac{2}{3}.$$

The sum $\Sigma_{n=0}^{\infty} a_n$ (or $\Sigma_{n=1}^{\infty} a_n$) which can also be read "the sum of a_n from $n = 0$ to infinity ($n = 1$ to infinity)" is not, of course, a sum in the usual sense (because there is an infinite number of terms). It is a generalization of the concept of sum and exists (by definition) only if the series is convergent.

The expression $\Sigma_n a_n$ is the name of a special kind of sequence (which may or may not converge) called a series; it exists for any sequence $\{a_n\}$.

On the other hand, $\Sigma_{n=0}^{\infty} a_n$, if it exists, is the name of a number, viz., the sum of the series $\Sigma_n a_n$; it exists if and only if the series is convergent. Many authors use one notation for both objects. Much confusion can be avoided by having the two notations.

Example 1. The *harmonic* series,

$$\sum_n \frac{1}{n} = 1 + \frac{1}{2} + \frac{1}{3} + \frac{1}{4} + \cdots,$$

diverges. This fact, which will be very useful to us later, can be proved as follows: If $k \in N$, and $k \geqslant 2$,

$$S_{2^k} = 1 + \frac{1}{2} + \frac{1}{3} + \frac{1}{4} + \frac{1}{5} + \frac{1}{6} + \frac{1}{7} + \frac{1}{8} + \frac{1}{9} + \cdots + \frac{1}{16} + \cdots + \frac{1}{2^k}$$

$$= 1 + \frac{1}{2} + \left(\frac{1}{3} + \frac{1}{4}\right) + \left(\frac{1}{5} + \frac{1}{6} + \frac{1}{7} + \frac{1}{8}\right)$$

$$+ \left(\frac{1}{9} + \cdots + \frac{1}{16}\right) + \cdots + \left(\frac{1}{2^{k-1}+1} + \cdots + \frac{1}{2^k}\right)$$

$$> 1 + \frac{1}{2} + \left(\frac{1}{4} + \frac{1}{4}\right) + \left(\frac{1}{8} + \frac{1}{8} + \frac{1}{8} + \frac{1}{8}\right)$$

$$+ 8\left(\frac{1}{16}\right) + \cdots + 2^{k-1}\left(\frac{1}{2^k}\right)$$

$$= 1 + \frac{1}{2} + \frac{1}{2} + \frac{1}{2} + \frac{1}{2} + \cdots + \frac{1}{2} = 1 + \frac{1}{2}k.$$

Since $\{k/2\}$ is unbounded, the set $\{S_{2^k}\}$ and the sequence of partial sums $\{S_n\}$ are unbounded, and the series diverges. In contrast to the harmonic series, we saw in Example 10 of 11.3 that the series $\Sigma_n(-1)^n/n = 1 - \frac{1}{2} + \frac{1}{3} - \frac{1}{4} + \cdots$ converges.

Example 2. Consider the series

$$\sum_n \frac{1}{n(n+1)}.$$

The partial sums of this series are:

$$S_n = \sum_{k=1}^{n} \frac{1}{k(k+1)}.$$

Now this sum is the same as (3) in 11.2 with

$$b_0 = \frac{1}{2}, \qquad b_{n-1} = \frac{1}{n(n+1)}.$$

Thus, by (5) in 11.2

$$s_n = \Delta^{-1} b_n = \Delta^{-1} \frac{1}{(n + 1)(n + 2)}.$$

But

$$\Delta \frac{1}{n + 1} = \frac{1}{n + 2} - \frac{1}{n + 1} = \frac{-1}{(n + 1)(n + 2)}.$$

Hence by Theorem 2 of 11.2

$$s_n = \Delta^{-1} \frac{1}{(n + 1)(n + 2)} = -\frac{1}{n + 1} + c.$$

Since $s_1 = b_0 = \frac{1}{2} = -\frac{1}{2} + c$, we find $c = 1$; hence

$$s_n = 1 - \frac{1}{n + 1}.$$

Alternatively by expression (4) in 11.4

$$s_n = -\sum_{k=1}^{n} \left(\frac{1}{k + 1} - \frac{1}{k} \right) = -\frac{1}{n + 1} + 1.$$

Thus

$$\lim_n s_n = 1$$

and so the series is convergent with limit 1. Thus we can also write

$$\sum_{n=1}^{\infty} \frac{1}{n(n + 1)} = 1.$$

A few words should be said about "finding" $\sum_{n=0}^{\infty} a_n$ if $\sum_n a_n$ converges. In this case $\sum_{n=0}^{\infty} a_n$ defines some real number. It may or may not be possible to give this real number a name in terms of already known real numbers such as, for example, $\pi^2/6$. Proving that $\sum_n a_n$ converges is usually easier than determining a value for $\sum_{n=1}^{\infty} a_n$. For example

$$\sum_{n=1}^{\infty} \frac{1}{n^2} = \frac{\pi^2}{6},$$

but the proof of this fact is beyond the level of our discussion. On the other hand, a similar "closed expression" or value is not known for $\sum_{n=1}^{\infty} 1/n^3$, although the corresponding series clearly converges.

Just as the convergence of the sequence $\{a_n\}$ depends only on its ultimate behavior, so the convergence of the series $\sum_n a_n$ depends only on the ultimate behavior of the sequence $\{a_n\}$. However, if the series $\sum_n a_n$ is convergent, its *sum* depends on the *entire* sequence $\{a_n\}$. More precisely, the series $\sum_n a_n$ has the sequence of partial sums

(5)
$$\left\{\sum_{k=0}^{n} a_k\right\}, \qquad n \geqslant 0.$$

Now by the comments in 11.3 following Example 3, sequence (5) converges if and only if the sequence

(6)
$$\left\{\sum_{k=0}^{n} a_k\right\}, \qquad n \geqslant q + 1$$

converges, where q is any fixed natural number. Furthermore if they both converge, they have the same limit.

The sequence (6) can be expressed as follows:

(7)
$$\left\{\sum_{k=0}^{n} a_k\right\} = \left\{\sum_{k=0}^{q} a_k + \sum_{k=q+1}^{n} a_k\right\} = \left\{\sum_{k=0}^{q} a_k\right\} + \left\{\sum_{k=q+1}^{n} a_k\right\}.$$

The sum $\sum_{k=0}^{q} a_k$ does not depend on n; it is equal to the constant sum of the first q terms even when n increases indefinitely. Therefore, if the various limits exist

(8)
$$\lim_{n} \sum_{k=0}^{n} a_k = \sum_{k=0}^{q} a_k + \lim_{n} \sum_{k=q+1}^{n} a_k.$$

or

(9)
$$\sum_{k=0}^{\infty} a_k = \sum_{k=0}^{q} a_k + \sum_{k=q+1}^{\infty} a_k.$$

By (8) and (9) we have the following useful lemma.

11.6 CONVERGENCE TESTS FOR SERIES OF POSITIVE TERMS

Lemma 1. *Let $\{a_n\}$ be any sequence; so that $\{a_{n+q+1}\}$, $q \in N$ is the sequence formed from $\{a_n\}$ by deleting the first $q + 1$ terms.*

(a) *The series $\Sigma_n a_n$ converges if and only if the series $\Sigma_n a_{n+q+1}$ converges, and*

(b) *if they converge, then*

$$\sum_{n=0}^{\infty} a_n = \sum_{n=0}^{q} a_n + \sum_{n=0}^{\infty} a_{n+q+1}.$$

We proceed now to prove some theorems which provide tests for convergence or divergence of series. The role of Lemma 1 is to supplement the tests in the following way. In testing whether $\Sigma_n a_n$ converges or diverges, we may find that one of the following theorems can be applied to $\Sigma_n \, a_{n+q+1}$ for some $q \in N$ although it does not apply to $\Sigma_n \, a_n$. From part (a) of Lemma 1, it follows that the series $\Sigma_n \, a_n$ converges if and only if the series $\Sigma_n \, a_{n+q+1}$ converges. Further, if we can evaluate the sum $\Sigma_{n=0}^{\infty} \, a_{n+q+1}$, then from part (b) $\Sigma_{n=0}^{\infty} \, a_n$ can be found.

The following Theorems 1 to 4 apply (directly) to series of nonnegative terms (that is, any series $\Sigma_n \, a_n$ such that $a_n \geqslant 0$ for $n \in N$). By Lemma 1, they apply also to series whose terms are ultimately nonnegative.

Observe that:

The partial sums of *a series* $\Sigma_n \, a_n$ *of nonnegative terms form a monotone nondecreasing sequence* $\{\Sigma_{k=0}^{n} \, a_k\}$.

Theorem 1. (Comparison Test).

If $\forall n \in N \; (0 \leqslant b_n \leqslant a_n)$, then

(a) $(\Sigma_n \, a_n \; converges) \Rightarrow (\Sigma_n \, b_n \; converges)$,

(b) $(\Sigma_n \, b_n \; diverges) \Rightarrow (\Sigma_n \, a_n \; diverges)$.

Proof. (a) Assume $\forall n \, (0 \leqslant b_n \leqslant a_n)$, then the series $\Sigma_n \, a_n$ and $\Sigma_n \, b_n$ are monotonic nondecreasing sequences such that

(1)
$$\sum_{k=0}^{n} b_k \leqslant \sum_{k=0}^{n} a_k \quad \text{for all } n \in N.$$

By Corollary 2 of 11.3, $\Sigma_n \, a_n$ converges to its l.u.b., $\Sigma_{n=0}^{\infty} \, a_n$, which must be an upper bound for $\Sigma_n \, b_n$ by (1). By Corollary 2 of 11.3 again $\Sigma_n \, b_n$ converges (to its l.u.b.).

(b) is the contrapositive of (a).

Corollary 1. *Theorem 1 holds with obvious modifications if* $\{a_n\}$, $\{b_n\}$ *are defined on* $N - \{0\}$ *rather than* N.

Proof. Use Lemma 1.

Corollary 2.

$$[\forall n \in N \; (b_n < a_n)] \Rightarrow \sum_{n=0}^{\infty} b_n < \sum_{n=0}^{\infty} a_n.$$

Proof. The proof will be left to the reader.

Example 1. As an example of the applications of Theorem 1, consider the series

$$\sum_n b_n = \sum_n \frac{1}{n^2}.$$

Now for each n

$$b_{n+1} = \frac{1}{(n+1)(n+1)} < \frac{1}{n(n+1)} = a_n, \text{ say.}$$

However, in Example 2 of 11.5 we showed that $\Sigma_{n=1}^{\infty} a_n = 1$, and so by Theorem 1, $\Sigma_n b_{n+1}$ converges. Therefore by Lemma 1, $\Sigma_n b_n$ also converges. The Lemma in conjunction with Corollary 2 yields an upper bound for $\Sigma_n b_n$. Thus

$$\sum_{n=1}^{\infty} \frac{1}{n^2} < 1 + \sum_{n=1}^{\infty} \frac{1}{n(n+1)} = 1 + 1 = 2.$$

Actually $\sum_{n=1}^{\infty} \frac{1}{n^2} = \frac{\pi^2}{6} = 1.6449...$, as we have already remarked.

Example 2. Consider the series $\sum_n \frac{1}{n^{2+k}}$, $k > 0$. Since

$$0 < \frac{1}{n^{2+k}} < \frac{1}{n^2} \quad \text{for all } n \geq 1,$$

and $\Sigma_n 1/n^2$ converges, Theorem 1 implies that $\Sigma_n 1/n^{2+k}$ is a convergent series.

Example 3. The series $\Sigma_n 1/n^{1-k}$ for $k > 0$, diverges by Corollary 1 and Example 1 of 11.5.

The inquiring reader may wonder about the behavior of the series

$$\sum_n 1/n^q \text{ where } 1 < q < 2.$$

This question is dealt with in Chapter 14.

Example 4. One important special application of Theorem 1 is the decimal representation of a real number

$$a = a_0 . a_1 a_2 a_3 \ldots$$

$$= a_0 + \sum_{n=1}^{\infty} \frac{a_n}{10^n},$$

where $a_i \in \{0, 1, 2, 3, 4, 5, 6, 7, 8, 9\}$ for all $i \in N - \{0\}$, and $a_0 \in I$.

The series $\Sigma_n a_n/10^n$ can be shown to converge by comparing it with the convergent geometric series $\Sigma_n 9/10^n$. This has already been discussed in 8.6.

We shall now examine two more tests for the convergence of series of nonnegative terms: the Ratio Test (Theorem 2) and the Root Test (Theorem 3).

Theorem 2. (Ratio Test). *Let* $\Sigma_n a_n$ *be a series of positive terms.*

(a) $\exists a \, \exists n_0 \in N \, \forall n > n_0 \, (a_{n+1}/a_n \leqslant a < 1) \Rightarrow (\Sigma_n a_n \text{ converges}).$
(b) $\exists n_0 \in N \, \forall n > n_0 \, (a_{n+1}/a_n > 1) \Rightarrow (\Sigma_n a_n \text{ diverges}).$

Theorem 3. (Root Test).   *Let* $\Sigma_n a_n$ *be a series of nonnegative terms.*

(a) $\exists a \, \exists n_0 \in N \, \forall n > n_0 \, (\sqrt[n]{a_n} \leqslant a < 1) \Rightarrow (\Sigma_n a_n \text{ converges}).$
(b) $\exists n_0 \in N \, \forall n > n_0 \, (\sqrt[n]{a_n} > 1) \Rightarrow (\Sigma_n a_n \text{ diverges}).$

We shall discuss the proof of these theorems presently.
The ratio test says: (a) If the ratio a_{n+1}/a_n ultimately becomes and remains less than some number that is less than 1, then $\Sigma_n a_n$ converges. In particular, if $\lim_n (a_{n+1}/a_n) < 1$, then $\Sigma_n a_n$ converges; or if $\lim \sup (a_{n+1}/a_n) < 1$, then $\Sigma_n a_n$ converges. (b) On the other hand, if the ratio a_{n+1}/a_n ultimately becomes bigger than 1, then $\Sigma_n a_n$ diverges.
If neither (a) nor (b) above obtains, for example, if $\lim_n (a_{n+1}/a_n) = 1$, *then the test provides no information.*
Analogous remarks apply to the root test. Before we give the proofs for these two tests, let us consider some examples.

Example 5.   The series

$$\sum_n a_n = \sum_n \frac{1}{n!} = \frac{1}{0!} + \frac{1}{1!} + \cdots$$

converges ($0! = 1$ by definition). Indeed

$$\frac{a_{n+1}}{a_n} = \frac{1}{(n+1)!} \bigg/ \frac{1}{n!} = \frac{1}{n+1} \leqslant \frac{1}{2}$$

for all $n > 1$. Therefore by the ratio test the series converges, and

$$\sum_{n=0}^{\infty} \frac{1}{n!} \text{ exists.}$$

The limit of $\Sigma_n 1/n!$ is an irrational number which is denoted by e. A proof that e is irrational is as follows:
Suppose that e is rational and that $e = m/n$ where $m, n \in N$, $n > 0$. Then

$$\frac{m}{n} = 1 + 1 + \frac{1}{2!} + \frac{1}{3!} + \cdots + \frac{1}{n!} + \cdots.$$

Multiply each side by $n!$ and define the two integers

$$p = m(n-1)!, \quad q = n!\left(1 + 1 + \frac{1}{2!} + \cdots + \frac{1}{n!}\right).$$

Then

$$p = q + \frac{1}{n+1} + \frac{1}{(n+1)(n+2)} + \frac{1}{(n+1)(n+2)(n+3)} + \cdots.$$

Since $p, q \in N$,

$$r = p - q = \frac{1}{n+1} + \frac{1}{(n+1)(n+2)} + \cdots \in N.$$

But

$$\frac{1}{n+1} < r < \frac{1}{n+1} + \frac{1}{(n+1)^2} + \frac{1}{(n+1)^3} + \cdots.$$

Now $\sum_{i=1}^{\infty} (n+1)^{-i}$ is a geometric series, and by Example 2 of 11.5, its sum is $1/n$. Thus

$$\frac{1}{n+1} < r < \frac{1}{n}$$

or

$$n < rn(n+1) < n+1.$$

But then $rn(n+1)$ is an integer between n and $n+1$ which is impossible, and consequently e must be irrational. Since e is irrational its decimal expansion neither terminates nor repeats. However, it can be computed as accurately as we wish from the series.

$$e = 2.7182818284\ldots.$$

The number e is one of the fundamental constants of mathematics and will appear again in 11.8 and Chapter 12.

Example 6.   The series $\sum_n 1/n^n$ converges, indeed

$$\sqrt[n]{a_n} = \sqrt[n]{(1/n^n)} = \frac{1}{n} \leqslant \frac{1}{2}$$

for $n \geqslant 2$. Therefore by the root test

$$\sum_{n=1}^{\infty} \frac{1}{n^n} \quad \text{exists.}$$

Example 7.   If we consider the series $\sum_n 1/n^2$ and apply the ratio test, we find that

$$\frac{a_{n+1}}{a_n} = \frac{1}{(n+1)^2}\bigg/\frac{1}{n^2} = \frac{n^2}{(n+1)^2},$$

and hence

$$\lim_n \frac{a_{n+1}}{a_n} = \lim_n \frac{n^2}{(n+1)^2} = 1.$$

Therefore the test fails to indicate whether $\Sigma_n 1/n^2$ converges or diverges. In fact we have shown in Example 1 that this series is convergent.

Example 8. In Example 7 we considered a convergent series for which the ratio test fails. If we consider $\Sigma_n 1/n$ which we know (by Example 4 of 11.5) to be divergent, the ratio of successive terms is

$$\frac{a_{n+1}}{a_n} = \frac{n}{n+1}.$$

Again

$$\lim_n \frac{n}{n+1} = 1$$

and so the test fails.

Examples of convergent and divergent series upon which the root test fails can also be produced.

Proof of Theorem 2a.   If for all $n > n_0$ we have

$$\frac{a_{n+1}}{a_n} \leqslant a < 1,$$

then

$$a_{n_0+2} \leqslant a_{n_0+1} \cdot a,$$
$$a_{n_0+3} \leqslant a_{n_0+2} \cdot a \leqslant a_{n_0+1} \cdot a^2,$$

etc., and in general

$$a_{n_0+k} \leqslant a_{n_0+1} a^{k-1},$$

for $k = 2, 3, \dots$. Since

$$\sum_{k=0}^n a_k = \sum_{k=0}^{n_0} a_k + \sum_{k=0}^{n-n_0-1} a_{k+n_0+1},$$

(2)
$$\sum_{k=0}^n a_k \leqslant \sum_{k=0}^{n_0} a_k + \sum_{k=0}^{n-n_0-1} a_{n_0+1} a^k.$$

Since the series $\Sigma_n a_{n_0+1} a^n$ is a convergent series of positive terms, $\Sigma_{n=0}^{\infty} a_{n_0+1} a^n$ is an upper bound of the sequence $\{\Sigma_{k=0}^{n-n_0-1} a_{k+n_0+1}\}$. Therefore $\{\Sigma_{k=0}^n a_k\}$ is a bounded increasing sequence by equation (2).

Hence

$$\lim_{n} \sum_{k=0}^{n} a_k = \sum_{n=0}^{\infty} a_n$$

exists.

The proofs of Theorems 2b, 3a, and 3b proceed in a similar fashion and are left to the reader.

The following theorem provides a simple test for divergence when applicable.

Theorem 4. *If* $\Sigma_n a_n$ *is a convergent series of nonnegative terms, then* $\lim_{n} a_n = 0$.

Proof. Suppose that $\sum_{n=1}^{\infty} a_n$ converges to a value S. Then

$$\lim_{n} \sum_{k=1}^{n} a_k = S$$

and

$$\lim_{n} \sum_{k=1}^{n-1} a_k = S.$$

Therefore

$$\lim_{n} \left(\sum_{k=1}^{n} a_k - \sum_{k=1}^{n-1} a_k \right) = \lim_{n} a_n = S - S = 0.$$

This proves that a necessary condition for the convergence of a series is that the limit of its nth term exist and equal 0. This is by no means sufficient for convergence, however, as we see from the divergent harmonic series $1 + \frac{1}{2} + \frac{1}{3} + \cdots$ (Example 1 of 11.5). The contrapositive statement of the theorem can be used in some cases to prove the divergence of a series.

Example 9. (a) By Theorem 4, the series $\Sigma_n 3n^2/(2n + 1)$ is divergent because $\{3n^2/(2n + 1)\}$ is divergent (Example 8 of 11.3).

(b) By Theorem 4, the series $\Sigma_n n/(n + 1)$ is divergent because $\lim_{n} [n/(n + 1)] = 1 \neq 0$.

11.7 CONVERGENCE OF SERIES OF TERMS WITH ARBITRARY SIGN

Theorems 1–4 of 11.6 apply to series of nonnegative terms (or by Lemma 1 of 11.6 to series whose terms are ultimately nonnegative).

Now let us consider series of arbitrary terms (such as $\Sigma_n (-1)^n/n$). First we show that certain of the arithmetic operations commute with the limit process.

Theorem 1. *If $\Sigma_n\, a_n$ and $\Sigma_n\, b_n$ are two convergent series, and if c is a real number, then*

(1)
$$\sum_{n=0}^{\infty} ca_n = c \sum_{n=0}^{\infty} a_n,$$

(2)
$$\sum_{n=0}^{\infty} (a_n + b_n) = \sum_{n=0}^{\infty} a_n + \sum_{n=0}^{\infty} b_n.$$

Proof. To prove (1), let $S_n = \sum_{k=0}^{n} a_k$ and $S_n' = \sum_{k=0}^{n} ca_k$.

From (5) of 11.4

$$\forall n \in N \; S_n' = cS_n,$$

and from Theorem 2 in 11.3,

$$\lim_n S_n' = \lim_n cS_n = c \lim_n S_n.$$

Therefore by Definition 2 of 11.5, $\sum_{n=0}^{\infty} ca_n$ exists and equals $c \sum_{n=0}^{\infty} a_n$.

We leave the proof of (2) to the reader.

Remark. The reader should note that Theorem 1 is also true in case we sum from $n = k$ (for some $k \in N$) to infinity instead of from $n = 0$ to infinity.

Example 1. As an example of the usefulness of Theorem 1 consider the series

$$\sum_n \frac{n^2 - 9n + 10}{n^4}.$$

This series converges by Theorem 1 because the series

$$\sum_n \frac{1}{n^2}, \quad \sum_n \frac{1}{n^3}, \quad \sum_n \frac{1}{n^4}$$

all converge (see Examples 1 and 2 of 11.6). Indeed, by Theorem 1,

$$\sum_{n=1}^{\infty} \frac{1}{n^2} - 9 \sum_{n=1}^{\infty} \frac{1}{n^3} + 10 \sum_{n=1}^{\infty} \frac{1}{n^4} = \sum_{n=1}^{\infty} \left(\frac{1}{n^2} - \frac{9}{n^3} + \frac{10}{n^4} \right)$$

$$= \sum_{n=1}^{\infty} \left(\frac{n^2 - 9n + 10}{n^4} \right).$$

Theorem 2.   *Let $\{a_n\}$ be a monotone decreasing sequence such that $\lim_n a_n = 0$. Then $\Sigma_n(-1)^n a_n$ (or $\Sigma_n(-1)^{n+1} a_n$) converges.*

Proof.   In Example 10 of 11.3 we proved that $\Sigma_n (-1)^n/n$ converges using Cauchy's convergence criterion. The proof of Theorem 2 proceeds in a similar manner and we leave it to the reader. Theorem 2 is sometimes called the Leibniz criterion. From Theorem 2 it follows that both

$$\sum_n (-1)^n/n \text{ and } \sum_n (-1)^n/n^2$$

are convergent series. However from Example 1 of 11.5 it is clear that $\Sigma_n |(-1)^n/n)| = \Sigma_n 1/n$ is divergent, while from Example 1 of 11.6 it follows that $\Sigma_n |(-1)^n/n^2| = \Sigma_n 1/n^2$ is convergent. The difference between these two situations is significant in a way which will now be discussed.

Definition 1.   *Let $\Sigma_n a_n$ be any series. Let $f : N \to N$ be a one-to-one mapping of N onto N. The series $\Sigma_n b_n$ where $b_n = a_{f(n)}$ is called a **rearrangement** of $\Sigma_n a_n$.*

There is no reason to expect that a rearrangement of a convergent series has the same limit as the original series or even that it is convergent. A series is, after all, equivalent to a sequence of partial sums and a rearrangement gives a new such sequence.

Example 2.   A simple rearrangement of $\Sigma_n (-1)^n/n$ is defined by taking the first term with negative sign, namely -1. Follow it by the next 10 terms with positive sign, namely $\frac{1}{2}, \frac{1}{4}, \frac{1}{6}, \ldots, \frac{1}{20}$, then follow with the second term with negative sign, namely $-\frac{1}{3}$, then follow that with the next 10^2 terms with positive sign and continue this *ad infinitum*. The result is a rearrangement of $\Sigma_n (-1)^n/n$ which diverges, since there is no upper bound upon its partial sums.

Example 3.   Another rearrangement of $\Sigma_n (-1)^n/n$ is achieved by the function f, tabled as follows:

n	1	2	3	4	5	6	7	8	9	10	11	$\cdots$
$f(n)$	1	3	2	5	7	4	9	11	6	13	15	$\cdots$

The procedure is to enumerate N by writing two successive odd numbers followed by an even number. The resulting rearrangement is

$$-1 - \frac{1}{3} + \frac{1}{2} - \frac{1}{5} - \frac{1}{7} + \frac{1}{4} - \frac{1}{9} - \frac{1}{11} + \frac{1}{6} - \frac{1}{13} - \frac{1}{15} + \frac{1}{8} - \cdots.$$

It can be shown (*Infinite Sequences and Series*, by K. Knopp, Dover

Publications, p. 77), that the rearrangement converges and its limit satisfies the equation

$$S = \frac{3}{2}\left(\sum_{n=1}^{\infty} \frac{(-1)^n}{n}\right).$$

Under some circumstances rearrangement of a series does not alter the value of its (generalized) sum as we shall soon see.

Definition 2.   *A series $\Sigma_n a_n$ is said to **converge absolutely** if $\Sigma_n |a_n|$ converges.*

It is clear that $\Sigma_n (-1)^n/n^2$ converges absolutely while $\Sigma_n (-1)^n/n$ does not converge absolutely.

Theorem 3.   *If $\Sigma_n a_n$ converges absolutely, then it converges.*

Proof.   Let $S_n = \Sigma_{k=0}^{n} a_n$. Let ε be a fixed positive real number. By the generalized triangle inequality (Expression (7) in 11.4), we have for $n < m$ (the case when $m < n$ can be treated in an identical fashion)

$$|S_n - S_m| = \left|\sum_{k=n+1}^{m} a_k\right| \le \sum_{k=n+1}^{m} |a_k| = S_m' - S_n',$$

where $S_j' = \Sigma_{k=0}^{j} |a_k|$. Since $\Sigma_n |a_n|$ is convergent, there is an n_0 such that $n > n_0$ and $m > n_0$ imply that

$$S_m' - S_n' < \varepsilon,$$

where ε is an arbitrarily chosen positive number, and so

$$|S_n - S_m| < \varepsilon$$

for $n, m > n_0$.

Therefore the requirements of the Cauchy Convergence Criterion are satisfied, so that $\{S_n\}$ is a convergent sequence and $\Sigma_n a_n$ is a convergent series.

A convergent series which is not absolutely convergent is called **condition-ally convergent**. The following theorem together with Examples 2 and 3 give some justification for this definition.

Theorem 4.   *A series $\Sigma_n a_n$ is absolutely convergent if and only if every rearrangement $\Sigma_n b_n$ of $\Sigma_n a_n$ is convergent to $\Sigma_{n=0}^{\infty} a_n$. Thus the limit of an absolutely convergent series is unchanged by rearrangement.*

For a proof of Theorem 4 see Knopp, pp. 79–81.

Example 4. In Example 6 in 11.6 it was shown that $\sum_n 1/n^n$ is convergent. Hence

$$\sum_n \frac{\sin (n\pi/4)}{n^n}$$

is absolutely convergent. Indeed

$$\left| \frac{\sin (n\pi/4)}{n^n} \right| \leqslant \frac{1}{n^n}$$

for all $n > 0$, and so

$$\sum_n \left| \frac{\sin (n\pi/4)}{n^n} \right|$$

is convergent by Theorem 1 of 11.6. By Theorem 3

$$\sum_n \frac{\sin (n\pi/4)}{n^n}$$

is convergent.

In view of Theorem 3 it should be clear that Theorems 2 and 3 of 11.6, can be generalized to apply to absolute convergence in an obvious way.

11.8 POWER SERIES

In this section we give a simple intuitive discussion of power series in general and the exponential function in particular. If the reader wishes to go into these matters more fully, he should consult *Infinite Sequences and Series*.

Definition 1. *A **power series** is a series of the form $\sum_n a_n x^n$, and is said to be **defined** for every real x such that $\sum_{n=0}^{\infty} a_n x^n$ exists.*

Thus $f: x \to \sum_{n=0}^{\infty} a_n x^n$ is a function whose domain is the set of values of $x \in R$ for which $\sum_{n=0}^{\infty} a_n x^n$ converges.

Power series can be used to define new functions, as we shall soon see, and also as an alternative form for known functions.

Example 1. Consider the series $\sum_n x^n/n!$. If we apply the ratio test to $\sum_n |x^n/n!|$ we find that

$$\left| \frac{x^{n+1}}{(n+1)!} \right| \Big/ \left| \frac{x^n}{n!} \right| = \frac{|x|}{n+1},$$

and so for $n > 2|x| - 1$, this ratio is less than $\frac{1}{2}$. Thus for every $x \in R$ there is an $n_x \in N$ such that if $n > n_x$ the corresponding ratio of successive terms is less than $\frac{1}{2}$. Therefore by Lemma 1 of 11.6 and Theorem 2 of 11.6, $\Sigma_n x^n/n!$ is absolutely convergent. Further, by Theorem 3 of 11.7, $\Sigma_{n=0}^{\infty} x^n/n!$ exists for all $x \in R$. Therefore the function $x \to \Sigma_{n=0}^{\infty} x^n/n!$ has the domain R. We call this function the *exponential function* and denote it by "exp". Thus

$$(1) \qquad\qquad \exp(x) = \sum_{n=0}^{\infty} \frac{x^n}{n!}.$$

Note that from Example 5 in 11.6 we have

$$\exp(1) = \sum_{n=0}^{\infty} \frac{1}{n!} = e.$$

Example 2.   From our work with the geometric series we know that

$$\sum_n x^n$$

converges for $|x| < 1$ and that it diverges for other real values of x. In addition we have found that

$$(2) \qquad\qquad \sum_{n=0}^{\infty} x^n = \frac{1}{1 - x}$$

when the sum of the series is defined. Thus $x \to \Sigma_{n=0}^{\infty} x^n$ defines a function with domain $\{x|\, |x| < 1\}$, and which has the value $1/(1 - x)$ there.

Example 3.   The series $\Sigma_n n!\, x^n$ converges only when $x = 0$. Indeed, forming the ratio of successive terms:

$$\frac{|(n + 1)!\, x^{n+1}|}{|n!\, x^n|} = (n + 1)|x|$$

we find that for any fixed $x \neq 0$ the ratio is ultimately greater than 1. Thus by Theorem 2 of 11.6, and Lemma 1 of 11.6, the series diverges. If $x = 0$, then each term is zero, and so the series converges in a trivial way.

Thus the function defined by $x \to \Sigma_{n=0}^{\infty} n!x^n$ has a domain containing only the point $0 \in R$.

Example 4.   Consider the two series

$$c(x) = \sum_{n=0}^{\infty} \frac{(-1)^n x^{2n}}{(2n)!}$$

and

$$s(x) = \sum_{n=0}^{\infty} \frac{(-1)^n x^{2n+1}}{(2n + 1)!}.$$

By proceeding as in Example 1, it is readily shown that these series are convergent for all x. In Chapter 12, we shall show that $c(x)$ and $s(x)$ are identical to the trigonometric functions cos x and sin x respectively. Later in this section we shall obtain some properties of $c(x)$ and $s(x)$ directly from their series definitions. By using these series, the values of sin x or cos x can be computed for any value of x. For example to find sin 1, that is, the sine of one *radian*, we have

$$s(1) = \sum_{n=0}^{\infty} \frac{(-1)^n}{(2n+1)!}.$$

Then sin 1 can be computed to any degree of accuracy, by taking a sufficient number of terms.

In order to continue our study of exp (x), $c(x)$ and $s(x)$, we shall need the following Lemma.

Lemma 1   *If $\Sigma_n a_n$ and $\Sigma_n b_n$ are absolutely convergent series, then*
(a) *the series $\Sigma_n c_n$ is absolutely convergent where*

$$c_n = \sum_{k=0}^{n} a_k b_{n-k},$$

and (b) *the following equation is valid,*

(3)
$$\left(\sum_{n=0}^{\infty} a_n \right) \left(\sum_{m=0}^{\infty} b_m \right) = \sum_{n=0}^{\infty} c_n.$$

Since our discussion of power series is mainly heuristic and expository and since the proof of this Lemma involves certain technical points not pertinent to the present discussion, we will not burden the reader with it here. If he is interested, he should see *Infinite Sequences and Series*, by K. Knopp, pp. 88–90.

We now discuss some of the properties of the exponential function introduced in Example 1. Indeed we shall ultimately show that for

$$q \in Q$$
$$\exp(q) = e^q,$$

that is, exp (q) is equal to the number e defined in Example 5 of 11.6 raised to the qth power. In Chapter 12 we shall return to power series in general and the exponential function in particular. There we shall show that

$$\exp(x) = e^x$$

for all $x \in R$.

Theorem 1.

$$(\exp x)(\exp y) = \exp(x + y)$$

for all real numbers x and y.

Proof. The product

$$(\exp x)(\exp y) = \left(\sum_{n=0}^{\infty} \frac{x^n}{n!}\right)\left(\sum_{m=0}^{\infty} \frac{y^m}{m!}\right)$$

can be expressed according to Lemma 1 in the form

$$\sum_{n=0}^{\infty} c_n = \sum_{n=0}^{\infty} \left(\sum_{k=0}^{n} \frac{x^k}{k!} \frac{y^{n-k}}{(n-k)!}\right).$$

Thus we can write

(4) $$(\exp x)(\exp y) = \sum_{n=0}^{\infty} \left[\frac{1}{n!}\left(\sum_{k=0}^{n} \frac{n!}{k!(n-k)!} x^k y^{n-k}\right)\right].$$

The reader should remember, however, that

$$\sum_{k=0}^{n} \frac{n!}{k!(n-k)!} x^k y^{n-k} = (x+y)^n$$

by the Binomial Theorem. Thus equation (4) becomes

$$(\exp x)(\exp y) = \sum_{n=0}^{\infty} \left[\frac{1}{n!}(x+y)^n\right], \text{ or}$$

(5) $$(\exp x)(\exp y) = \exp(x+y).$$

Note that equation (5) is a first step toward showing that $\exp x = e^x$ because, in general, exponentiation gives $a^x a^y = a^{x+y}$.

Corollary 1. *For $x \in R$,*

(6) $$\exp(-x) = \frac{1}{\exp x}.$$

Proof. By Theorem 1,

$$(\exp x)\left(\exp(-x)\right) = \exp\left(x + (-x)\right)$$

$$= \exp 0 = \sum_{n=0}^{\infty} \frac{0^n}{n!}$$

$$= \lim_{n}\left(1 + \frac{0}{1!} + \frac{0^2}{2!} + \cdots + \frac{0^n}{n!}\right) = 1.$$

Therefore

$$\exp(-x) = \frac{1}{\exp x}.$$

Remark. As a by-product of Corollary 1 we have established

(7) $\exp(0) = 1.$

In addition, note that since $x > 0$ implies $\exp x > 0$, equation (6) ensures that $\exp(x) > 0$ for all $x \in R$.

Theorem 2. *If $x \geqslant 0$, then*

(8) $\exp x \geqslant 1,$

and if $x < 0$,

(9) $0 < \exp x < 1.$

Proof. If $x \geqslant 0$, then

$$\exp x = \sum_{n=0}^{\infty} \frac{x^n}{n!} = \lim_n \left(1 + \frac{x}{1!} + \frac{x^2}{2!} + \cdots + \frac{x^n}{n!}\right) \geqslant 1.$$

Thus expression (8) is verified. On the other hand if $x < 0$, then $-x > 0$ and so

$$\exp(-x) > 1.$$

From Corollary 1 it follows that

$$\frac{1}{\exp x} > 1,$$

and so, if $x < 0$,

$$\exp x < 1.$$

From the Remark following the proof of Corollary 1, it is clear that $0 < \exp x < 1$ for all $x < 0$.

Corollary 2. *The exponential function is a monotone increasing function, that is,*

$$\forall x \in R \ \forall y \in R \ (x < y \Rightarrow \exp x < \exp y).$$

Proof. If $x < y$, then $y - x > 0$, and so

$$\exp(y - x) = \frac{\exp y}{\exp x} > 1$$

by Theorems 1, 2 and Corollary 1. Therefore (using Theorem 2)

$$\exp x < \exp y.$$

By using the preceding Theorems and Corollaries we can make a sketch of the graph of exp (see Figure 11–4).

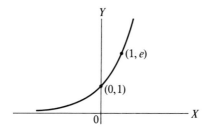

Figure 11–4 Graph of the function $x \to \exp x$

Theorem 3. *If $q \in Q$, then*
$$\exp(q) = e^q,$$
where
$$e = \exp 1 = \sum_{n=0}^{\infty} \frac{1}{n!}.$$

Proof. We proceed in three steps; we show that: (i) $\exp(n) = e^n$ for $n \in N$, then (ii) $\exp(i) = e^i$ for $i \in I$, and finally (iii) $\exp(q) = e^q$ for $q \in Q$.

(i) First note that $\exp 2 = \exp(1+1) = \exp 1 \cdot \exp 1$
$$= (\exp 1)^2 = e^2.$$
If we proceed by mathematical induction, we can show that
$$\exp n = (\exp 1)^n = e^n$$
for all $n \in N$.

In general, if $x \in R$, we can easily show by an argument similar to that employed in the proof of (i) that

(10)
$$\exp(nx) = (\exp x)^n$$

for $n \in N$.

(ii) Since by (6)
$$\exp(-1) = \frac{1}{\exp 1} = \frac{1}{e},$$
we can use equation (10) to produce the result
$$\exp(-n) = \exp(n(-1)) = (\exp(-1))^n$$
$$= e^{-n}.$$

If we observe that $\exp 0 = 1$, then we have shown that for $i \in I$,
$$\exp(i) = e^i.$$

(iii) Since for any $n \in N$, $e^{1/n} = s$ if and only if $s^n = e$ and since $(\exp(1/n))^n = \exp(n(1/n)) = \exp 1 = e$ by equation (10), it follows that

$$\exp\left(\frac{1}{n}\right) = e^{1/n}.$$

Now since any $q \in Q$ can be written $q = m/n$,

$$\exp q = \exp\left(\frac{m}{n}\right) = \left(\exp\frac{1}{n}\right)^m = (e^{1/n})^m = e^{m/n} = e^q$$

by equation (10) and the foregoing remark. Thus the theorem is proved.

In Chapter 12, we show that

$$\exp x = e^x$$

for all $x \in R$, even when x is irrational.

We now return to the functions $c(x)$ and $s(x)$ defined in Example 4 and deduce some of their properties. The methods used are similar to the methods used previously for $\exp x$.

Theorem 4.   *For all real numbers x and y*

$$c(x)\, c(y) - s(x)\, s(y) = c(x + y)$$

Proof.   Using Lemma 1, we have

$$(11) \hspace{4cm} c(x)\, c(y) = \sum_{n=0}^{\infty} c_n$$

where

$$c_n = \sum_{k=0}^{n} \frac{(-1)^k x^{2k}}{(2k)!} (-1)^{n-k} \frac{y^{2(n-k)}}{[2(n-k)]!}$$

$$= (-1)^n \sum_{k=0}^{n} \frac{x^{2k} y^{2n-2k}}{(2k)!\,(2n-2k)!}$$

and

$$(12) \hspace{4cm} s(x)\, s(y) = \sum_{n=0}^{\infty} c_n'$$

where

$$c_n' = (-1)^n \sum_{k=0}^{n} \frac{x^{2k+1} y^{2n-2k+1}}{(2k+1)!\,(2n-2k+1)!}.$$

Also

(13)
$$c(x + y) = \sum_{n=0}^{\infty} (-1)^n \frac{(x + y)^{2n}}{(2n)!}$$

$$= \sum_{n=0}^{\infty} \frac{(-1)^n}{(2n)!} \sum_{k=0}^{2n} \frac{(2n)! \, x^k y^{2n-k}}{k! \, (2n - k)!}$$

by the Binomial Theorem applied to $(x + y)^{2n}$. Now the sum over k in (13) can be written as two sums, one over even k, and the other over odd k. Thus

$$c(x + y) = \sum_{n=0}^{\infty} (-1)^n \sum_{k=0,2\ldots}^{2n} \frac{x^k y^{2n-k}}{k! \, (2n - k)!} + \sum_{n=1}^{\infty} (-1)^n \sum_{k=1,3\ldots}^{2n-1} \frac{x^k y^{2n-k}}{k! \, (2n - k)!}$$

In writing the second series, we should note that the term corresponding to $n = 0$ is absent since it has been included in the first series. Now in the first series set $k = 2p$, and in the second set $k = 2p + 1$, $n = m + 1$. We find

$$c(x + y) = \sum_{n=0}^{\infty} (-1)^n \sum_{p=0}^{n} \frac{x^{2p} y^{2n-2p}}{(2p)! \, (2n - 2p)!}$$

$$+ \sum_{m=0}^{\infty} (-1)^{1+m} \sum_{p=0}^{m} \frac{x^{2p+1} y^{2m-2p+1}}{(2p + 1)! \, (2m - 2p + 1)!}$$

$$= c(x) \, c(y) - s(x) \, s(y),$$

from (11) and (12).

Theorem 5.   *The functions $c(x)$, $s(x)$ have the properties that, if $x \in R$,*

$$c(-x) = c(x),$$
$$s(-x) = -s(x),$$
$$c(0) = 1,$$
$$s(0) = 0.$$

Proof.   The proofs will be left to the reader.

Corollary 3. $c^2(x) + s^2(x) = 1$ *for all* $x \in R$.

Proof.   Set $y = -x$ in Theorem 4, and apply Theorem 5; the result follows.

The properties of $c(x)$, $s(x)$ proved in Theorem 4, Theorem 5, and Corollary 3 are, of course, well-known properties of cos x and sin x.

11.9 EXERCISES

1. In each of the following cases, decide whether the sequence is bounded or unbounded if $n \in N$:

(a) $\left| \dfrac{2n+1}{n} \right|, \; n > 0,$ (b) $\left| \dfrac{n^2+n+1}{n^2} \right|, \; n > 0,$

(c) $\{a_n\}$ where $a_0 = a_1 = 1$ and $a_n = 2a_{n-1} - a^2_{n-2}$ for $n \geqslant 2,$

(d) $\{a_n\}$ where $a_0 = 0$, $a_1 = 1$ and $a_n = 2a_{n-1} - a^2_{n-2}$ for $n \geqslant 2,$

(e) $\left| \dfrac{1 + 2 + \cdots + n}{n^2} \right|,$

(f) $\left| \dfrac{1 + 2 + \cdots + n}{n} \right|.$

2. Find Δa_n in each of the following cases:

(a) $a_n = n^3,$ (b) $a_n = 2^n,$

(c) $a_n = \dfrac{n}{n+1},$ (d) $a_n = \begin{cases} \dfrac{n!}{(n-r)!\,r!} & n \geqslant r, \\ 0 & n < r, \end{cases}$

(e) $a_n = \dfrac{1}{n},$ (f) $a_n = n^4.$

3. Show that each of the following statements is valid.

(a) $\Delta(a_n b_n) = b_n \Delta a_n + a_{n+1} \Delta b_n.$

(b) $\Delta(a_n/b_n) = \dfrac{b_n \Delta a_n - a_n \Delta b_n}{b_{n+1} b_n}.$

(c) $\Delta(\sin na) = 2 \sin \dfrac{a}{2} \cos \dfrac{(2n+1)a}{2}.$

4. Using Exercise 3, find Δa_n when

(a) $a_n = n \sin na,$

(b) $a_n = \dfrac{n^2 + n}{2^n}.$

5. Find the following sums as simple functions of n.

(a) $1 + 4 + 9 + \cdots + n^2.$

(b) $1 \cdot 2 + 2 \cdot 4 + 3 \cdot 8 + \cdots + n \cdot 2^n.$

   (Hint: Show that for $n > 2$, $\Delta[(n-2)\,2^n] = n\,2^n$.)

(c) $\cos \dfrac{a}{2} + \cos \dfrac{3a}{2} + \cdots + \cos \dfrac{2n+1}{2} a$

   (Hint: Use Exercise 3c.)

(d) $\dfrac{1}{2} + \dfrac{1}{6} + \cdots + \dfrac{1}{n(n+1)}$

   (Hint: Use Exercise 2e.)

6. In the following cases ascertain whether the sequence converges or diverges. Find the limit if it exists.

(a) $\left\{1 + \dfrac{1}{n}\right\}$,

(b) $\left\{\dfrac{n^2 + n + 1}{n^2}\right\}$,

(c) $\left\{\dfrac{1 + 2 + \cdots + n}{n}\right\}$,

(d) $\left\{\dfrac{1 + 2 + 3 + \cdots + n}{n^2}\right\}$,

(e) $\left\{\dfrac{1}{2} + \dfrac{1}{6} + \cdots + \dfrac{1}{n(n+1)}\right\}$.

7. A Fibonacci sequence is defined by the following recursion equations:
$$a_1 = a_2 = 1, \quad a_n = a_{n-1} + a_{n-2} \quad \text{for } n \geqslant 3.$$
If $\{a_n\}$ is the Fibonacci sequence, prove by mathematical induction that
$$a_{n+m} = a_{n-1}a_m + a_{m+1}a_n, \quad m > 1, \ n > 1.$$

8. Consider the sequence $\{b_n\}$ where
$$b_0 = 1, \quad b_{n+1} = \frac{b_n^2 + 1}{2b_n - 1}$$

for $n = 0, 1, 2, \ldots$

(a) Prove that the sequence $\{b_{n+1}\}$ is monotone decreasing with a lower bound, and hence that it is convergent.

(b) Prove that $\displaystyle\lim_n b_n = \frac{\sqrt{5} + 1}{2}$

(c) If $\{a_n\}$ is the Fibonacci sequence, prove using mathematical induction that
$$b_n = \frac{a_{2^n + 1}}{a_{2^n}}.$$

(Hint: Use the results of Exercise 7.)

(d) Prove that
$$\lim_k \frac{a_{k+1}}{a_k} = \frac{\sqrt{5} + 1}{2}.$$

9. (a) If a and b are arbitrary real numbers, prove that
$$a^2 + b^2 \geqslant 2ab.$$

(b) By substituting
$$a = \frac{|a_j|}{\left[\displaystyle\sum_{i=p}^{q} a_i^2\right]^{\frac{1}{2}}} \quad \text{and} \quad b = \frac{|b_j|}{\left[\displaystyle\sum_{i=p}^{q} b_i^2\right]^{\frac{1}{2}}}$$

with $j = p, p+1, \ldots, q$ in the formula of 9(a) and adding the resulting inequalities, show that
$$\left(\sum_{i=p}^{q} a_i^2\right)\left(\sum_{i=p}^{q} b_i^2\right) \geqslant \left(\sum_{i=p}^{q} a_i b_i\right)^2$$

(The Cauchy–Schwarz inequality.)

10. Suppose that a_i satisfies the conditions

(i) $a_i \in R$,

(ii) $a_i > -1$,

(iii) all a_i are of the same sign for $i = 1, 2, \ldots n$.

(a) Prove by mathematical induction that if $a_i \geqslant 0$,

$$(1 + a_1)(1 + a_2) \ldots (1 + a_n) \geqslant 1 + a_1 + a_2 + \cdots + a_n.$$

(b) If $a_1 = a_2 = \cdots = a_n = x$ prove that $(1 + x)^n \geqslant 1 + nx$ (the Bernoulli inequality).

11. (a) Prove that

$$\binom{n+1}{k} = \binom{n}{k-1} + \binom{n}{k}.$$

(b) Prove that

$$\binom{n}{k} + \binom{n}{k+1} = \binom{n+1}{k+1}.$$

(c) From 11(a) and mathematical induction deduce the Binomial Theorem.

12. Suppose that $\{a_n\}$ is the Fibonacci sequence defined in Exercise 7. Prove that

(a) $\qquad \sum_{i=1}^{n} a_i = a_{n+2} - 1.$

(See "The Fibonacci Numbers", by N. N. Vorobyov, in *Topics in Mathematics*, Heath and Co. 1963.)

(b) $\qquad \sum_{i=1}^{n} a_{2i-1} = a_{2n}.$

Define

$$\sigma_n = \sum_{i=0}^{\left[\frac{n-1}{2}\right]} \binom{n-i-1}{i}$$

(recall Theorem 3 of 8.3).

(c) Now prove that

$$\sigma_{n-1} + \sigma_{n-2} = a_n.$$

(Hint: Use Exercise 11 (b).)

13. Show that:

(a) $\lim\limits_{n} x^n = 0$ for $|x| < 1.$

(b) $\lim\limits_{n} |a_n - a| = 0$ if and only if $\lim\limits_{n} a_n = a.$

(c) $\lim\limits_{n} \left(\dfrac{1}{n} \sin n\pi x \right) = 0$ for all $x \in R.$

14. Consider the sequence $\{a_n\}$ and consider the sequence $\{b_n\}$ formed from $\{a_n\}$ by letting

$$b_n = \frac{1}{n+1} \sum_{k=0}^{n} a_k$$

(a) Show that the function $f: \{a_n\} \to \{b_n\}$ carries null sequences into null sequences; that is if $\lim\limits_{n} a_n = 0$, then $\lim\limits_{n} b_n = 0.$

(b) Show, using part (a), that if $\lim\limits_{n} a_n = \xi$ then $\lim\limits_{n} b_n = \xi$. Thus f carries convergent sequences into convergent sequences with the same limit.

(c) It is the case that f carries some divergent sequences into convergent

sequences. Show this for
$$a_n = (-1)^n.$$

15. (a) Prove the following:

If $a_n \leqslant b_n$ for all n and if $\lim_n a_n = a$ and $\lim_n b_n = b$, then $a \leqslant b$.

   (b) Prove that $\lim_n a_n = 0$ if and only if $\lim_n |a_n| = 0$.

   (c) Prove that if $|a_n| \leqslant b_n$, then $\lim_n |a_n| \leqslant \lim_n b_n$.

16. (a) Prove that if $a,b \in R$, $|a - b| \geqslant \big| |a| - |b| \big|$.

   (b) Prove that $\big|\lim_n a_n\big| = \lim_n |a_n|$.

17. Which of the following series converge?

   (a) $\displaystyle\sum_n \frac{\sin(\frac{1}{4} n\pi)}{n^3}$ (b) $\displaystyle\sum_n \frac{n^3 + 2n^2 - 1}{n^5}$

   (c) $\displaystyle\sum_n n^2(\frac{1}{2})^n$ (d) $\displaystyle\sum_n n^3/n!$

18. (a) Suppose that a power series $\sum_n a_n x^n$ is absolutely convergent at $x = x_0$. Show that the series is absolutely convergent for all $x \in R$ such that
$$|x| < |x_0|.$$

   The l.u.b. of the set $\{|x| \mid \sum_n a_n x^n$ is absolutely convergent$\}$ if it exists is called the *radius of convergence* of $\sum_n a_n x^n$; if this l.u.b. does not exist, then $\sum_n a_n x^n$ is said to have infinite radius of convergence.

   Using the ratio test show that

   (b) $\sum_n n! x^n$ has zero radius of convergence,

   (c) $\sum_n x^n$ has radius of convergence 1,

   (d) $\sum_n x^n/n!$ has infinite radius of convergence.

12

Limits and Continuity of Real Functions

12.1 LIMITS

In this chapter we examine functions whose domain and range are arbitrary subsets of the real numbers. Unless some statement is made to the contrary, it will be assumed that every number used is real and that all functions are limited to real values. We shall find that the concepts of limit, difference, sum, and series introduced in Chapter 11 for sequences have analogs for functions of a real variable. The analog of difference, the derivative, is considered in Chapter 13; the analog of sum, the integral, and the analog of series, the improper integral, are considered in Chapter 14. First, however, limits and continuity must be discussed.

Because the real numbers form a continuum, that is, the number line contains no gaps, we can say, for example, that the function defined by

(1) $$f(x) = \frac{x^2 - 1}{x - 1}$$

has a limit 2 as x approaches 1, even though $f(1)$ is undefined since $0/0$ is undefined. To verify this, note that $x^2 - 1 = (x - 1)(x + 1)$ and, provided $x \neq 1$,

$$\frac{x^2 - 1}{x - 1} = x + 1.$$

If the concept of limit (which is defined below) is to behave as our intuition tells us it should, then the limit of $x + 1$ as x approaches 1 should be 2. To aid us in our discussion of limits, we need the following ideas.

Definition 1. *An **open interval** is a set of the form*

(2) $$\{x \in R \,|\, a < x < b\}.$$

We call the set defined in (2) the **open interval with end points** a and b and denote it by (a, b).

The **closed interval** with end points a, b $(a < b)$ is the set

$$[a, b] = \{x \in R \,|\, a \leqslant x \leqslant b\}.$$

The following pair of intervals are called **semi-open**

$$[a, b) = \{x \in R \,|\, a \leqslant x < b\},$$
$$(a, b] = \{x \in R \,|\, a < x \leqslant b\}.$$

Note that the intersection of two open intervals is either the empty set or an open interval. This fact will prove useful in the sequel. Note also that the end points of intervals are assumed to be real numbers. Some texts use $(-\infty, \infty)$ to represent the infinite interval containing all the numbers in R; that is, the number line. We will avoid this notation.

The definitions of limit used in the present context depend on the idea of a **neighborhood of a point**.

Definition 2. *A **neighborhood** N_x **of a point** x is an open interval containing x.*

Thus $(-1, 1)$ is a neighborhood of, for example, $\frac{1}{2}$, $-\frac{1}{2}\sqrt{2}$, 0, and $\frac{3}{4}$ but is not a neighborhood of $+1$, -1, 25, or π.

Sometimes when there is no possibility of confusion the subscript x may be omitted from N_x.

The reader should observe that the intersection of two neighborhoods of a point $x \in R$ is also a neighborhood of x. It will also prove useful to have the following restricted class of neighborhoods at our disposal.

Definition 3. *For any real number $\varepsilon > 0$, the open interval $N_{x,\varepsilon} = (x - \varepsilon, x + \varepsilon)$ of length 2ε with x at its center is called the **symmetric neighborhood of** x **with radius** ε, or, more briefly, the ε neighborhood of x.*

Clearly every point $y \in R$ such that $|x - y| < \varepsilon$ belongs to $N_{x,\varepsilon}$ and conversely. Thus $N_{x,\varepsilon}$ contains exactly those real numbers whose distance from x is less than ε.

It is also clear that *every neighborhood of x contains, as a subset, a symmetric neighborhood for some ε*. Indeed, suppose $N_x = (a, b)$. Since N_x is a neighborhood of x the distances $|a - x|$ and $|b - x|$ are both positive. Let ε be equal to the smaller of the two, then

$$N_{x,\varepsilon} \subseteq (a, b).$$

Neighborhoods are used in our work on limits since the concept of limit, which we shall define shortly, involves a discussion of the behavior of a function *near* a particular point a, but not *at a*. In particular we are concerned with distinguishing the cases where the values of the function, as, for example, the function defined in (1), tend to a fixed number l as x nears a. Another way of saying that $f(x)$ tends to l as x tends to a is to say: for any neighborhood N_l of l, no matter how small, there is a neighborhood N_a of a such that if $x \in N_a$ and $x \neq a$ (remember we are not interested in $x = a$), then $f(x) \in N_l$. An example of the situation is given in Figure 12–1. A neighbor-

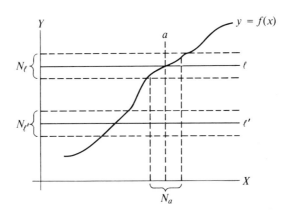

Figure 12–1   See Text. Symmetric neighborhoods are shown here; they need not be.

hood N_l is given and a corresponding N_a is shown which insures that $(x \in N_a$ and $x \neq a) \Rightarrow f(x) \in N_l$. It should be clear that the function indicated in Figure 12–1 is such that no matter how "small" N_l is, a corresponding N_a can be formed. It is for this reason that we say that in this case $f(x)$ tends to l as x tends to a. On the other hand we know that $f(x)$ in Figure 12–1 does not tend to l' as x tends to a. This can be seen easily, for there is no neighborhood N_a' of a with the property that $(x \in N_a'$ and $x \neq a) \Rightarrow f(x) \in N_{l'}$.

    Since we are not interested in the value of $f(x)$ at a but only in the values of $f(x)$ near a, we introduce the notations $N_a^* = N_a - \{a\}$ and $N_{a,\varepsilon}^* = N_{a,\varepsilon} - \{a\}$ and refer to N_a^* and $N_{a,\varepsilon}^*$ as a **punctured** (*or* **deleted**) **neighborhood** *and a* **punctured** (*or* **deleted**) **symmetric neighborhood**, respectively, of a.

    After these preliminary remarks we are now ready to define the concept of limit.

Definition 4.   *Let f be a function defined on some set which includes a punctured neighborhood N_a^* of a. Then $f(x)$ is said to approach the limit l as x tends toward a, or* $\lim_{x \to a} f(x) = l$, *if, corresponding to every neighborhood M of l, there is a punctured neighborhood $N^*(M)$ of a (depending on M) such that $N^*(M) \subset N_a^*$ and*

$$(x \in N^*(M) \Rightarrow f(x) \in M).$$

Remark 1.   Note that $\lim_{x \to a} f(x)$, if it exists, does not depend on the value of f at a. Indeed, as indicated above, the function need not even be defined at a. Our concern is only with the behavior of $f(x)$ for x near a and *not* with the value (if it exists) of $f(x)$ at $x = a$.

Remark 2.   The definition demands that in order for f to have a limit as $x \to a$ (x approaches a) it is necessary that for any neighborhood M of l (*no matter how small*) a neighborhood $N(M)$ of a (*sufficiently small*) exist so that whenever $x \in N^*(M)$, we have $f(x) \in M$.

Example 1.   The function defined by

(1) $$f(x) = \frac{x^2 - 1}{x - 1}$$

has limit 2 as $x \to 1$. Indeed, for any neighborhood N_1 of 1, f is defined in $N_1{}^*$ and

$$f(x) = x + 1$$

in $N_1{}^*$. To show that $\lim_{x \to 1} f(x) = 2$, consider an arbitrary neighborhood M of 2 (see Figure 12–2), and let d be some positive number such that if $|x - 2| < d$ then $x \in M$. We observe that if N_1 is a neighborhood of 1 and if $x \in N_1{}^*$,

$$|f(x) - 2| = |(x + 1) - 2| = |x - 1|.$$

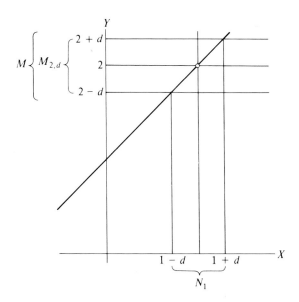

Figure 12–2   Graph of f: $x \to (x^2 - 1)/(x - 1)$.
$M_{2,d}$ is a symmetric neighborhood of 2

As N_1 we take the neighborhood $(1 - d, 1 + d)$. Then if $x \in N_1{}^*$, then $|x - 1| < d, |f(x) - 2| < d$, and $f(x) \in M$. By the definition of limit, it follows that $\lim_{x \to 1} f(x) = 2$.

In example 1 we had an instance where $\lim_{x \to a} f(x)$ exists. Now we give an example where it does not exist.

Example 2. Consider the step function s where

(3)
$$s(x) = \begin{cases} 0, & x < 0, \\ 1, & x \geqslant 0. \end{cases}$$

In the case of s we shall show that $\lim_{x \to 0} s(x)$ does not exist (although $s(0)$ is defined). If $\lim_{x \to 0} s(x)$ is to exist, there can be only three possible cases

    (i) $\lim_{x \to 0} s(x) = 1$,

    (ii) $\lim_{x \to 0} s(x) = 0$,

    (iii) $\lim_{x \to 0} s(x) = a$, where a is different from both 1 and 0.

In each of these cases the number on the right can be shown to be inadmissible as a value of $\lim_{x \to 0} s(x)$; and thus $\lim_{x \to 0} s(x)$ does not exist. Consider first Case (i): The set $(\frac{1}{2}, \frac{3}{2}) = M$ is a neighborhood of 1 (see Figure 12–3). We now show that there is no punctured neighborhood $N^*(M)$ of 0 such that

(4)
$$x \in N^*(M) \Rightarrow s(x) \in M.$$

Indeed any neighborhood $N_0{}^*$ of 0 contains both positive and negative real numbers and so there is at least one $x \in N_0{}^*$ such that $s(x) = 0$; therefore no neighborhood $N^*(M)$ can be found such that (4) holds.

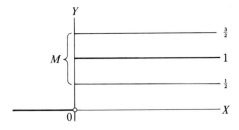

Figure 12–3 Graph of s; M is a neighborhood of 1

A similar argument can be used to rule out Case (ii).

Now consider Case (iii). Since $a \neq 1$ and $a \neq 0$, the distances $|1 - a|$ and $|-a|$ are both positive. Let $\varepsilon > 0$ be the smaller of the two values. Then $N_{a,\varepsilon}$ contains neither 0 nor 1. Since for every punctured neighborhood N_0^*, $x \in N_0^* \Rightarrow s(x) = +1$ or 0, it is clear that there is no punctured neighborhood $N^*(M)$ such that

$$x \in N^*(M) \Rightarrow s(x) \in N_{a,\varepsilon}.$$

Thus $\lim_{x \to 0} s(x) \neq a$, and Case (iii) is ruled out. Since no real number satisfies the conditions of Definition 4, $\lim_{x \to 0} s(x)$ does not exist.

Although the function defined in Example 2 has no limit at $x = 0$, it does exhibit some regularity as $x \to 0$ through values of $x < 0$ and also as $x \to 0$ through values of $x > 0$.

Definition 5. $\lim_{x \to a-} f(x) = l$ *if for each neighborhood M of l there is a neighborhood $N(M)$ such that* $(x \in N(M) \wedge x < a) \Rightarrow f(x) \in M$.

Definition 6. $\lim_{x \to a+} f(x) = l$ *if for each neighborhood M of l there is a neighborhood $N(M)$ such that* $(x \in N(M) \wedge x > a) \Rightarrow f(x) \in M$.

If $\lim_{x \to a-} f(x) = l$, we say "the limit of $f(x)$ as x approaches a from below is l", and if $\lim_{x \to a+} f(x) = l$, we say "the limit of $f(x)$ as x approaches a from above is l".

It is easy to see that for $s(x)$ in Example 2,

$$\lim_{x \to 0-} s(x) = 0 \quad \text{and} \quad \lim_{x \to 0+} s(x) = 1.$$

Example 3. The function f, with

$$f(x) = \frac{1}{x^2},$$

is defined throughout the set $R - \{0\}$, but does not have a limit as $x \to 0$. Indeed for any neighborhood N_0, values of $x \in N_0^*$ can be found for which $f(x)$ is arbitrarily large. Thus the conditions of Definition 4 (or 5 and 6 for that matter) cannot be satisfied. Therefore $f(x)$ has no limit as $x \to 0$.

Example 4. $\lim_{x \to 0} \sin (1/x)$ does not exist even though $\sin (1/x)$ is defined and

(5) $$-1 \leqslant \sin \frac{1}{x} \leqslant 1 \text{ for all } x \in R - \{0\}.$$

By (5), there is no need to look for a limit outside the interval $[-1, 1]$. This can be verified using the argument employed in Example 2. Suppose $\lim_{x \to 0} \sin (1/x) = a$ for $-1 \leqslant a \leqslant 1$ (see Figure 12–4). Let M be the symmetric neighborhood $N_{a,q}$ where q is the smaller of the two numbers $1 - a$ and $a + 1$, if $a \neq \pm 1$, and $q = \frac{1}{2}$ otherwise. Since $2/n\pi$ ultimately lies in any neighborhood of 0 for n large enough, $\sin (1/x)$ takes on the values 1, -1, and 0 in every neighborhood of zero. Thus for $N_{a,q}$ there is no punctured neighborhood $N^*(N_{a,q})$ of 0 such that

$$x \in N^*(N_{a,q}) \Rightarrow f(x) \in N_{a,q}.$$

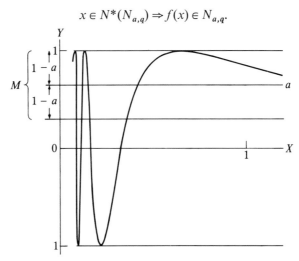

Figure 12–4   Graph of $x \to \sin (1/x)$; $\lim_{x \to 0} \sin (1/x)$ does not exist

Examples 2, 3, and 4 exhibit three different situations in which $\lim_{x \to a} f(x)$ does not exist. Any situation in which $\lim_{x \to a} f(x)$ does not exist will bear some resemblance to one or more of these examples.

Example 5. $\lim_{x \to 0} x \sin (1/x) = 0$. To show this note that

(6) $$\left| x \sin \frac{1}{x} \right| \leqslant |x| \text{(see Figure 12–5)}$$

for all $x \neq 0$. If M is an arbitrary symmetric neighborhood of 0, let $N(M) = N_{0,\varepsilon}$ where ε is chosen so that $N_{0,\varepsilon} \subseteq M$. Then $x \in N^*(M) \Rightarrow \pm |x| \in N_{0,\varepsilon}^*$ and by (6)

$$x \sin \frac{1}{x} \in N_{0,\varepsilon} = N(M).$$

Therefore $x \sin (1/x) \in M$. Thus for $N(M) = N_{0,\varepsilon}$,

$$x \in N^*(M) \Rightarrow x \sin \frac{1}{x} \in M.$$

Since M was arbitrary symmetric (see Theorem 1, p. 258),

$$\lim_{x \to 0} x \sin \frac{1}{x} = 0.$$

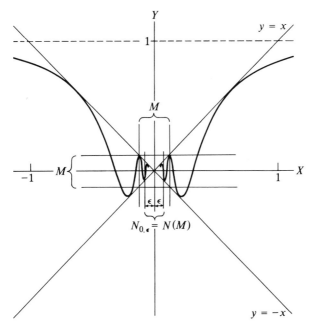

Figure 12–5   Graph of $x \to x \sin (1/x)$; $\lim_{x \to 0} x \sin (1/x)$ exists

Example 6. $\lim_{x \to 0} x^2 = 0$. To prove this we must show that for an arbitrary neighborhood M_0 of 0 we can find a corresponding deleted neighborhood N_0^* of 0 such that if $x \in N_0^*$ then $x^2 \in M_0$. Given $M_0 = (a, b)$ (as in Figure 12–6) where $a < 0 < b$, let $c = \sqrt{b}$. Now, as N_0 take the interval $(-c, c)$. Then

$$x \in N_0^* \Rightarrow |x| < c \Rightarrow x^2 < c^2 = b,$$

that is, $x \in N_0^* \Rightarrow x^2 \in M_0$.

The following results will help to clarify the concept of limit defined in Definition 4.

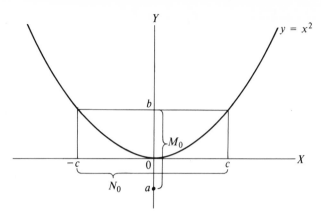

Figure 12–6 $\lim\limits_{x\to 0} x^2 = 0$

Theorem 1. $\lim\limits_{x\to a} f(x) = l$ *if and only if for every symmetric neighborhood*
$M = N_{l,\varepsilon}$ *of l there is a symmetric neighborhood $N = N_{a,\delta}$ of a such that*

$$x \in N_{a,\delta}{}^* \Rightarrow f(x) \in M.$$

Proof. The proof follows immediately from Definition 4 and will be left to the reader.

Corollary 1. $\lim\limits_{x\to a} f(x) = l$ *if and only if*

$$\forall \varepsilon > 0 \; \exists \delta > 0 \; [0 < |x - a| < \delta \Rightarrow |f(x) - l| < \varepsilon].$$

Corollary 1 gives an alternative definition of $\lim\limits_{x\to a} f(x) = l$ and is perhaps the one in most common use in textbooks.

We have employed Definition 4 for our basic definition because of its greater intuitive appeal.

Proof of Corollary 1. First note that $0 < |x - a| < \delta$ means that

$$x \neq a \text{ and } a - \delta < x < a + \delta,$$

which in turn means that

$$x \in N_{a,\delta}{}^*.$$

Similarly,

$$|f(x) - l| < \varepsilon \text{ means that } f(x) \in N_{l,\varepsilon}.$$

It should now be clear that Corollary 1 is nothing more than a restatement of Theorem 1.

Using the triangle inequality and Corollary 1, it is easy to prove the following theorem.

Theorem 2. *If* $\lim\limits_{x \to a} f(x) = l$ *and* $\lim\limits_{x \to a} g(x) = m$ *and if b is a real number, then*

(i) $\lim\limits_{x \to a} b\, f(x) = bl,$

(ii) $\lim\limits_{x \to a} \left(f(x) + g(x) \right) = l + m,$

(iii) $\lim\limits_{x \to a} \left(f(x) \cdot g(x) \right) = l \cdot m.$

If, in addition, $m \neq 0$, then

(iv) $\lim\limits_{x \to a} f(x)/g(x) = l/m.$

Proof. The proofs of (i) and (ii) are similar to the proofs given in 8.4 for the analogous statements regarding rational sequences. The proof of (iii) resembles the proof of Theorem 2 in 11.3 for the analogous statement regarding real sequences. Here we prove (iv).

Since $\lim\limits_{x \to a} g(x) = m \neq 0$, there is a symmetric neighborhood $N_{a,\delta'}$ of a such that if $x \in N_{a,\delta'}{}^*$, then

$$g(x) \in N_m = \left(m - \frac{|m|}{2}, m + \frac{|m|}{2} \right),$$

that is, $|g(x) - m| < \frac{1}{2}|m|$.

Thus, if $x \in N_{a,\delta'}{}^*, |g(x)| > \frac{1}{2}|m|$ and so $f(x)/g(x)$ is defined. Moreover, if $x \in N_{a,\delta'}{}^*$

$$\left| \frac{f(x)}{g(x)} - \frac{l}{m} \right| = \left| \frac{f(x)m - lg(x)}{g(x)m} \right|$$

$$= \frac{1}{|g(x)|\,|m|} \cdot |f(x)m - lg(x)|$$

$$\leq \frac{1}{\frac{1}{2}|m| \cdot |m|} \cdot |f(x)m - lg(x)|,$$

because $|g(x)| > \frac{1}{2}|m|$. Proceeding as in the proof of (3) in Theorem 2 of 11.3, we have

(7) $$\left| \frac{f(x)}{g(x)} - \frac{l}{m} \right| \leq \frac{2}{|m|^2} |f(x)m - lm + lm - lg(x)|$$

$$\leq \frac{2}{|m|^2} \cdot [|m|\,|f(x) - l| + |l|\,|m - g(x)|].$$

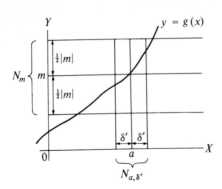

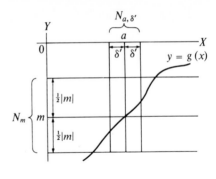

Figure 12–7a   Case of $m > 0$:
$x \in N_{a,\delta'}{}^* \Rightarrow g(x) \in N_m$

Figure 12–7b   Case of $m < 0$:
$x \in N_{a,\delta'}{}^* \Rightarrow g(x) \in N_m$

Now since $\lim_{x \to a} f(x) = l$ (assume $l \neq 0$) and $\lim_{x \to a} g(x) = m$, for an arbitrary but fixed $\varepsilon > 0$ there exist δ_1 and δ_2 such that

$$(8) \qquad 0 < |x - a| < \delta_1 \Rightarrow |f(x) - l| < \frac{\varepsilon |m|}{4}$$

and

$$(9) \qquad 0 < |x - a| < \delta_2 \Rightarrow |g(x) - m| < \frac{\varepsilon |m|^2}{4|l|} \qquad \text{(by Corollary 1).}$$

If δ is taken to be the smallest of the three numbers δ', δ_1, and δ_2, then from statements (7), (8), and (9),

$$0 < |x - a| < \delta \Rightarrow \left| \frac{f(x)}{g(x)} - \frac{l}{m} \right| < \frac{2}{|m|^2} \left(|m| \frac{\varepsilon |m|}{4} + |l| \frac{\varepsilon |m|^2}{4|l|} \right).$$

Thus

$$0 < |x - a| < \delta \Rightarrow \left| \frac{f(x)}{g(x)} - \frac{l}{m} \right| < \varepsilon,$$

and $\lim_{x \to a} \dfrac{f(x)}{g(x)} = \dfrac{l}{m}$, provided $l \neq 0$.

If $\lim_{x \to a} f(x) = 0$, then when $x \in N_{a,\delta'}{}^*$,

$$\left| \frac{f(x)}{g(x)} \right| = \frac{|f(x)|}{|g(x)|} \leqslant \frac{2}{|m|} |f(x)|.$$

Now, for an arbitrary $\varepsilon > 0$, there exists a $\delta > 0$ such that

$$0 < |x - a| < \delta \Rightarrow |f(x)| < \varepsilon \frac{|m|}{2}.$$

Then

$$0 < |x - a| < \min(\delta, \delta') \Rightarrow \left|\frac{f(x)}{g(x)}\right| < \varepsilon.$$

Thus

$$\lim_{x \to a} \frac{f(x)}{g(x)} = 0,$$

and the proof is completed.

Corollary 2. *If* $\lim_{x \to a} f_1(x) = l_1$, $\lim_{x \to a} f_2(x) = l_2$, ..., $\lim_{x \to a} f_n(x) = l_n$, *then*

(i) $\lim_{x \to a} (f_1(x) + f_2(x) + \cdots + f_n(x)) = l_1 + l_2 + \cdots + l_n$

(ii) $\lim_{x \to a} (f_1(x)f_2(x) \ldots f_n(x)) = l_1 l_2 \ldots l_n$.

This result can be proved by induction using Theorem 2.

Theorem 2 and Corollary 2 simplify computation of limits for algebraic combinations of functions with known limits.

Example 7. Consider

$$f(x) = \frac{x^2 + 3x + 2}{x^3 - 3}.$$

To find $\lim_{x \to 1} f(x)$, note that

$$\lim_{x \to 1} (x^3 - 3) = \lim_{x \to 1} x^3 - \lim_{x \to 1} 3$$

$$= \left(\lim_{x \to 1} x\right)^3 - \lim_{x \to 1} 3 = -2$$

by Theorem 2. Thus $\lim_{x \to 1} (x^3 - 3) = -2 \neq 0$ and so we can apply (iv) of Theorem 2:

$$\lim_{x \to 1} \frac{x^2 + 3x + 2}{x^3 - 3} = \frac{\lim_{x \to 1} (x^2 + 3x + 2)}{\lim_{x \to 1} (x^3 - 3)}.$$

Now applying (i), (ii), and (iii) of Theorem 2 and Corollary 2, we have

$$\lim_{x \to 1} \frac{x^2 + 3x + 2}{x^3 - 3} = \frac{\lim_{x \to 1} x^2 + 3 \lim_{x \to 1} x + \lim_{x \to 1} 2}{-2}.$$

Hence by Theorem 2,

$$\lim_{x \to 1} \frac{x^2 + 3x + 2}{x^3 - 3} = \frac{1 + 3 + 2}{-2} = -3.$$

Theorem 3.   *If* $\lim_{x \to a} f_1(x) = l_1$ *and* $\lim_{x \to a} f_2(x) = l_2$ *exist and if* $f_1(x) \leqslant f_2(x)$
for all x in some neighborhood of a, then

$$\lim_{x \to a} f_1(x) \leqslant \lim_{x \to a} f_2(x).$$

Proof.   Suppose the conclusion is not valid. Then

$$l_2 = \lim_{x \to a} f_2(x) < \lim_{x \to a} f_1(x) = l_1.$$

Now consider $\varepsilon = (l_1 - l_2)/2$. Let N_i $(i = 1, 2)$ be a neighborhood of a with
the property:

$$x \in N_i{}^* \Rightarrow f_i(x) \in N_{l_i, \varepsilon}$$

(since $\lim_{x \to a} f_i(x)$ exists, we know that N_i exists). The set $N_1 \cap N_2$ is a neighborhood of a and the reader can verify that

$$(N_1 \cap N_2)^* = N_1{}^* \cap N_2{}^*.$$

Now $x \in (N_1 \cap N_2)^*$ implies

$$f_1(x) \in N_{l_1, \varepsilon} \text{ and } f_2(x) \in N_{l_2, \varepsilon},$$

or

$$|f_1(x) - l_1| < \varepsilon \text{ and } |f_2(x) - l_2| < \varepsilon. \text{ (See Figure 12–8.)}$$

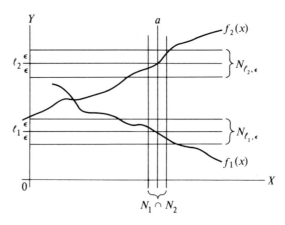

Figure 12–8

Thus by the definition of ε, if $x \in (N_1 \cap N_2)^*$, then

$$f_1(x) > l_1 - \varepsilon = l_1 - \frac{l_1 - l_2}{2} = \frac{l_1 + l_2}{2}$$

and

$$f_2(x) < l_2 + \varepsilon = l_2 + \frac{l_1 - l_2}{2} = \frac{l_1 + l_2}{2}.$$

Therefore

$$f_2(x) < \frac{l_1 + l_2}{2} < f_1(x).$$

Since this is contrary to the hypothesis of the theorem, the theorem is proved.

Remark 3. The hypothesis of this theorem, that $f_1(x) \leqslant f_2(x)$, clearly includes the case in which $f_1(x) < f_2(x)$.

12.2 INFINITE LIMITS

If we examine the graph of the function $x \to 1/x$ we see that as x becomes very large in absolute value and positive then $1/x$ becomes very small. In fact, merely by taking x sufficiently large, we can make $1/x$ as small as we please. We write in this case $\lim_{x \to \infty} 1/x = 0$. More generally, if $f(x)$ is arbitrarily close to a number l whenever x exceeds a properly chosen number, then we write $\lim_{x \to \infty} f(x) = l$. We say that the limit as $x \to \infty$ of $f(x)$ is l.

A similar situation arises for the function $x \to 1/x$ if we take x very large in absolute value and negative. We see then that $1/x$ becomes very small, and that the larger we take x in absolute value, the smaller $1/x$ becomes in absolute value. We write $\lim_{x \to -\infty} 1/x = 0$. More generally, if $f(x)$ becomes arbitrarily close to a number l as x is taken negative but sufficiently large in absolute value, then we write $\lim_{x \to -\infty} f(x) = l$. We say that the limit as $x \to -\infty$ of $f(x)$ is l.

The symbols ∞, $-\infty$ do not stand for numbers in R; they are parts of the expression "$\lim_{x \to \infty} f(x)$" which is being defined. We define these limits more precisely as follows:

Definition 1. $\lim_{x \to \infty} f(x) = l$ *if for every neighborhood M of l there is a natural number n such that* $x > n \Rightarrow f(x) \in M$.

Definition 2. $\lim_{x \to -\infty} f(x) = l$ *if for every neighborhood M of l there is a natural number n such that*

$$x < -n \Rightarrow f(x) \in M.$$

(see Figures 12–9(a) and 12–9(b) for a graphical interpretation of Definitions 1 and 2.)

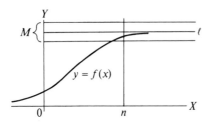
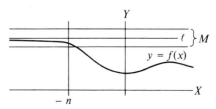

Figure 12–9a $\lim\limits_{x \to \infty} f(x) = l$;   **Figure 12–9b** $\lim\limits_{x \to -\infty} f(x) = l$

It is easy to show that Theorems 2 and 3 in 12.1 remain valid when a is replaced by ∞ or $-\infty$. The reader should attempt to show this as an exercise.
If

(1) $\forall m \in N \; \exists n \in N \; [x > n \Rightarrow f(x) > m]$,

then we say $\lim\limits_{x \to \infty} f(x) = \infty$.
If, on the other hand,

(2) $\forall m \in N \; \exists n \in N \; [x > n \Rightarrow -f(x) > m]$,

we say $\lim\limits_{x \to \infty} f(x) = -\infty$.
Similar definitions can be made for

$$\lim_{x \to -\infty} f(x) = \pm \infty \quad \text{and} \quad \lim_{x \to a} f(x) = \pm \infty.$$

Example 1. $\lim\limits_{x \to \infty} x^2 = \infty$. Indeed, if m is an arbitrary natural number, then for $n = [\sqrt{m}] + 1$

$$x > n \Rightarrow x > \sqrt{m} \Rightarrow x^2 > m.$$

Thus
$$\forall m \in N \; \exists n \in N \; [x > n \Rightarrow x^2 > m]$$

so we can say $\lim\limits_{x \to \infty} x^2 = \infty$.

Example 2. $\lim\limits_{x \to \infty} \exp x = \infty$. Indeed since

$$\exp x = \sum_{n=0}^{\infty} \frac{x^n}{n!},$$

for each $x > 0$
$$\exp x > 1 + x.$$

Now given $m \in N$ we choose $n = m - 1$. Then if $x > n$, $\exp x > 1 + x > 1 + n = m$, so that if $x > n$, $\exp x > m$. Thus by (1) $\lim\limits_{x \to \infty} \exp x = \infty$. This agrees with the graph in Figure 11–4.

The reader should note that:

$$(3) \qquad\qquad \text{if} \quad \lim_{x \to +\infty} f(x) = \pm\, \infty, \quad \text{then} \quad \lim_{x \to \infty} \frac{1}{f(x)} = 0.$$

Indeed, if $\lim\limits_{x \to +\infty} f(x) = \pm\, \infty$, then for any $m > 0$ (no matter how large) there is an n such that

$$(4) \qquad\qquad x > n \Rightarrow |f(x)| > m.$$

Now for an arbitrary $\varepsilon > 0$ choose $m = [1/\varepsilon] + 1$. Then, by the above remarks there exists an n such that $x > n \Rightarrow |f(x)| > [1/\varepsilon] + 1$. That is,

$$x > n \Rightarrow \frac{1}{|f(x)|} = \left| \frac{1}{f(x)} - 0 \right| < \frac{1}{[1/\varepsilon] + 1} < \frac{1}{1/\varepsilon} = \varepsilon,$$

and so

$$\lim_{x \to +\infty} \frac{1}{f(x)} = 0.$$

As an extension of Theorem 2 of 12.1 to the cases in which $|x|$ increases without limit, we can state the following theorem. The reader should supply the proofs, which are similar to the proofs of Theorem 2 of 12.1.

Theorem 1. *If* $\lim\limits_{x \to \infty} f(x) = l$ *and* $\lim\limits_{x \to \infty} g(x) = m$ *and* b *is some real number, then*

(i) $\lim\limits_{x \to \infty} bf(x) = bl,$

(ii) $\lim\limits_{x \to \infty} (f(x) + g(x)) = l + m,$

(iii) $\lim\limits_{x \to \infty} (f(x) \cdot g(x)) = l \cdot m.$

(iv) *If, in addition to the preceding hypotheses,* $m \neq 0$, *then*

$$\lim_{x \to \infty} \left(\frac{f(x)}{g(x)} \right) = \frac{l}{m}.$$

The theorem is equally valid if $+\infty$ is replaced by $-\infty$ throughout.

Example 3. $\lim\limits_{x \to \infty} 1/x^2 = 0$. We know from Example 1 that $\lim\limits_{x \to \infty} x^2 = \infty$, and so by (3) $\lim\limits_{x \to \infty} 1/x^2 = 0$.

The reader can easily verify by arguments similar to Examples 1 and 3 that

(5)
$$\forall n \in N - \{0\} \left[\lim_{x \to \infty} \frac{1}{x^n} = 0 \right].$$

Example 4. $\lim\limits_{x \to -\infty} \exp x = 0$. From Chapter 11, we have

$$\exp x = \frac{1}{\exp(-x)}.$$

Thus

$$\lim_{x \to -\infty} \exp x = \lim_{y \to +\infty} \frac{1}{\exp y}.$$

Therefore by (3)

$$\lim_{x \to -\infty} \exp x = 0. \quad \text{(cf. Figure 11–4)}$$

Example 5. Consider the behavior of

$$f(x) = \frac{2x^3 + x^2 - x + 1}{5x^3 - 1} \quad \text{as } x \to +\infty.$$

Note that both numerator and denominator become large as $x \to +\infty$ so that it is not clear whether the limit exists or not. However, for $x \neq 0$,

$$f(x) = \frac{2 + 1/x - 1/x^2 + 1/x^3}{5 - 1/x^3},$$

and so

$$\lim_{x \to \infty} f(x) = \lim_{x \to \infty} \frac{2 + 1/x - 1/x^2 + 1/x^3}{5 - 1/x^3}.$$

By Theorem 1,

$$\lim_{x \to \infty} f(x) = \frac{\lim\limits_{x \to \infty} 2 + \lim\limits_{x \to \infty} 1/x - \lim\limits_{x \to \infty} 1/x^2 + \lim\limits_{x \to \infty} 1/x^3}{\lim\limits_{x \to \infty} 5 - \lim\limits_{x \to \infty} 1/x^3},$$

and by (5)

$$\lim_{x \to \infty} f(x) = \frac{2 + 0 - 0 + 0}{5 - 0} = \frac{2}{5}.$$

In general, as the reader can show,

$$\lim_{x \to \infty} \frac{f(x)}{g(x)} = 0 \quad \text{and} \quad \lim_{x \to -\infty} \frac{f(x)}{g(x)} = 0$$

whenever

$$f(x) = a_0 + a_1 x + \cdots + a_{n-1} x^{n-1} + x^n,$$
$$g(x) = b_0 + b_1 x + \cdots + b_{m-1} x^{m-1} + x^m$$

and $m > n$, that is, whenever the numerator and denominator of the fraction $f(x)/g(x)$ are polynomials and the degree of the denominator is greater than that of the numerator.

12.3   LIMIT PROPERTIES OF exp x, $c(x)$, $s(x)$

The following theorem will prove useful in the discussion of the derivative of exp x in Chapter 13.

Theorem 1.

$$\lim_{x \to 0} \frac{\exp x - 1}{x} = 1.$$

Proof.

$$\left(\frac{\exp x - 1}{x} - 1\right) = \lim_n \left(\frac{1 + x + x^2/2! + \cdots + x^n/n! - 1}{x} - 1\right)$$

$$= \lim_n \left(\frac{x}{2!} + \frac{x^2}{3!} + \cdots + \frac{x^n}{(n + 1)!}\right).$$

Now

$$\left|\frac{\exp x - 1}{x} - 1\right| = \left|\lim_n \left(\frac{x}{2!} + \frac{x^2}{3!} + \cdots + \frac{x^n}{(n + 1)!}\right)\right|$$

$$= \lim_n \left|\frac{x}{2!} + \frac{x^2}{3!} + \cdots + \frac{x^n}{(n + 1)!}\right|.$$

(Exercise 16 of 11.9).
Now for $|x| < 1$, by the triangle inequality,

$$\left|\frac{x}{2!} + \frac{x^2}{3!} + \cdots + \frac{x^n}{(n + 1)!}\right| \leq \frac{|x|}{2!} + \frac{|x|^2}{3!} + \cdots + \frac{|x|^n}{(n + 1)!}$$

$$\leq |x| + |x|^2 + \cdots + |x|^n$$

$$= \frac{|x|}{1 - |x|} - \frac{|x|^{n+1}}{1 - |x|}$$

$$\leq \frac{|x|}{1 - |x|}.$$

Hence, by Exercise 15(c) of 11.9, if $|x| < 1$,

$$\left| \frac{\exp x - 1}{x} - 1 \right| \leqslant \lim_n \frac{|x|}{1 - |x|} = \frac{|x|}{1 - |x|}.$$

Thus

$$\lim_{x \to 0} \left| \frac{\exp x - 1}{x} - 1 \right| = 0,$$

so that

$$\lim_{x \to 0} \left(\frac{\exp x - 1}{x} - 1 \right) = 0,$$

and Theorem 1 follows.

A corresponding theorem for the function $c(x)$ is also valid:

Theorem 2.

$$\lim_{x \to 0} \frac{c(x) - 1}{x} = 0.$$

Proof.

$$\frac{c(x) - 1}{x} = \lim_n \left(-\frac{x}{2!} + \frac{x^3}{4!} - \cdots + \frac{(-1)^n x^{2n-1}}{(2n)!} \right).$$

Hence

$$\left| \frac{c(x) - 1}{x} \right| = \left| \lim_n \left(-\frac{x}{2!} + \frac{x^3}{4!} - \cdots + \frac{(-1)^n x^{2n-1}}{(2n)!} \right) \right|$$

$$= \lim_n \left| -\frac{x}{2!} + \frac{x^3}{4!} - \cdots + \frac{(-1)^n x^{2n-1}}{(2n)!} \right|,$$

(Exercise 16 of 11.9).
Now

$$\left| -\frac{x}{2!} + \frac{x^3}{4!} - \cdots + \frac{(-1)^n x^{2n-1}}{(2n!)} \right| \leqslant \frac{|x|}{2!} + \frac{|x|^3}{4!} + \cdots + \frac{|x|^{2n-1}}{(2n)!}$$

$$\text{(triangle inequality)}$$

$$\leqslant |x| + |x|^3 + \cdots + |x|^{2n-1}$$

$$= \frac{|x|}{1 - |x|^2} - \frac{|x|^{2n+1}}{1 - |x|^2}$$

$$\leqslant \frac{|x|}{1 - |x|^2}$$

for all n, $|x| < 1$.

Hence, by Exercise 15(c) of 11.9 we have

$$\left| \frac{c(x) - 1}{x} \right| \leq \lim_n \frac{|x|}{1 - |x|^2} = \frac{|x|}{1 - |x|^2}$$

for $|x| < 1$.
Thus

$$\lim_{x \to 0} \left| \frac{c(x) - 1}{x} \right| = 0, \quad \text{and so} \quad \lim_{x \to 0} \frac{c(x) - 1}{x} = 0.$$

The following theorem will be left as an exercise for the reader.

Theorem 3. *For $s(x)$ defined in 11.8*

$$\lim_{x \to 0} \frac{s(x)}{x} = 1.$$

12.4 CONTINUITY

Intuitively speaking, we think of a function as continuous if its graph has no breaks in it. In most instances, this is easy to ascertain by inspecting the graph of the function, but in some cases this method is not easy to apply. Consider, for example, the graph of f where

$$(1) \qquad f(x) = \begin{cases} x \sin 1/x & \text{for } x \neq 0, \\ 0 & \text{for } x = 0 \text{ (see Figure 12--5)}. \end{cases}$$

Most readers would feel intuitively that $x \sin 1/x$ is continuous so long as $x \neq 0$, but there might be some question concerning the continuity of f at $x = 0$. In order that no break in the curve occur at $x = 0$, we would wish $f(x)$ to approach $f(0) = 0$ as x approaches 0. By 12.1, Example 5, this is in fact the case, so we might intuitively assert that $f(x)$ is continuous at $x = 0$, as well as elsewhere.

These preliminary considerations lead to the following definition.

Definition 1. *A function f defined in a neighborhood N_a of a point a is continuous at a if*

$$\lim_{x \to a} f(x) = f(a),$$

or, equivalently, if

$$\lim_{h \to 0} f(a + h) = f(a).$$

Note that Definition 1 demands (i) that f is defined at a, and (ii) that

$\lim\limits_{x \to a} f(x)$ exists and is equal to $f(a)$. For example, the function f, where

$$f(x) = \frac{x^2 - 1}{x - 1},$$

discussed in 12.1 Example 1, is not continuous at $x = 1$ because $f(1)$ is undefined. However, the function f with

$$f(x) = \frac{x^2 - 1}{x - 1}, \quad x \ne 1,$$

$$f(x) = 2, \qquad\qquad x = 1,$$

is continuous at $x = 1$ since

$$\lim\limits_{x \to 1} f(x) = 2 = f(1).$$

The function s where

(2)
$$s(x) = \begin{cases} 0, & x < 0, \\ 1, & x \geqslant 0 \end{cases}$$

of Example 2 in 12.1 is not continuous because $\lim\limits_{x \to 0} s(x)$ does not exist.

Let us return now to $x \sin 1/x$. From Example 5 of 12.1, it follows that

$$\lim\limits_{x \to 0} x \sin \frac{1}{x} = 0,$$

and so by Definition 1 the function f defined by (1) is continuous at $x = 0$.
However, note that the function g where

$$g(x) = \begin{cases} x \sin 1/x & \text{for } x \ne 0 \\ 2 & \text{for } x = 0 \end{cases}$$

is not continuous at $x = 0$, since $\lim\limits_{x \to 0} g(x) = 0$ but $\lim\limits_{x \to 0} g(x) \ne g(0)$.

A discontinuity like that of g which can be removed by merely redefining the function at the point is called a **removable discontinuity**. The discontinuity of the function s in (2) is obviously not removable.

Theorem 1. *A polynomial is continuous at every point of R and a rational function is continuous at every point of its domain.*

Proof. First note that for any $n \in N$

$$\lim\limits_{x \to a} x^n = a^n$$

for any $a \in R$ by Corollary 2 in 12.1. Similarly

$$\lim_{x \to 0} \sum_{k=0}^{n} a_k x^k = \sum_{k=0}^{n} a_k \lim_{x \to a} x^k \quad \text{(by Corollary 2 in 12.1)}$$

$$= \sum_{k=0}^{n} a_k a^k.$$

Thus for any polynomial $f(x)$

$$\lim_{x \to a} f(x) = f(a)$$

for any $a \in R$, and $f(x)$ is continuous.

Denote the value of the rational function by $f(x)/g(x)$ where f and g are polynomials. Its domain is the set R with the zeroes of $g(x)$ deleted. Suppose that a belongs to this domain. Then $g(a) \neq 0$, and so, by Theorem 2 part (iv) of 12.1,

$$\lim_{x \to a} \frac{f(x)}{g(x)} = \frac{\lim_{x \to a} f(x)}{\lim_{x \to a} g(x)}.$$

Then, since polynomials are continuous at every point of R,

$$\lim_{x \to a} \frac{f(x)}{g(x)} = \frac{f(a)}{g(a)}.$$

Therefore any rational function is continuous except at zeros of the denominator. An example of a function which is discontinuous everywhere is the following.

Example 1. Let f be the function defined by

$$f(x) = 1, \quad x \in Q,$$
$$f(x) = 0, \quad x \notin Q. \quad \text{(see Figure 12–10)}$$

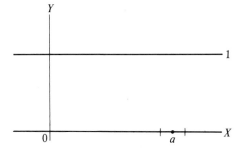

Figure 12–10 The function f where $f(x) = 1$, $x \in Q$, $f(x) = 0$, $x \notin Q$

This function is discontinuous everywhere. We prove this as follows:

Let a be any real number and let ε be such that $0 < \varepsilon < \frac{1}{2}$. Since both the rationals and the irrationals are everywhere dense *any* neighborhood of a contains some $x \neq a$ such that $f(x) = 0$ and some such that $f(x) = 1$. Therefore whether $a \in Q$ or $a \notin Q$ any punctured neighborhood of a contains points for which $|f(x) - f(a)| \not< \varepsilon$. Consequently $f(x)$ is not continuous at a.

12.5 CONTINUITY ON AN INTERVAL

Definition 1. *A function f is **continuous** on an open interval* (a, b) *if it is continuous at each point of* (a, b).

Example 1. As an example of this definition let us prove that the function f defined by

$$f(x) = \sqrt{x}, \quad x \in (0, 1)$$

is continuous on $(0, 1)$.

Let $a \in (0, 1)$, $x \in (0, 1)$; then

$$|f(x) - f(a)| = |\sqrt{x} - \sqrt{a}| = \frac{|x - a|}{|\sqrt{x} + \sqrt{a}|} < \frac{|x - a|}{\sqrt{a}}.$$

Suppose $\varepsilon > 0$, and let $\delta = \varepsilon\sqrt{a}$. Then if $0 < |x - a| < \delta$,

$$|f(x) - f(a)| < \frac{\varepsilon\sqrt{a}}{\sqrt{a}} = \varepsilon.$$

Consequently $\lim_{x \to a} f(x) = f(a)$, and $f(x)$ is continuous at $x = a$. Hence by Definition 1, f is continuous on $(0, 1)$.

Suppose we wished to investigate the continuity of f (in Example 1) on the closed interval $[0, 1]$. We would naturally try to extend the proof given in Example 1 by proving that f is continuous at $x = 0$ and $x = 1$. However, Definition 1 of 12.4 cannot be applied at $x = 0$ since it involves neighborhoods that extend to the right and left of $x = 0$, and f is not defined if $x < 0$. Thus, in general, Definition 1 of 12.4 cannot be applied at the end point of the domain of a function. To handle situations like this we introduce the concepts of left-hand and right-hand continuity.

Definition 2. *Let $x = a$ be in the domain of a function f. The function f is **continuous to the right** at $x = a$ if* $\lim_{x \to a^+} f(x) = f(a)$; *the function f is **continuous to the left** at $x = a$ if* $\lim_{x \to a^-} f(x) = f(a)$.

It seems reasonable, from Definitions 2 and 1 of 12.4, that if a function is continuous to the right and continuous to the left at a point then the function is continuous at the point. We shall prove this shortly.

Definition 3. *A function f is **continuous** on a closed interval $[a, b]$ if it is continuous on the open interval (a, b), continuous to the right at $x = a$, and continuous to the left at $x = b$.*

We shall now apply Definition 3 to prove that the function f where $f(x) = \sqrt{x}$ is continuous on $[0, 1]$. Let us first of all prove that $\lim\limits_{x \to 0+} \sqrt{x} = f(0)$. We have $|f(x) - f(0)| = |\sqrt{x} - \sqrt{0}| = \sqrt{x}$. If $0 < \varepsilon < 1$, $\varepsilon^2 < \varepsilon$.

Take $\delta = \varepsilon^2$; then if $|x| < \delta = \varepsilon^2$, $\sqrt{x} < \sqrt{\delta} = \varepsilon$.
Therefore $f(x) < \varepsilon$ if $0 \leqslant x \leqslant \delta$, so that

(1) $$\lim_{x \to 0+} \sqrt{x} = 0 = f(0).$$

Similarly it may be proved that

(2) $$\lim_{x \to 1-} \sqrt{x} = 1 = f(1).$$

Results (1) and (2) and the result of Example 1, that is, that f is continuous on $(0, 1)$, prove that f is continuous on $[0, 1]$. (See Figure 12–11.)

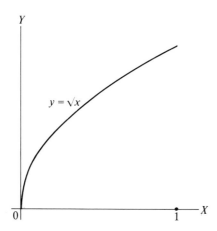

Figure 12–11   The function $\sqrt{x}$ is continuous on $[0, 1]$

The step function s defined as equal to 0 when $x < 0$ and equal to 1 when $x \geqslant 0$ is continuous on the interval $[0, 1]$ but not continuous on $[-1, 0]$

because $\lim\limits_{x \to 0^-} s(x) = 0 \neq s(0)$. In addition, s is not continuous on R because s is not continuous at $x = 0 \in R$.

We now prove the following Lemma.

Lemma.   *A function f is continuous at $x = a$ if and only if f is continuous to the left and continuous to the right at $x = a$.*

Proof.   Suppose f is continuous at $x = a$.

$$\text{Then } |f(x) - f(a)| < \varepsilon \text{ if } 0 < |x - a| < \delta,$$

that is, if $a < x < a + \delta$ and $a - \delta < x < a$. Hence f is continuous to the right and to the left at $x = a$. Similarly if f is continuous to the right and to the left at $x = a$ then f is continuous at $x = a$.

12.6 CONTINUITY AND LIMIT PROPERTIES OF SINE AND COSINE

Theorem 1.   *The sine function is continuous at every point of R.*

Proof.   We first show that the sine and cosine are continuous at $x = 0$. The circle in Figure 12–12 has unit radius; suppose the angles are measured in radians.

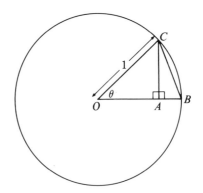

Figure 12–12   Proof of continuity of sine

If $|\theta| < \tfrac{1}{2}\pi$ we have

$$|\sin \theta| = AC < BC < \text{arc } BC = |\theta|.$$

By Theorem 3 of 12.1 and the result $\lim_{\theta \to 0} \theta = 0$, we have

$$\lim_{\theta \to 0} |\sin \theta| \leqslant \lim_{\theta \to 0} |\theta| = 0$$

and

(1) $$\lim_{\theta \to 0} \sin \theta = 0.$$

Similarly, if $|\theta| < \frac{1}{2}\pi$, we have

$$|\cos \theta - 1| = AB < BC < \text{arc } BC = |\theta|.$$

Hence   ·

$$\lim_{\theta \to 0} |\cos \theta - 1| \leqslant \lim_{\theta \to 0} |\theta| = 0$$

and

(2) $$\lim_{\theta \to 0} \cos \theta = 1.$$

In this proof we must appeal to geometry since we have not as yet strictly defined the sine and cosine functions.

Now let x_0 be an arbitrary real number.

Set $\theta = x - x_0$, then

$$\lim_{x \to x_0} (x - x_0) = \lim_{\theta \to 0} \theta = 0$$

and

$$\lim_{x \to x_0} \sin \theta = 0, \; \lim_{x \to x_0} \cos \theta = 1.$$

Hence

$$\lim_{x \to x_0} \sin x = \lim_{x \to x_0} \sin [(x - x_0) + x_0]$$

$$= \lim_{x \to x_0} \sin (\theta + x_0)$$

$$= \lim_{x \to x_0} [\sin \theta \cos x_0 + \cos \theta \sin x_0]$$

$$= 0 \cdot \cos x_0 + 1 \cdot \sin x_0$$

$$= \sin x_0.$$

Therefore sine is continuous at x_0, and since x_0 is arbitrary, sine is continuous on R.

A similar proof shows that cosine is continuous on R.

Corollary 1. *Cosine is continuous on R.*

Proof. For an arbitrary point $x_0 \in R$,

$$\cos x = \cos (x - x_0 + x_0) = \cos (x - x_0) \cos x_0 - \sin (x - x_0) \sin x_0.$$

Therefore from Theorem 2 in 12.1 and (1) and (2),

$$\lim_{x \to x_0} \cos x = \lim_{x \to x_0} \cos (x - x_0) \cos x_0 - \lim_{x \to x_0} \sin (x - x_0) \sin x_0$$

$$= 1 \cdot \cos x_0 + 0 \cdot \sin x_0$$

$$= \cos x_0.$$

The following result is useful for finding the derivatives of trigonometric functions in Chapter 13.

Theorem 2.

$$\lim_{x \to 0} \frac{\sin x}{x} = 1.$$

Proof. If the circle in Figure 12–13 is a unit circle, then it is clear that for $0 < |x| < \pi/2$:

(3)   Area of triangle $COB <$ area of sector $COA <$ area of triangle DOA,

(4)   Area of triangle $COB = \frac{1}{2}|\sin x| \cos x$,

(5)   Area of sector $COA = \frac{1}{2}|x|$,

(6)   Area of triangle $DOA = \frac{1}{2}|\tan x|$.

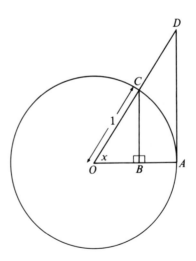

Figure 12–13

Thus

(7) $$\tfrac{1}{2}|\sin x|\cos x < \tfrac{1}{2}|x| < \tfrac{1}{2}|\tan x|.$$

Multiplying (7) by $2/|\sin x|$ we have

(8) $$\cos x < \frac{|x|}{|\sin x|} < \frac{1}{\cos x}.$$

Since $0 < a < b \Rightarrow 1/b < 1/a$, (8) becomes

(9) $$\frac{1}{\cos x} > \left|\frac{\sin x}{x}\right| > \cos x.$$

However, for $0 < |x| < \pi/2$,

$$\left|\frac{\sin x}{x}\right| = \frac{\sin x}{x}, \quad \text{so} \quad \frac{1}{\cos x} > \frac{\sin x}{x} > \cos x.$$

Now, by Corollary 1, $\lim_{x \to 0} \cos x = 1$, and so by Theorems 2 and 3 of 12.1,

$$1 = \lim_{x \to 0} \frac{1}{\cos x} \geqslant \lim_{x \to 0} \frac{\sin x}{x} \geqslant \lim_{x \to 0} \cos x = 1.$$

Thus

$$\lim_{x \to 0} \frac{\sin x}{x} = 1.$$

12.7   SOME PROPERTIES OF CONTINUOUS FUNCTIONS

In this section we establish some properties of continuous functions that will be useful in the sequel. We first show that a continuous function of a continuous function is continuous or, in other words, that the composition of continuous functions is a continuous function.

Theorem 1.   *If f is a continuous function with domain $[a, b]$ and range $[c, d]$ and if g is a continuous function with domain containing $[c, d]$, then $g \circ f$ is a continuous function.*

Proof.   Let $x_0 \in [a, b]$. By hypothesis there is a $y_0 \in [c, d]$ such that $f(x_0) = y_0$. Since $y_0 \in [c, d]$, for every neighborhood M of $g(y_0)$, there is a neighborhood $N(M)$ of y_0 such that $[y \in (N(M) \cap (c, d))] \Rightarrow [g(y) \in M]$.

Now $N(M)$ is a neighborhood of $f(x_0)$, and so there is a neighborhood V of x_0 such that by the continuity of f and g

$$[x \in V^* \cap (a, b)] \Rightarrow [y_0 = f(x_0) \in (N(M) \cap (c, d))]$$
$$\Rightarrow g \circ f(x_0) = g(y_0) = g(f(x_0)) \in M.$$

Therefore by Exercise 22 in 12.10, $g \circ f$ is continuous on $[a, b]$.

From this theorem it is clear that $f(x) = \sin 1/x$ is continuous on every closed interval that excludes 0 because $\sin x$ is continuous everywhere on R and $1/x$ is continuous on every interval excluding 0 (Theorem 1 of 12.4). Similarly $\sqrt{(1 - x^2)}$ is continuous on $[-1, 1]$ since we have proved in Example 1 of 12.5 that $\sqrt{x}$ is continuous on $[0, 1]$, and by Theorem 1 of 12.4, $(1 - x^2)$ is continuous on R.

The following pair of theorems tie in with some of the results of Chapter 9.

Theorem 2.   *The set $C[a, b]$ of all continuous functions on the interval $[a, b]$ forms a ring with respect to the operations*

(1) $$(f + g)(x) = f(x) + g(x),$$

(2) $$(f \cdot g)(x) = f(x) \cdot g(x),$$

defined in Chapter 9.

The proof follows easily from Theorem 2 of 12.1.

Theorem 3.   *The set C composed of all continuous functions defined on R forms a ring with respect to the pair of operations "$+$", as defined in (1), and "composition"*

(3) $$f \circ g(x) = f(g(x)).$$

The proof follows from Theorem 2 of 12.1 and from Theorem 1.

Note that $C[a, b]$, which is the set of all functions continuous on the bounded set $[a, b]$, does *not* form a ring with respect to "$+$" defined in (1) and the composition operation, because $C[a, b]$ is not closed under the composition operation. Indeed, if $c \notin [a, b]$, then if $f(x) = c$ for all $x \in [a, b]$, $f \in C[a, b]$, but $f \circ f(x) = f(f(x))$ is undefined.

Note also that C provides another example of a noncommutative ring. Indeed, both f and g belong to C, where

$$f(x) = \sin x \quad \text{and} \quad g(x) = x^2$$

and
$$f \circ g(x) = f(g(x)) = \sin x^2$$
while
$$g \circ f(x) = g(f(x)) = (\sin x)^2.$$

Since $(\sin x)^2 \neq \sin x^2$, $f \circ g \neq g \circ f$.

The following theorem indicates how the density of the rational numbers in the real numbers interacts with the concept of continuity. Intuitively the theorem says that a continuous function is determined completely by its values on the rational numbers of its domain.

Theorem 4. *Suppose that the two functions f and g are continuous on (a, b). If $f(r) = g(r)$ for all $r \in (a, b) \cap Q$, then*
$$f(x) = g(x)$$
for all $x \in (a, b)$.

Proof. Suppose that there is an $x_0 \in (a, b)$ such that $f(x_0) \neq g(x_0)$. The density of Q in R implies that there is a sequence $\{r_n\}$ of rationals such that $\lim_n r_n = x_0$. Since f and g are continuous,
$$\lim_{x \to x_0} f(x) = f(x_0) \quad \text{and} \quad \lim_{x \to x_0} g(x) = g(x_0).$$

Thus we have the following three statements: If $R^+ = \{x \in R \,|\, x > 0\}$,

(4) $\forall \delta \in R^+ \; \exists n_0 \in N \; [n > n_0 \Rightarrow |r_n - x_0| < \delta]$,

(5) $\forall \varepsilon \in R^+ \; \exists \delta \in R^+ \; [|x - x_0| < \delta \Rightarrow |f(x) - f(x_0)| < \varepsilon]$,

(6) $\forall \varepsilon \in R^+ \; \exists \delta \in R^+ \; [|x - x_0| < \delta \Rightarrow |g(x) - g(x_0)| < \varepsilon]$.

Now let ε be chosen so that
$$0 < \varepsilon < \tfrac{1}{2}|f(x_0) - g(x_0)|$$
and let δ be chosen so that

(7) $|x - x_0| < \delta \Rightarrow (|f(x) - f(x_0)| < \varepsilon) \wedge (|g(x) - g(x_0)| < \varepsilon)$.

Now by (4) there are terms r_{n_1} in the sequence $\{r_n\}$ such that $|r_{n_1} - x_0| < \delta$, and so, by (7),
$$|f(r_{n_1}) - f(x_0)| < \varepsilon \quad \text{and} \quad |g(r_{n_1}) - g(x_0)| < \varepsilon.$$

Thus by the triangle inequality
$$|f(x_0) - f(r_{n_1}) + g(r_{n_1}) - g(x_0)| \leqslant |f(r_{n_1}) - f(x_0)| + |g(r_{n_1}) - g(x_0)| < 2\varepsilon.$$

However, $f(r_{n_1}) = g(r_{n_1})$, and so

$$|f(x_0) - g(x_0)| < 2\varepsilon < |f(x_0) - g(x_0)|.$$

This contradiction indicates that the assumption $g(x_0) \neq f(x_0)$ for some $x_0 \in (a, b)$ is inadmissible, and the theorem follows.

In Example 1 of 12.4, we showed that the function f such that

$$f(x) = \begin{cases} 1, x \in Q \\ 0, x \notin Q \end{cases}$$

is discontinuous; actually this result follows immediately from Theorem 4. Indeed, the constant function $g\colon g(x) = 1$ for $x \in R$ is continuous and $f(x) = g(x)$ if $x \in Q$. Now if f were continuous, we would have $f(x) = g(x)$ for all $x \in R$, which is not so since at an irrational value of x, $f(x) = 0$ and $g(x) = 1$. Thus f is not continuous.

Now suppose that $[a, b]$ is a closed interval and f is a function defined on $Q \cap [a, b]$ such that for every $\varepsilon > 0$ there is a $\delta > 0$ with the property

(8) $(|r_1 - r_2| < \delta) \wedge (r_1, r_2 \in Q \cap [a, b]) \Rightarrow |f(r_1) - f(r_2)| < \varepsilon.$

By arguments similar to those employed in the proof of Theorem 4, we can now show that there is a unique g with domain $[a, b]$ which is equal to f on $Q \cap [a, b]$ and is continuous on $[a, b]$. Roughly speaking the proof goes as follows: if x is an irrational point in $[a, b]$, then there is a rational sequence $\{r_n\}$ such that $r_n \to x$. Then we can show that $\{f(r_n)\} = \{g(r_n)\}$ is a convergent sequence by the Cauchy Convergence Criterion. We define $g(x) = \lim_n g(r_n)$.

Of course it must be shown that $g(x)$ is independent of the choice of sequence $\{r_n\}$ of rationals chosen such that $r_n \to x$. If x is rational, we have by definition $g(x) = f(x)$. Finally using (8), we can show g is continuous on $[a, b]$.

12.8   EXPONENTIAL AND LOGARITHM

Let us now apply the process discussed at the end of the preceding section to the exponential function. This function, which will be of the form $f\colon x \to a^x$ for some positive real number a, can easily be defined for all rational x (see Chapter 8 and 11.8). The problem arises, however, to define a^x for irrational values of x.

Since irrational numbers can be obtained from rationals only as limits, we must employ a limiting process. It can be shown that, for rational x in any closed interval $[a, b]$, property (8) of 12.7 holds for $f\colon x \to a^x$. Therefore using the development described above, we can, by a natural limiting process, extend the function f to a function g so that g agrees with f on $[a, b] \cap Q$ for any $[a, b]$, *and* g is continuous on any closed interval $[a, b]$ in R.

This extension g of f is what we define to be the exponential function $x \rightarrow a^x$.

In Theorem 3 of 11.8, we showed that for every $q \in Q$,

$$\exp q = e^q.$$

By the above argument, we can take the function $x \rightarrow e^x$ to be a function which is continuous in any interval $[a, b]$ (and hence at every point) in R. In Exercise 21 of 12.10, you are asked to show that $\exp x$ is continuous on R, and so by Theorem 4 of 12.7, we can conclude that for every $x \in R$

$$\exp x = e^x.$$

Since the exponential function a^x ($a > 0$) with domain R has now been defined, we shall establish the following theorem.

Theorem 1. *If* $a > 1$, $x, y \in R$, *then* $x < y \Rightarrow a^x < a^y$.

Proof. It has been established in Corollary 2 of 11.8, that

(1) $$x < y \Rightarrow \exp x < \exp y.$$

We now show that a^x can be written as e^{px} for some $p \in R$ as follows: By Exercise 18 of 12.10, there exists a real number $p > 0$ such that $a = e^p$. Now if $x \in Q$, $(e^p)^x = e^{px}$ (Exercise 17 of 12.10). Thus $a^x = (e^p)^x = e^{px}$ where $x \in Q$.

Since both $x \rightarrow a^x$ and $x \rightarrow e^{px}$ are continuous functions which agree on Q and $\exp x = e^x$ for all $x \in R$, we have $a^x = e^{px} = \exp(px)$ for all $x \in R$.

Now (1) can be written as

$$px < py \Rightarrow e^{px} < e^{py}.$$

Hence

$$x < y \Rightarrow a^x < a^y.$$

Thus the exponential function is monotone increasing, and the mapping $x \rightarrow a^x$ is one-to-one.

The result of Theorem 1 and of Definition 4 of 7.8 enables us to say that the exponential function $x \rightarrow a^x$, where $x \in R$, has an inverse function which we call the logarithm to the base a, and we write $x = \log_a (a^x)$.

Definition 1. *The exponential function* $x \rightarrow a^x$, *where* $x \in R$ *and* $a > 1$, *has an inverse function called the* **logarithm to the base** a, *such that*

$$y = a^x \Leftrightarrow x = \log_a y.$$

There are two common choices for a: $a = 10$ which gives common or Briggsian logarithms, and $a = e = 2.71828...$ which gives natural or Napierian logarithms.

Usually if the base is unspecified, natural logarithms are meant, although sometimes these are written as $\ln x$. By Lemma 1 of 9.10 and Figure 11–4, the graph of the logarithm to the base a can be obtained and is shown in Figure 12–14 for three different values of a. It is seen that the domain is the set $x \in R$, $x > 0$ and the range is the set $y \in R$.

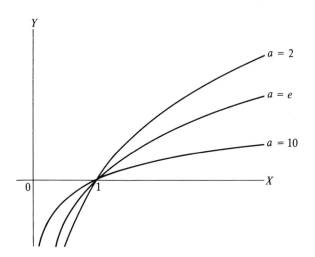

Figure 12–14   Graph of $x \rightarrow \log_a x$ for $a = 10$, $a = e$, $a = 2$

Some of the simplest properties of the logarithm are given in the following theorem.

Theorem 2.   *The logarithm to the base a has the following properties: If a > 1,*

(2) $\log_a a = 1$,

(3) $\log_a 1 = 0$,

(4) $\lim_{x \to \infty} \log_a x = +\infty$,

(5) $\lim_{x \to 0+} \log_a x = -\infty$,

(6) $\log_a (xy) = \log_a x + \log_a y$, $x, y \in R$, $x > 0$, $y > 0$,

(7) $\log_a x^y = y \log_a x$, $x, y \in R$, $x > 0$,

(8) $\log_a x = \dfrac{\log_b x}{\log_b a}$, $b > 1$,

(9) $\log_a x > \log_a y$ if $x > y$.

Proof. We prove (6) and (8) and leave the rest as exercises for the reader.

Proof of (6):

Since $a^p \cdot a^q = a^{p+q}$ for $a > 1$ and $p, q \in R$, if we let $x = a^p$, $y = a^q$, then

$$x = a^p \Rightarrow p = \log_a x,$$

$$y = a^q \Rightarrow q = \log_a y,$$

$$xy = a^{p+q} \Rightarrow p + q = \log_a (xy).$$

Hence $\log_a x + \log_a y = \log_a (xy)$.

Proof of (8): Since $a = b^{\log_b a}$, we have

$$a^{\log_b x / \log_b a} = (b^{\log_b a})^{\log_b x / \log_b a}.$$

Hence $\quad a^{\log_b x / \log_b a} = b^{\log_b x} = x$, and $\quad \dfrac{\log_b x}{\log_b a} = \log_a x.$

12.9 THE INTERMEDIATE VALUE THEOREM

The following important property of continuous functions (sometimes called Bolzano's theorem) was referred to in 8.5. The proof was deferred until this section since the concept of continuity was needed.

Theorem 1. *If $P(x)$ is a polynomial with real coefficients and if x_1, x_2 and c are real numbers such that*

$$P(x_1) < c < P(x_2),$$

then there exists $x_3 \in R$ between x_1 and x_2 such that $P(x_3) = c$.

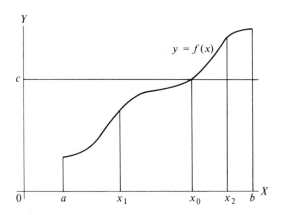

Figure 12–15 f is continuous on (a, b): $f(x_0) = c$

Theorem 1 is a special case of Theorem 2. (See Figure 12–15.)

Theorem 2. *If f is a function continuous on (a, b) with $f(x_1) < c < f(x_2)$ for $x_1, x_2 \in (a, b)$, then there is an $x_0 \in R$ between x_1 and x_2 such that $f(x_0) = c$.*

Proof. First let us assume $x_1 < x_2$. The proof if $x_2 < x_1$ is analogous. Consider the set S of points x such that $f(x) < c$ for $x_1 \leqslant x < x_2$. Now S is nonempty since $x_1 \in S$, and S has an upper bound, x_2; hence by OC S has a l.u.b. We denote l.u.b. S by x_0. Let us now prove that $f(x_0) = c$. Assume $f(x_0) > c$. Since f is continuous at $x = x_0$,

$$\forall \varepsilon > 0 \ \exists \delta > 0 \ [x_0 - \delta < x < x_0 + \delta \Rightarrow f(x_0) - \varepsilon < f(x) < f(x_0) + \varepsilon].$$

In particular, if $\varepsilon = f(x_0) - c > 0$, then for some $\delta > 0$

$$x_0 - \delta < x < x_0 + \delta \Rightarrow c < f(x) < 2f(x_0) - c.$$

That is, there is a neighborhood N_{x_0} such that for $x \in N_{x_0}, f(x) > c$. But N_{x_0} has a nonempty intersection with S, and in this intersection we have both $f(x) > c$ and $f(x) < c$. Since this is impossible we conclude that $f(x_0) \not> c$.

Assume $f(x_0) < c$; then, as above, there is a neighborhood N_{x_0}' such that for $x \in N_{x_0}', f(x) < c$. But N_{x_0}' has a nonempty intersection with the set (x_0, x_2) and in this intersection $f(x) < c$ and $f(x) > c$ (otherwise x_0 would not be an upper bound of S). Since this is impossible we conclude that $f(x_0) \not< c$. Thus we can only conclude that $f(x_0) = c$.

Theorem 1 follows immediately as a corollary to Theorem 2 because polynomials are continuous on R (see Theorem 1 of 12.4). The requirement of continuity in Theorem 2 is essential, as the following example demonstrates.

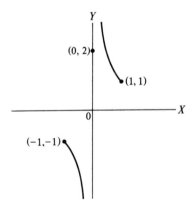

Figure 12–16   Graph of $f(x) = 1/x, \quad x \in [-1, 0) \cup (0, 1], \quad f(0) = 2$

Example 1. Let f be the function

$$f(x) = \begin{cases} 1/x, & x \in [-1, 0) \cup (0, 1] \\ 2, & x = 0 \end{cases}$$

(See Figure 12–16)

It is clear that $f(-1) < 0 < f(1)$, but that there is no x_0, $-1 < x_0 < 1$, such that $f(x_0) = 0$. Since this function is not continuous on $[-1, 1]$ the conditions of Theorem 2 are not satisfied.

12.10 EXERCISES

1. Find the limit of $f(x)$ as $x \to a$ in each of the following cases:

(a) $f(x) = x^2$, $a = 1$.

(b) $f(x) = \dfrac{\sin x}{x^2}$, $a = 0$.

(c) $f(x) = |x|$, $a = 0$.

(d) $f(x) = \dfrac{x^3 - 1}{x - 1}$, $a = 1$.

(e) $f(x) = x \sin^2 \dfrac{1}{x}$, $a = 0$.

2. (a) Show that the intersection of two closed intervals is either a closed interval, a singleton or empty.

(b) Show that the intersection of two semi-open intervals of the type $[a,b)$ is either void or a semi-open interval of the same type.

3. Find $\lim\limits_{x \to \infty} f(x)$ in each of the following cases:

(a) $f(x) = \dfrac{1}{x}$;

(b) $f(x) = \dfrac{\sin x}{x}$;

(c) $f(x) = \sin \dfrac{1}{x}$;

(d) $f(x) = e^{-x}$;

(e) $f(x) = \dfrac{x^2 - 1}{x^2 - 2}$;

(f) $f(x) = \dfrac{x^3 + x^2 + 1}{x^4 + x^3}$;

(g) $f(x) = \dfrac{x^3 + x}{x^2}$.

4. Find the indicated limits if they exist; otherwise show that the limit does not exist.

(a) $\lim\limits_{x \to 0+} \sqrt{x}$;

(b) $\lim\limits_{x \to 0-} \sqrt{(1 - x^2)}$;

(c) $\lim\limits_{x \to -1+} \sqrt{(1-x^2)};$ (d) $\lim\limits_{x \to 0+} x \sin \dfrac{1}{x};$

(e) $\lim\limits_{x \to 0+} \sin \dfrac{1}{x};$ (f) $\lim\limits_{x \to 0-} x \sin \dfrac{1}{x}.$

5. Show that $\lim\limits_{x \to a} f(x) = L$ if and only if both $\lim\limits_{x \to a-} f(x) = L$ and $\lim\limits_{x \to a+} f(x) = L$.

6. Show that if $|f(x)| < |g(x)|$ throughout a deleted neighborhood of a and if $\lim\limits_{x \to a} g(x) = 0$, then $\lim\limits_{x \to a} f(x) = 0$.

7. Prove that if $f(x) \leqslant g(x) \leqslant h(x)$ for x in some neighborhood M of a and if $\lim\limits_{x \to a} f(x) = F$, $\lim\limits_{x \to a} g(x) = G$ and $\lim\limits_{x \to a} h(x) = H$, then

$$F \leqslant G \leqslant H.$$

8. Prove that $\lim\limits_{x \to 0} (1+x)^{1/x} = e$.

9. Prove (i) and (ii) of Theorem 2 in 12.1.

10. Prove (iii) and (iv) of Theorem 1 in 12.2.

11. Prove Theorem 3 in 12.3.

12. Suppose f and g are two functions defined on the open interval (a, b). If they are continuous at $c \in (a,b)$, then show that

(a) $f + g$,   (b) fg,

(c) $f - g$

are continuous functions at c, and if $\lim\limits_{x \to c} g(x) \neq 0$, then

(d) f/g

is also continuous at c.

13. Prove that the tangent function is continuous at every point of its domain.

14. Prove parts (2) through (5), (7), and (9) in Theorem 2 of 12.8.

15. Prove the following statements:

(a) If $\lim\limits_{x \to \infty} f(x) = \infty$ and $f(x) \leqslant g(x)$ for all $x > M$ for some integer M, then

$$\lim\limits_{x \to \infty} g(x) = + \infty.$$

(b) If $\lim\limits_{x \to \infty} f(x) = F$ and $\lim\limits_{x \to \infty} g(x) = G$ and c is a real number then

(i) $\lim\limits_{x \to \infty} c f(x) = cF$,

(ii) $\lim\limits_{x \to \infty} (f(x) + g(x)) = F + G$,

(iii) $\lim\limits_{x \to \infty} f(x) \cdot g(x) = F \cdot G$,

(iv)   if $G \neq 0$, then $\lim\limits_{x \to \infty} f(x)/g(x) = F/G$.

16. Prove that if f is a continuous function in $[a, b]$, then $1/f$ is continuous except for the set of $x \in [a, b]$ such that $f(x) = 0$.

17. If $x \in Q$, $y \in R$, prove that $(e^y)^x = e^{yx}$.

18. (a) By using Theorem 2 of 12.9, show that there exists $x \in R$ such that $e^x = a$, where $a > 0$.

    (b) If $a > 1$, show that $x > 0$.

19. Prove that the following statement is correct:

    If $\lim_{x \to a} |f(x)| = 0$, then $\lim_{x \to a} f(x) = 0$. (Hint: See Lemma 2 of 8.5.)

20. Show that for a and b greater than zero,

$$\log_b a = \frac{1}{\log_a b} .$$

21. (a) Prove that $\lim_{x \to 0} \exp x = 1$. (*Hint*: Use the technique used in proof of Theorem 1 of 12.3.)

    (b) Use (a) to show that

$$\lim_{h \to 0} \exp (x + h) = \exp x$$

    and hence show that exp is continuous for all $x \in R$.

22. Prove that f is continuous on $[a, b]$ if for each $x_0 \in [a, b]$ and each neighborhood M of $f(x_0)$, there is a neighborhood $N(M)$ of x_0 such that

$$x \in N(M) \cap (a, b) \Rightarrow f(x) \in M.$$

13

Differentiation

13.1 THE DERIVATIVE

In a first course in calculus one of the principal topics is the derivative. Like
the difference discussed in Chapter 11, the derivative indicates the trend of
the function values $f(x)$ as x increases. Intuitively we consider the derivative
of a function f at x as a measure of the rate of change of $f(x)$ at x. For
example, suppose that a car is traveling along a straight road and that the
distance, $s(t)$, that it has covered in time t is known; then the derivative of
$s(t)$ is the velocity of the car at time t. This is what is read on the speedometer
at time t. If we plot the graph s, against t, we may have a graph like that
appearing in Figure 13–1. If we wish to determine the velocity at time t_0,
we might take a point $t_0 + h$ which is a short interval (either $h > 0$ or $h < 0$)
from t_0, measure the distance covered between t_0 and $t_0 + h$,

$$\Delta_h s = s(t_0 + h) - s(t_0),$$

and divide by h to get the average velocity over the interval $[t_0, t_0 + h]$
(or $[t_0 + h, t_0]$ depending on whether $h > 0$ or $h < 0$). This average is

$$\frac{\Delta_h s}{h} = \frac{s(t_0 + h) - s(t_0)}{h}.$$

It is clear, intuitively, that the smaller h is, the better estimate we get for
what we think of as the *instantaneous* velocity at t_0. The ideal value would
appear to be

$$\lim_{h \to 0} \frac{\Delta_h s}{h} = \lim_{h \to 0} \frac{s(t_0 + h) - s(t_0)}{h},$$

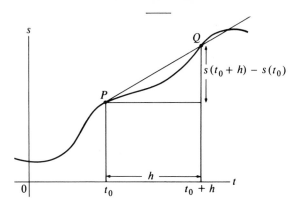

Figure 13–1   The average velocity over $[t_0, t_0 + h]$ is

$$\frac{s(t_0 + h) - s(t_0)}{h}$$

if it exists. This limit is called the **derivative** of s at $t = t_0$. The general definition of the derivative follows:

Definition 1. *Let f be a function defined in some neighbourhood N_a of a point a. Then the **derivative of f at** a is*

(1)
$$f'(a) = \lim_{h \to 0} \frac{f(a + h) - f(a)}{h}.$$

*If $f'(a)$ exists, then f is said to be **differentiable** at a.*

In Definition 1 if we put

$$F(h) = \frac{f(a + h) - f(a)}{h}$$

then $f'(a) = \lim_{h \to 0} F(h)$. It is important to notice that in general $F(0)$ is not defined. However (provided $f'(a)$ exists), $\lim_{h \to 0} F(h)$ is defined. The derivative of a function has an important geometrical interpretation. For in Figure 13–1 observe that

$$\frac{s(t_0 + h) - s(t_0)}{h}$$

is the slope of the secant PQ. As $h \to 0$, Q moves along the curve toward P, and (secant PQ) $\to$ (the tangent to the curve at P). Since this is the case in general the slope of the tangent to the graph of f at a is by definition the derivative of f at a.

Note that the relation f' defined by $y = f'(x)$ is a function, the domain of which is the set of values where the real-valued function f possesses a derivative.

There are a number of notations for the derivative function besides that used in (1). The following are in current use:

(i) $\dfrac{df}{dx}$ to stand for the derivative function of f and

$\left(\dfrac{df}{dx}\right)_{x=a}$ for the value of $\dfrac{df}{dx}$ at $x = a$.

(ii) Df to stand for the derivative function of f and $Df(a)$ for the value of Df at a.

(iii) $\dfrac{dy}{dx}$ to stand for the derivative function and $\left(\dfrac{dy}{dx}\right)_{x=a}$ to stand for the value of $\dfrac{dy}{dx}$ at a when $y = f(x)$.

(iv) $\dfrac{df(x)}{dx}$ to stand for the value of the derivative at x.

Let us now apply Definition (1) to calculate the derivatives of some specific functions.

Example 1. If $f(x) = x^2$, then

$$f'(x) = \lim_{h \to 0} \frac{f(x + h) - f(x)}{h}$$

$$= \lim_{h \to 0} \frac{(x + h)^2 - x^2}{h}$$

$$= \lim_{h \to 0} \frac{x^2 + 2xh + h^2 - x^2}{h}$$

$$= \lim_{h \to 0} (2x + h)$$

$$= 2x.$$

If we suppose that the domain of f is R, then the domain of f' is also R. In other words f is differentiable for all $x \in R$. By the geometric interpretation of the derivative we see that the slope of the tangent at any point (x, x^2) on the graph of f is $2x$.

Example 2. Consider the function f with

$$f(x) = |x|.$$

If $x > 0$, then $f(x) = x$. Let $x_0 > 0$.

Then

$$f'(x_0) = \lim_{h \to 0} \frac{f(x_0 + h) - f(x_0)}{h}$$

$$= \lim_{h \to 0} \frac{x_0 + h - x_0}{h}$$

$$= \lim_{h \to 0} \frac{h}{h} = \lim_{h \to 0} 1$$

$$= 1.$$

On the other hand, if $x < 0$, then $f(x) = -x$ and so, if $x_0 < 0$,

$$f'(x_0) = \lim_{h \to 0} \frac{f(x_0 + h) - f(x_0)}{h}$$

$$= \lim_{h \to 0} \frac{-(x_0 + h) - (-x_0)}{h} = \lim_{h \to 0} (-1)$$

$$= -1.$$

However, if we consider $x_0 = 0$, $f(x_0 + h) = |h| = h$, for all $h > 0$, while $f(x_0 + h) = |h| = -h$ for $h < 0$. Thus

$$F(h) = \frac{f(0 + h) - f(0)}{h} \qquad \begin{cases} 1 \text{ for } h > 0, \\ -1 \text{ for } h < 0, \end{cases}$$

but, for $h = 0$, $F(h)$ is undefined.

This function $F(h)$ clearly has no limit as $h \to 0$. That is, $\lim_{h \to 0} F(h)$ does not exist, and so f' is undefined at 0. In fact,

$$f'(x) = \begin{cases} -1 \text{ for } x < 0, \\ 1 \text{ for } x > 0, \\ \text{is undefined at } x = 0. \end{cases}$$

The function f is consequently not differentiable at $x = 0$. The geometrical interpretation of f' reinforces this result since it is clear by an examination of a graph of f, that there is no tangent to f at $x = 0$.

Example 3.   Consider the function f, where

$$f(x) = 2x^3 + 3x - 7$$

Now

$$f'(x) = \lim_{h \to 0} \{[2(x + h)^3 + 3(x + h) - 7] - [2x^3 + 3x - 7]\}/h$$

But

$$2(x + h)^3 + 3(x + h) - 7 = 2x^3 + 6x^2h + 6xh^2 + 2h^3 + 3x + 3h - 7$$

and

$$[2(x + h)^3 + 3(x + h) - 7] - [2x^3 + 3x - 7]$$
$$= 6x^2h + 6xh^2 + 2h^3 + 3h$$

Hence

$$f'(x) = \lim_{h \to 0} (6x^2h + 6xh^2 + 2h^3 + 3h)/h$$

$$= \lim_{h \to 0} (6x^2 + 3 + 6xh + 2h^2)$$

$$= 6x^2 + 3.$$

Example 4.   If $f(x) = \sqrt{x}$, then

$$f'(x) = \lim_{h \to 0} \frac{\sqrt{(x + h)} - \sqrt{x}}{h}$$

$$= \lim_{h \to 0} \frac{\sqrt{(x + h)} - \sqrt{x}}{h} \cdot \frac{\sqrt{(x + h)} + \sqrt{x}}{\sqrt{(x + h)} + \sqrt{x}}$$

$$= \lim_{h \to 0} \frac{x + h - x}{h[\sqrt{(x + h)} + \sqrt{x}]}$$

$$= \lim_{h \to 0} \frac{1}{\sqrt{(x + h)} + \sqrt{x}} = \frac{1}{2\sqrt{x}} = \frac{1}{2} x^{-\frac{1}{2}}$$

provided $x > 0$.

13.2 DERIVATIVE OF SINE, COSINE, EXPONENTIAL

Theorem 1. $(\sin x)' = \cos x$, $(\cos x)' = -\sin x$.

Proof. We have $(\sin x)' = \lim_{h \to 0} \dfrac{\sin(x + h) - \sin x}{h}$.

Using the trigonometric formula

$$\sin A - \sin B = 2 \cos \tfrac{1}{2}(A + B) \sin \tfrac{1}{2}(A - B),$$

we have

$$\frac{\sin(x + h) - \sin x}{h} = \frac{2 \cos \tfrac{1}{2}(2x + h) \sin \tfrac{1}{2}h}{h}$$

$$= \cos (x + \tfrac{1}{2}h) \frac{\sin \tfrac{1}{2}h}{\tfrac{1}{2}h}.$$

Since cosine is a continuous function (Corollary 1 of 12.6)

$$\lim_{h \to 0} \cos (x + \tfrac{1}{2}h) = \cos x,$$

and by Theorem 2 of 12.6

$$\lim_{h \to 0} \frac{\sin \tfrac{1}{2}h}{\tfrac{1}{2}h} = 1.$$

Hence

$$(\sin x)' = \cos x.$$

Similarly

$$(\cos x)' = \lim_{h \to 0} \frac{\cos (x + h) - \cos x}{h}$$

$$= \lim_{h \to 0} \left[-\sin \left(x + \frac{h}{2} \right) \frac{\sin \tfrac{1}{2}h}{\tfrac{1}{2}h} \right]$$

$$= -\sin x,$$

using the continuity of the sine function.

Example 1.   If $f(x) = x \sin x$ then

$$f'(x) = x \cos x + \sin x.$$

$$f'(x) = \lim_{h \to 0} \frac{(x + h) \sin (x + h) - x \sin x}{h}$$

$$= \lim_{h \to 0} x \left[\frac{\sin (x + h) - \sin x}{h} \right] + \lim_{h \to 0} \sin (x + h)$$

$$= x \lim_{h \to 0} \frac{\sin (x + h) - \sin x}{h} + \lim_{h \to 0} \sin (x + h)$$

(Theorem 2 of 12.1)

$$= x \cos x + \lim_{h \to 0} \sin (x + h) \qquad \text{(Theorem 1)}$$

$$= x \cos x + \sin x \qquad \text{(Theorem 1 of 12.6).}$$

Theorem 2. $(\exp x)' = \exp x.$

Proof. We have

$$(\exp x)' = \lim_{h \to 0} \frac{\exp (x + h) - \exp x}{h}$$

$$= \lim_{h \to 0} \frac{\exp x \exp h - \exp x}{h} \qquad \text{(Theorem 1 of 11.8)}$$

$$= \exp x \lim_{h \to 0} \frac{\exp h - 1}{h} \qquad \text{(Theorem 2 of 12.1)}$$

$$= \exp x \qquad \text{(Theorem 1 of 12.3).}$$

Example 2. Suppose that $f(x) = (\exp x)^2$. We prove that

$$f'(x) = 2(\exp x)^2.$$

Now

$$f'(x) = \lim_{h \to 0} \frac{[\exp (x + h)]^2 - (\exp x)^2}{h}$$

$$= \lim_{h \to 0} \frac{(\exp x)^2 \cdot (\exp h)^2 - (\exp x)^2}{h} \qquad \text{(Theorem 1 of 11.8)}$$

$$= (\exp x)^2 \lim_{h \to 0} \frac{(\exp h)^2 - 1}{h}$$

$$= (\exp x)^2 \lim_{h \to 0} \frac{\exp 2h - 1}{h} \qquad \begin{array}{l}\text{(Theorem 1 of 11.8} \\ \text{with } x = y = h)\end{array}$$

$$= (\exp x)^2 \, 2 \lim_{h \to 0} \frac{\exp 2h - 1}{2h}$$

$$= 2(\exp x)^2 \lim_{\theta \to 0} \frac{\exp \theta - 1}{\theta}$$

$$= 2(\exp x)^2 \qquad \text{(Theorem 1 of 12.3).}$$

The following theorem establishes a connection between differentiability and continuity. With the aid of this theorem and Theorem 2 we can show that the function exp is continuous in R. We can then draw the important conclusion, from Theorem 4 of 12.7 and the discussion following it, applied to exp x, that

$$\exp x = e^x, \quad \text{for all } x \in R.$$

Theorem 3. *If $f'(a)$ exists, then f is continuous at a.*

Proof. If h is any non-zero real number we have

(1)
$$f(a + h) = \left[\frac{f(a + h) - f(a)}{h} \right] h + f(a).$$

Now
$$\lim_{h \to 0} \frac{f(a + h) - f(a)}{h} = f'(a).$$

By Theorem 2 of 12.1 we have

$$\lim_{h \to 0} f(a + h) = \lim_{h \to 0} \left[\frac{f(a + h) - f(a)}{h} \right] \lim_{h \to 0} h + \lim_{h \to 0} f(a).$$

That is
$$\lim_{h \to 0} f(a + h) = f'(a) \cdot 0 + f(a),$$

or
$$\lim_{h \to 0} f(a + h) = f(a).$$

Hence, by Definition 1 of 12.4, f is continuous at a.

Remark 1. The converse of this theorem is not true. To show this, consider the function defined by $f(x) = |x|$. This function is continuous on R but not differentiable at $x = 0$ (as shown in Example 2 of 13.1).

Corollary 1.   *The function* exp *is continuous for every $x \in R$.*

Corollary 2.   *For all $x \in R$,*

$$\exp x = e^x.$$

Proof. This follows from Corollary 1 (see 11.8 and 12.8).

Remark 2. The number e is the only positive real value of a for which

(2)
$$Da^x = a^x.$$

Indeed, for $a > 0$, consider the function for which $f(x) = a^x$. It can be shown (see Exercise 18 of 12.10) that $a = e^p$ for some positive real number p, and so

$$a^x = e^{px}.$$

Hence

$$\frac{da^x}{dx} = \frac{de^{px}}{dx} = \lim_{h \to 0} \frac{e^{p(x+h)} - e^{px}}{h} = \lim_{h \to 0} \frac{e^{px}e^{ph} - e^{px}}{h}$$

$$= \lim_{h \to 0} e^{px} \left(\frac{e^{ph} - 1}{h} \right) = e^{px} \lim_{h \to 0} p \frac{e^{ph} - 1}{ph}$$

$$= p\, e^{px} \lim_{h \to 0} \frac{e^{ph} - 1}{ph}.$$

Since $ph \to 0$ if and only if $h \to 0$

$$\lim_{h \to 0} \frac{e^{ph} - 1}{ph} = \lim_{ph \to 0} \frac{e^{ph} - 1}{ph} = \lim_{q \to 0} \frac{e^q - 1}{q} = 1$$

by Theorem 1 in 12.3. Therefore

$$\frac{da^x}{dx} = pe^{px} = pa^x.$$

Since $e^p = a$, $p = \log_e a$. Thus

(3) $$\frac{da^x}{dx} = (\log_e a)\, a^x,$$

and since $\log_e a \neq 1$ if $a \neq e$, e is the only positive value of a for which (2) is valid.

Theorem 3 provides an alternative proof of Theorem 1 of 12.6.

Corollary 3. *The functions sine and cosine are continuous.*

Proof. In Theorem 1 we found the derivatives of sine and cosine. By Theorem 3, since these derivatives exist these functions are continuous.

13.3 DIFFERENTIATION THEOREMS

The following theorem can be used to simplify the process of differentiation.

Theorem 1. *Let f and g be two functions whose derivatives exist at $x = a$.*

(1) $$(f + g)'(a) = f'(a) + g'(a).$$

(2) $$(f \cdot g)'(a) = f'(a) g(a) + f(a) g'(a),$$

(3) $$(cf)'(a) = cf'(a) \text{ where } c \text{ is a fixed real number};$$

(4) $$(f/g)'(a) = \frac{f'(a) g(a) - g'(a) f(a)}{(g(a))^2} \text{ for } g(a) \neq 0.$$

Proof. (1):

$$(f + g)'(a) = \lim_{h \to 0} \frac{(f + g)(a + h) - (f + g)(a)}{h}$$

$$= \lim_{h \to 0} \frac{f(a + h) + g(a + h) - (f(a) + g(a))}{h}$$

$$= \lim_{h \to 0} \left[\frac{f(a + h) - f(a)}{h} + \frac{g(a + h) - g(a)}{h} \right]$$

$$= f'(a) + g'(a).$$

(2):

$$(f \cdot g)'(a) = \lim_{h \to 0} \frac{f(a + h) g(a + h) - f(a) g(a)}{h}.$$

Now

$$f(a + h) g(a + h) - f(a) g(a)$$

$$= f(a + h) g(a + h) - f(a) g(a + h) + f(a) g(a + h) - f(a) g(a)$$

$$= [f(a + h) - f(a)] g(a + h) + f(a) [g(a + h) - g(a)],$$

and so

$$(f \cdot g)'(a) = \lim_{h \to 0} \left\{ \left(\frac{f(a + h) - f(a)}{h} \right) g(a + h) + f(a) \left(\frac{g(a + h) - g(a)}{h} \right) \right\}$$

$$= \lim_{h \to 0} \left(\frac{f(a + h) - f(a)}{h} \right) \lim_{h \to 0} g(a + h)$$

$$+ f(a) \lim_{h \to 0} \left(\frac{g(a + h) - g(a)}{h} \right).$$

Since both $f'(a)$ and $g'(a)$ exist,

$$\lim_{h \to 0} \frac{f(a + h) - f(a)}{h} = f'(a),$$

$$\lim_{h \to 0} \frac{g(a + h) - g(a)}{h} = g'(a);$$

and by Theorem 3 of 13.2, g is continuous, thus

$$\lim_{h \to 0} g(a + h) = g(a).$$

Therefore $(f \cdot g)'(a) = f'(a) g(a) + f(a) g'(a)$.

(3):

$$\frac{d[cf(x)]}{dx} = \lim_{h \to 0} \frac{cf(x + h) - cf(x)}{h}$$

$$= c \lim_{h \to 0} \frac{f(x + h) - f(x)}{h}$$

$$= c \frac{df(x)}{dx}.$$

Equation (4) follows from (2) and (5) below. The proof of (4) is left to the reader.

(5)

$$\frac{d}{dx}\left(\frac{1}{f(x)}\right) = -\frac{f'(x)}{(f(x))^2}$$

is valid for all x where $f(x) \neq 0$.

To demonstrate (5) note that

$$\frac{d}{dx}\left(\frac{1}{f(x)}\right) = \lim_{h \to 0} \left(\frac{1}{f(x + h)} - \frac{1}{f(x)}\right)\frac{1}{h}$$

$$= \lim_{h \to 0} \frac{f(x) - f(x + h)}{hf(x + h)f(x)}$$

$$= \lim_{h \to 0} \frac{1}{f(x + h)} \cdot \frac{1}{f(x)} \cdot \frac{f(x) - f(x + h)}{h}$$

$$= \frac{1}{f(x)} \lim_{h \to 0} \frac{1}{f(x + h)} \lim_{h \to 0} \left[-\frac{f(x + h) - f(x)}{h}\right].$$

Since f is continuous,

$$\lim_{h \to 0} \frac{1}{f(x + h)} = \frac{1}{f(x)},$$

and therefore

$$\frac{d}{dx}\left[\frac{1}{f(x)}\right] = -\frac{f'(x)}{(f(x))^2}.$$

Remark 1. By successive application of (1), we can prove

$$(f_1 + f_2 + \cdots + f_n)'(a) = f_1'(a) + \cdots + f_n'(a)$$

and by successive application of (2), we can prove (cf. proof of Corollary 1)

$$(f^n)'(a) = nf^{n-1}(a)f'(a).$$

Corollary 1. *For all integers $n \geqslant 0$,*

(6)
$$\frac{dx^n}{dx} = nx^{n-1}.$$

Proof. We prove this by induction. First consider the case where $n = 0$. In this case we have to prove

$$\frac{d1}{dx} = 0$$

To show this, note that

$$\frac{d1}{dx} = \lim_{h \to 0} \frac{1-1}{h} = \lim_{h \to 0} \frac{0}{h} = 0.$$

Now suppose that

(7)
$$\frac{dx^k}{dx} = kx^{k-1},$$

and then try to prove that

(8)
$$\frac{dx^{k+1}}{dx} = (k+1)x^k.$$

Given (7) we note that

$$\frac{dx^{k+1}}{dx} = \frac{d(x \cdot x^k)}{dx} = \frac{dx}{dx}x^k + x\frac{dx^k}{dx} \quad \text{(by Theorem 1)}$$

$$= \frac{dx}{dx}x^k + x\,kx^{k-1} \quad\quad\quad \text{(by (7))}$$

$$= x^k\left(\frac{dx}{dx} + k\right).$$

Now

$$\frac{dx}{dx} = \lim_{h \to 0} \frac{x+h-x}{h} = \lim_{h \to 0} 1 = 1,$$

so that

$$\frac{dx^{k+1}}{dx} = x^k(1 + k) = (k+1)\,x^k.$$

Thus (8) follows from (7), the induction proof is complete, and (6) is valid for all $n \in N$.

The reader should compare these results to those obtained for the difference operation in 11.2 and 11.9. Let us illustrate Corollary 1 and Theorem 1 by some examples.

Example 1. Evaluate $D_x(3x^3 + 7x^2 - 2x + 3)$.

By Theorem 1, we have

$$D_x(3x^3 + 7x^2 - 2x + 3) = D_x(3x^3) + D_x(7x^2) - D_x(2x) + D_x 3$$

$$= 3D_x x^3 + 7D_x x^2 - 2D_x x + 3D_x 1$$

$$= 3 \cdot 3x^2 + 7 \cdot 2x - 2 + 0 \text{ (Corollary 1)}$$

$$= 9x^2 + 14x - 2.$$

Example 2. Find

$$D_x\left[\frac{x^2 - 2}{x^3 + 4x}\right].$$

By Theorem 1(4) we have

$$D_x\left[\frac{x^2 - 2}{x^3 + 4x}\right] = \frac{(x^3 + 4x)\,D_x(x^2 - 2) - (x^2 - 2)\,D_x(x^3 + 4x)}{(x^3 + 4x)^2}$$

$$= \frac{(x^3 + 4x)\,(D_x x^2 - D_x 2) - (x^2 - 2)\,(D_x x^3 + 4D_x x)}{(x^3 + 4x)^2}$$

$$= \frac{(x^3 + 4x)\,2x - (x^2 - 2)\,(3x^2 + 4)}{(x^3 + 4x)^2}$$

$$= \frac{(-x^4 + 10x^2 + 8)}{(x^3 + 4x)^2}.$$

The derivatives of $c(x)$, $s(x)$ defined in Example 4 of 11.8 are obtained in the following theorem and corollary.

Theorem 2. *If* $f(x) = c(x)$, $f'(x) = -s(x)$.

Proof.

$$\frac{c(x + h) - c(x)}{h} = c(x)\frac{c(h) - 1}{h} - s(x)\frac{s(h)}{h} \quad \text{(Theorem 4 of 11.8)}.$$

Now

$$f'(x) = \lim_{h \to 0} \left\{ c(x)\frac{c(h) - 1}{h} - s(x)\frac{s(h)}{h} \right\}$$

$$= c(x)\lim_{h \to 0}\frac{c(h) - 1}{h} - s(x)\lim_{h \to 0}\frac{s(h)}{h}$$

$$= c(x)\cdot 0 - s(x)\cdot 1 = -s(x) \quad \text{(Theorems 2 and 3 of 12.3)}.$$

Corollary 2. *If* $f(x) = s(x)$, $f'(x) = c(x)$ *provided* $s(x) \neq 0$.

Proof. We use a procedure called implicit differentiation. Differentiating each side of the equation

$$c^2(x) + s^2(x) = 1 \qquad \text{(Corollary 3 of 11.8)}$$

we obtain

$$2c(x)\,c'(x) + 2s(x)\,s'(x) = 0 \qquad \text{(Remark 1)}.$$

Hence

$$s(x)\,s'(x) = -c(x)\,(-s(x)) \qquad \text{(Theorem 2)}$$

and

$$s'(x) = c(x) \qquad \text{if } s(x) \neq 0.$$

(But see Theorem 1 of 3.2 and Theorem 6 of 13.6.)

We are now in a position to find derivatives of more complicated expressions.

Example 3. If $f(x) = x^3 \exp x$, then

$$f'(x) = \frac{dx^3}{dx}\exp x + x^3\frac{d\exp x}{dx}$$

by (2). Thus by (6) and Theorem 2 of 13.2

$$f'(x) = 3x^2 \exp x + x^3 \exp x$$

$$= (x^3 + 3x^2)\exp x.$$

Example 4. If

$$f(x) = \frac{\sin x}{\exp(x + 1)},$$

then by an application of (3), (4), Theorems 1 and 2 of 13.2,

$$f'(x) = \frac{d}{dx}\left(\frac{\sin x}{\exp x \exp 1}\right) = \frac{1}{\exp 1}\left(\frac{d\sin x}{dx}\exp x - \frac{d\exp x}{dx}\sin x\right)\frac{1}{(\exp x)^2}$$

$$= e^{-1}\frac{\cos x \exp x - \exp x \sin x}{(\exp x)^2}$$

$$= e^{-1}\left(\frac{\cos x - \sin x}{\exp x}\right).$$

The following additional Corollary to Theorem 2 is an extension of Corollary 1.

Corollary 3.   *The formula*

(9)
$$\frac{dx^n}{dx} = nx^{n-1}$$

is true for all $n \in I$.

Proof.   We have already shown that (9) is valid for $n \in N$; now we must show it for $n = -k$ where $k \in N$. Thus

$$\frac{dx^n}{dx} = \frac{dx^{-k}}{dx} = \frac{d}{dx}\left(\frac{1}{x^k}\right) = \left(\frac{d1}{dx}x^k - \frac{dx^k}{dx}\cdot 1\right)\frac{1}{x^{2k}}$$

$$= \frac{0 \cdot x^k - kx^{k-1}}{x^{2k}} = (-k)x^{-k-1};$$

$$= nx^{n-1}.$$

13.4 THE CHAIN RULE AND APPLICATIONS

In the following theorem called the Chain Rule, we develop a method for finding the derivative of the composition of two functions in terms of the derivatives of the functions. This theorem is important since it enables us to obtain the derivatives of such functions as

$$\sin(2x^3 + 3), \exp(\cos x^5), \text{ and } \sqrt{(5x^4 + 7x^{-1})}.$$

Theorem 1.   **The Chain Rule.** *If $g'(a)$ exists and $f'(g(a))$ exists, then*

$$(f \circ g)'(a)$$

exists and

(1) $$(f \circ g)'(a) = g'(a)f'(g(a)).$$

Proof. We have $(f \circ g)(x) = f(g(x))$.

Hence

$$(f \circ g)'(a) = \lim_{h \to 0} \frac{f(g(a + h)) - f(g(a))}{h}$$

Now define a new function F, by setting

(2) $$F(k) = \begin{cases} \dfrac{f(b + k) - f(b)}{k}, & k \neq 0, \\ f'(b), & k = 0. \end{cases}$$

In this definition of F, b and $(b + k)$ are any numbers in the range of g. With the aid of F, we can write

$$(f \circ g)'(a) = \lim_{h \to 0} F(k) \cdot \frac{k}{h}$$

where $k = g(a + h) - g(a)$ and $b = g(a)$. Since g is differentiable at a it is continuous at a. Whence

$$\lim_{h \to 0} k = \lim_{h \to 0} [g(a + h) - g(a)] = 0.$$

Also

$$\lim_{k \to 0} F(k) = f'(b) = f'(g(a)).$$

Thus

$$(f \circ g)'(a) = \lim_{h \to 0} F(k) \cdot \lim_{h \to 0} \frac{k}{h}$$

$$= \lim_{k \to 0} F(k) \cdot \lim_{h \to 0} \frac{k}{h}$$

$$= f'(g(a)) g'(a).$$

Remark 1. It is evident from (2) that $F(k)$ is defined even if $k = 0$. The value of $F(0)$ was arbitrarily chosen, but the choice $F(0) = f'(b)$ makes F continuous. It is necessary to introduce F and to define $F(0)$ since, for some functions g, $k = g(a + h) - g(a)$ may vanish for values of h other than $h = 0$.

Remark 2. If we write $y = f(u)$, $u = g(x)$, then the Chain Rule can be written as

$$\frac{df(g(x))}{dx} = \frac{dy}{dx} = \frac{dy}{du} \cdot \frac{du}{dx}.$$

In this notation if we *think* of dy/dx, etc. as quotients of numbers then the Chain Rule appears to be an arithmetic triviality.

The following Corollary is a direct deduction from Theorem 1.

Corollary 1. *If f^{-1} is the function inverse to f, then*

$$(f^{-1})'(x) = \frac{1}{f'(f^{-1}(x))}$$

Proof. We have $f \circ f^{-1}(x) = f(f^{-1}(x)) = x$.

Differentiating we have

$$D_x x = 1 = (f^{-1})'(x)f'(f^{-1}(x)).$$

That is

$$(f^{-1})'(x) = \frac{1}{f'(f^{-1}(x))}$$

Remark 3. If we write $y = f^{-1}(x)$, then, since $y = f^{-1}(x) \Leftrightarrow x = f(y)$, we can write the result of Corollary 1 as

$$\frac{dy}{dx} = 1 \bigg/ \frac{dx}{dy}$$

In this form Corollary 1 appears to be an obvious arithmetical result.

Example 1.   Let us use Corollary 1 to find $(x^{1/p})'$, where $p \in I - \{0\}$.

If $f^{-1}(x) = x^{1/p}$, then $f(x) = x^p$.

Now by Corollary 3 of 13.3, $f'(x) = px^{p-1}$.
Hence by Corollary 1

$$(x^{1/p})' = \frac{1}{p(x^{1/p})^{p-1}} = \frac{1}{px^{1-1/p}} = \frac{1}{p}x^{1/p-1}$$

We can use the result of Example 1 to generalize Corollary 3 of 13.3.

Corollary 2.   *The formula*

$$\frac{dx^n}{dx} = nx^{n-1}$$

is true for all $n \in Q$.

Proof. Let $n = p/q$, then $x^{p/q} = f \circ g(x)$ where

$$f(x) = x^p, \qquad g(x) = x^{1/q}.$$

By Theorem 1 $(f \circ g)'(x) = (x^{1/q})' f'(x^{1/q})$.

But

$$(x^{1/q})' = \frac{1}{q} x^{1/q - 1}$$

by Example 1, and

$$f'(x^{1/q}) = p(x^{1/q})^{p-1} = p x^{p/q - 1/q}.$$

Thus

$$(f \circ g)'(x) = (x^{p/q})' = \left(\frac{1}{q} x^{1/q-1} \right) (p x^{p/q - 1/q})$$

$$= \frac{p}{q} x^{p/q - 1} = n x^{n-1}.$$

By an argument involving the definition of x^n for n irrational, Corollary 2 can be further generalized so that it is valid for all $n \in R$.

Example 2. Find the derivative of $\exp(\cos x^5)$. This can be found by successive applications of the Chain Rule. We have

$$[\exp(\cos x^5)]' = \exp(\cos x^5)(\cos x^5)'$$

$$= \exp(\cos x^5)(-\sin x^5)(x^5)'$$

$$= -5x^4 \sin x^5 \exp(\cos x^5).$$

Theorem 2. *If $f(x) = \log_a x$, where $a > 0$, then*

(3)
$$f'(x) = \frac{\log_a e}{x}.$$

Proof. Let $y = \log_a x$.

Now

$$y = \log_a x \Leftrightarrow x = a^y.$$

Let

$$p = \log_e a, \qquad \text{that is, } a = e^p.$$

Then

$$x = (e^p)^y = e^{py}.$$

Differentiating with respect to x, we have by Theorem 1

$$1 = \frac{d}{dx} e^{py} = e^{py} \frac{d}{dx}(py) = p e^{py} \frac{dy}{dx}.$$

Hence

$$\frac{dy}{dx} = \frac{1}{p e^{py}} = \frac{1}{px} = \frac{1}{x \log_e a}$$

But

$$\log_e a = \frac{1}{\log_a e} \qquad \text{(Exercise 20 of 12.10)}$$

Hence

$$f'(x) = \frac{\log_a e}{x}$$

Now the naturalness of the natural logarithm can be justified. For if $a = e$, then $\log_e e = 1$, and (3) becomes

(4) $$\frac{d \log_e x}{dx} = \frac{1}{x}.$$

Expression (4) and Theorem 3 of 13.2 show that the function $\log_a x$ is continuous if $x > 0$.

13.5   THE INVERSE TRIGONOMETRIC FUNCTIONS

It is clear that the functions sin, cos, and tan, since they are not one-to-one onto functions, have no inverses. However, by redefining these functions with new restricted domains one can produce one-to-one onto functions that have inverses. These inverse functions are called inverse trigonometric functions and they have numerous important applications. Before we define them we need to state the following lemma.

Lemma 1.   *If* $x, y \in [-\frac{1}{2}\pi, \frac{1}{2}\pi]$ *and* $x > y$, *then* $\sin x > \sin y$.

Proof.   By using

$$\sin x - \sin y = 2 \cos \left(\frac{x + y}{2} \right) \sin \left(\frac{x - y}{2} \right)$$

the result can be proved readily by the reader.

By Lemma 1, the mapping $x \rightarrow \sin x$ for $x \in [-\frac{1}{2}\pi, \frac{1}{2}\pi]$ is one-to-one and hence, by Definition 4 of 7.8, for $x \in [-\frac{1}{2}\pi, \frac{1}{2}\pi]$ the function $f: x \rightarrow \sin x$ has an inverse function which we shall call the inverse sine.

Definition 1.   *The* sine *function* $x \rightarrow \sin x$, $x \in [-\frac{1}{2}\pi, \frac{1}{2}\pi]$ *has an inverse function called the* **inverse sine** *or* **arcsine.** *There are alternative notations: if* $-\frac{1}{2}\pi \leqslant x \leqslant \frac{1}{2}\pi$, $y = \sin x \Leftrightarrow x = \sin^{-1} y$ *or* $y = \sin x \Leftrightarrow x = \arcsin y$.

The graphs of the sine and inverse sine are shown in Figure 13–2 and Figure 13–3.

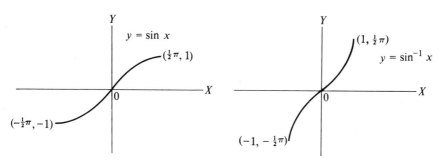

Figure 13–2   Figure 13–3

The function $y = \sin^{-1}x$ has domain $[-1, 1]$. It is not defined for $|x| > 1$.

In a similar way the inverse cosine and inverse tangent can be defined.

Definition 2.   *The cosine function* $x \rightarrow \cos x$, $x \in [0, \pi]$ *has an inverse function called the* **inverse cosine** *such that* $y = \cos x \Leftrightarrow x = \cos^{-1}y$ *($\cos^{-1}y$ is often written* arcos y).

Definition 3.   *The tangent function* $x \rightarrow \tan x$, $x \in (-\frac{1}{2}\pi, \frac{1}{2}\pi)$ *has an inverse function called the* **inverse tangent** *such that* $y = \tan x \Leftrightarrow x = \tan^{-1}y$ *($\tan^{-1}y$ is often written* arctan y).

These definitions are justified, since it is readily established that the mapping $x \rightarrow \cos x$, $x \in [0, \pi]$ and the mapping $x \rightarrow \tan x$, $x \in (-\frac{1}{2}\pi, \frac{1}{2}\pi)$ are both one-to-one. The graphs of these functions are shown in Figures 13–4 to 13–7.

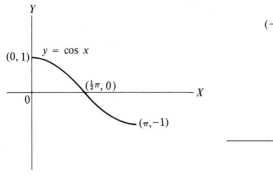

Figure 13–4

$f: x \rightarrow \cos x$, Dom $f = [0, \pi]$

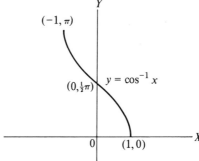

Figure 13–5

$f: x \rightarrow \cos^{-1} x$, Dom $f = [-1, 1]$

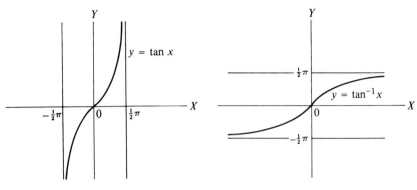

Figure 13–6

$f: x \to \tan x,\ \mathrm{Dom}\, f = (-\pi/2,\ \pi/2)$

Figure 13–7

$f: x \to \tan^{-1} x,\ \mathrm{Dom}\, f = R$

In the case of $y = \cos x$, it was necessary to choose the domain $[0, \pi]$ since for the domain, $[-\tfrac{1}{2}\pi, \tfrac{1}{2}\pi]$ the mapping $x \to \cos x$ is not one-to-one, and consequently no inverse function exists. The properties of these three inverse trigonometric functions are easily found from the properties of the corresponding trigonometric functions.

Theorem 1. *If $x \in (-1, 1)$,*

$$\frac{d \sin^{-1} x}{dx} = \frac{1}{\sqrt{(1 - x^2)}}.$$

Proof. Let $y = \sin^{-1} x$. Then for $x \in (-1, 1)$, $y = \sin^{-1} x \Leftrightarrow x = \sin y$ where $y \in (-\tfrac{1}{2}\pi, \tfrac{1}{2}\pi)$.
Differentiating $x = \sin y$ with respect to x, we have

$$\frac{dx}{dx} = 1 = \cos y \, \frac{dy}{dx}.$$

Hence

$$\frac{dy}{dx} = \frac{1}{\cos y} = \frac{1}{\sqrt{(1 - \sin^2 y)}} = \frac{1}{\sqrt{(1 - x^2)}}.$$

Note: Since $y \in (-\tfrac{1}{2}\pi, \tfrac{1}{2}\pi)$, $\cos y = \sqrt{(1 - \sin^2 y)}$, that is, we choose the positive square root.

Corollary 1(a).

$$\frac{d \cos^{-1} x}{dx} = \frac{-1}{\sqrt{(1 - x^2)}},$$

(b).

$$\frac{d \tan^{-1} x}{dx} = \frac{1}{1 + x^2}.$$

The proof is left to the reader.

13.6 ROLLE'S THEOREM AND THE MEAN VALUE THEOREM

In this section we prove some theorems which appear obvious from a geometrical point of view, but the proofs of these theorems are rather intricate, and they depend on considerations of continuity developed in Chapter 12. In Theorem 6 of this section we finally establish (by the application of the earlier theorems of the section) the identification of the functions $s(x)$, $\sin x$, and $c(x)$, $\cos x$.

In order to demonstrate that the derivative of a function (like the difference in Chapter 11) indicates the trend of the function, we begin this section with the following theorems.

Theorem 1.   *Let f be differentiable at every point in the interval (a, b). If f is monotone nondecreasing (nonincreasing) in (a, b), then $f'(x) \geqslant 0$ $(f'(x) \leqslant 0)$ for $x \in (a, b)$.*

Proof.   First assume f monotone nondecreasing. Then for every pair $x_1 < x_2$, we have $f(x_1) \leqslant f(x_2)$.

Now since

$$f'(x) = \lim_{h \to 0} \frac{f(x + h) - f(x)}{h}$$

exists for any $x \in (a, b)$, it follows that

$$f'(x) = \lim_{h \to 0+} \frac{f(x + h) - f(x)}{h}.$$

Since f is nondecreasing, $f(x + h) - f(x) \geqslant 0$ for $h > 0$, and so

$$\frac{f(x + h) - f(x)}{h} \geqslant 0 \text{ for } h > 0.$$

Thus

$$f'(x) = \lim_{h \to 0+} \frac{f(x + h) - f(x)}{h} \geqslant 0.$$

For f nonincreasing the proof is analogous.

To prove the converse of Theorem 1 we need both Rolle's Theorem and the Mean Value Theorem. The proof of these requires the following lemmas.

Lemma 1.   *If $f(x)$ is continuous on $[a, b]$ then $f(x)$ is bounded on $[a, b]$.*

Proof. Since $f(x)$ is continuous at $x = a$, for any $\varepsilon > 0$ there is a $\delta' > 0$ and a $\delta < \delta'$ such that

$$|f(x) - f(a)| < \varepsilon \text{ or } f(a) - \varepsilon < f(x) < f(a) + \varepsilon \text{ if } a \leqslant x \leqslant a + \delta.$$

Thus, for $x \in [a, a + \delta]$, $f(x)$ is bounded. Now consider the set of points $c \in [a, b]$ such that $f(x)$ is bounded in $[a, c]$. It is nonempty ($a + \delta$ is a member) and bounded. Hence, by OC it possesses a l.u.b.—call it c_1 and assume $c_1 \neq b$. Now $f(x)$ is bounded in $[a, c_1 - \varepsilon']$ and unbounded in $[a, c_1 + \varepsilon']$, for any $\varepsilon' > 0$ (if this were not true c_1 would not be a l.u.b.). But $f(x)$ is continuous at $x = c_1$, thus for any $\varepsilon_1 > 0$ there exists $\delta_1 > 0$ such that $f(c_1) - \varepsilon_1 < f(x) < f(c_1) + \varepsilon_1$ if $c_1 - \delta_1 < x < c_1 + \delta_1$, that is $f(x)$ is bounded if $c_1 - \delta_1 < x < c_1 + \delta_1$. But this implies that $f(x)$ is both bounded and unbounded in the nonempty set $[a, c_1 + \varepsilon'] \cap [c_1, c_1 + \delta_1]$ which is impossible (see Figure 13–8). Thus $c_1 = b$ and $f(x)$ is bounded on $[a, b]$.

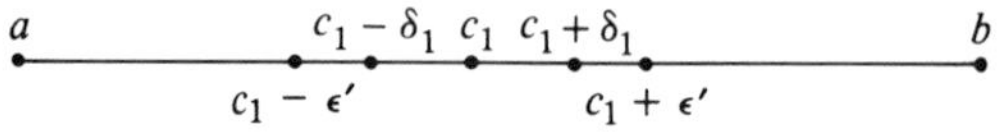

Figure 13-8   The point c_1 is the l.u.b. of $\{c | f(x) \text{ is bounded in } [a, c]\}$

The necessity of assuming that we are dealing with a *closed* interval $[a, b]$ in this lemma becomes apparent when one considers the function $1/x$ which is continuous on $(0, 1]$ but not bounded on $(0, 1]$.

Lemma 2.   *If $f(x)$ is continuous on $[a, b]$ there exists at least one point $c \in [a, b]$ and at least one point $d \in [a, b]$ such that $f(d) \leqslant f(x) \leqslant f(c)$ for all $x \in [a, b]$.*

Proof.   Since, by Lemma 1, $\{f(x) | x \in [a, b]\}$ has an upper bound, by axiom OC it must have a l.u.b. which we call M. Now, given any positive number n, no matter how large, we can find x_n such that $M - f(x_n) < 1/n$ (by the definition of l.u.b.). Hence $1/(M - f(x_n)) > n$, for these x_n; this means that $1/(M - f(x))$ is unbounded. Thus, by the contrapositive of Lemma 1, $1/(M - f(x))$ is not continuous in $[a, b]$. But $M - f(x)$ is continuous in $[a, b]$, and by Exercise 16 of 12.10 this implies that there is at least one $c \in [a, b]$ such that $f(c) = M$. Hence $f(x) \leqslant f(c)$ for all $x \in [a, b]$.
The proof of the existence of $d \in [a, b]$ is analogous to the above.

We now state and prove Rolle's Theorem.

Theorem 2. Rolle's Theorem.   *If $f(x)$ is continuous on $[a, b]$ and differenti-able at every point of (a, b) and if $f(a) = f(b)$, there exists $c \in (a, b)$ such that $f'(c) = 0$.*

Proof.   From Lemma 2 we know that there exist $c, d \in [a, b]$ such that $f(d) \leqslant f(x) \leqslant f(c)$ for all $x \in [a, b]$. However, unless $f(x) = f(a)$ for all $x \in [a, b]$—in which case the theorem is trivial—there will be $x \in (a, b)$ such that one of the following holds: either $f(x) > f(a)$, or $f(x) < f(a)$. If we deal with the former case (the latter can be similarly dealt with), we see by Lemma 2 that there is such a $c \in (a, b)$ and

$$f(c + h) \leqslant f(c)   \text{or}   \frac{f(c + h) - f(c)}{h} \leqslant 0$$

for all $h > 0$. But

$$\lim_{h \to 0} \frac{f(c + h) - f(c)}{h}$$

exists since $c \in (a, b)$ and by Theorem 3 of 12.1, $f'(c) \leqslant 0$. By taking $h < 0$ we find $f'(c) \geqslant 0$. Hence $f'(c) = 0$, establishing the result.

Figure 13–9 shows that the result is intuitively obvious.

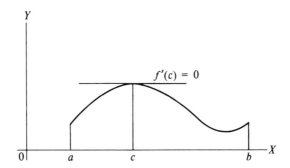

Figure 13–9   Rolle's Theorem establishes that $f'(c) = 0$

Remark 1.   In Rolle's Theorem differentiability at every point of (a, b) implies, of course, continuity on that interval. In addition, the conditions of the theorem require continuity at $x = a$ and $x = b$. If these conditions are not satisfied the theorem may not be true. Consider, for example:

$$f(x) = \frac{1}{x - 1},   1 < x \leqslant 2$$

$$f(1) = 1.$$

This function is differentiable on $(1, 2)$ and satisfies all the other conditions of the theorem except that it is not continuous at $x = 1$. For this example there is no $c \in (1, 2)$ such that $f'(c) = 0$. In the case of the function f for which

$$f(x) = \begin{cases} x - 1, & 1 \leqslant x \leqslant 2 \\ (7 - x)/5, & 2 \leqslant x \leqslant 7 \end{cases}$$

the requirement of differentiability is not satisfied on $(1, 7)$ and again there is no $c \in (1, 7)$ such that $f'(c) = 0$.

Theorem 3.   The Mean Value Theorem. *If $f(x)$ is continuous on $[a, b]$ and if $f'(x)$ exists for $x \in (a, b)$ then there is some $c \in (a, b)$ such that*

(1) $$f(b) - f(a) = f'(c)(b - a).$$

Proof.   Let

$$F(x) = f(x) - f(a) - (x - a)\frac{f(b) - f(a)}{b - a}.$$

By the hypothesis of the Mean Value Theorem, $F(x)$ is continuous on $[a, b]$ and differentiable on (a, b), and $F(b) = F(a) = 0$. Therefore the hypothesis of Rolle's Theorem is satisfied, and so there is a $c \in (a, b)$ such that $F'(c) = 0$ or

$$F'(c) = f'(c) - \frac{f(b) - f(a)}{b - a} = 0.$$

Thus

$$f(b) - f(a) = f'(c)(b - a).$$

Theorem 4.   *If $f'(x) \geqslant 0$ ($f'(x) \leqslant 0$) for $x \in (a, b)$, then $f(x)$ is monotone nondecreasing (nonincreasing) for $x \in (a, b)$.*

Proof.   Let $x_1 < x_2$ be two points in (a, b) and let $f'(x) \geqslant 0$. Then by the Mean Value Theorem applied to $[x_1, x_2]$, there is a $c \in [x_1, x_2]$ such that

$$f(x_2) - f(x_1) = (x_2 - x_1)f'(c).$$

Since $f'(c) \geqslant 0$, it follows that $f(x_2) \geqslant f(x_1)$. The case when $f'(x) \leqslant 0$ is handled analogously.

Another deduction from the Mean Value Theorem is the following.

Theorem 5. $(\forall x \in R \, f'(x) = 0) \Rightarrow (\forall x \in R \, \exists k \in R \, f(x) = k)$.

Proof.   Let x_1, x_2 $(x_1 \neq x_2)$ be any two elements of R. By the Mean Value Theorem $f(x_2) - f(x_1) = f'(c)(x_2 - x_1) = 0$ where $c \in (x_1, x_2)$ Hence $f(x_1) = f(x_2)$ for any two elements of R and $f(x) = f(x_1) = k$ for all $x \in R$.

If $f'(x) = 0$ on (a, b) and $f(x)$ is continuous on $[a, b]$, the same result holds on (a, b).

From a geometrical point of view Theorem 5 is obvious—it merely states that a curve with zero slope is a horizontal straight line. Theorem 5 can be used to identify $c(x)$ with $\cos x$ and $s(x)$ with $\sin x$.

Theorem 6. $s(x) = \sin x$, $c(x) = \cos x$ *for all* $x \in R$.

Proof.   Consider the function f defined by

$$f(x) = (s(x) - \sin x)^2 + (c(x) - \cos x)^2, \quad x \in R.$$

Now

$$f'(x) = 2(s(x) - \sin x)(s'(x) - (\sin x)') + 2(c(x) - \cos x)(c'(x) - (\cos x)')$$

$$= 2(s(x) - \sin x)(c(x) - \cos x) + 2(c(x) - \cos x)(-s(x) + \sin x)$$

$$= 0.$$

Hence $f(x) = k$ for all $x \in R$ (Theorem 5). Now

$$f(0) = (s(0) - \sin 0)^2 + (c(0) - \cos 0)^2 = (0 - 0)^2 + (1-1)^2 = 0$$

(Theorem 5 of 11.8).

Thus $(s(x) - \sin x)^2 + (c(x) - \cos x)^2 = 0$ for all $x \in R$, and therefore $s(x) = \sin x$, $c(x) = \cos x$ for all $x \in R$.

The following theorem is useful in establishing inequalities.

Theorem 7.   *Consider two functions f, g such that $f(a) = g(a)$, $f'(x) > g'(x)$ for all $x > a$. Then $f(x) > g(x)$ for all $x > a$.*

Proof.   (See Figure 13–10.) Consider $F(x) = f(x) - g(x)$. Now $F(a) = 0$ and $F'(x) = f'(x) - g'(x) > 0$ for all $x > a$. By the Mean Value Theorem, if $x > a$, there is a $c, a < c < x$ such that

$$F(x) - F(a) = F'(c)(x - a).$$

Since $F'(c) > 0$ and $x - a > 0$, $F(x) > 0$ for all $x > 0$, and so $f(x) > g(x)$ for all $x > 0$.

Example 1.   Using Theorem 7 we prove that

$$\log_a(1 + x) < x \text{ if } x > 0 \text{ and } 1 < a < e.$$

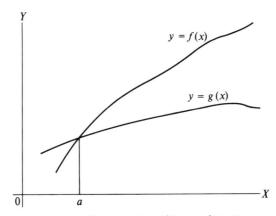

Figure 13–10 $f(a) = g(a)$, $f'(x) > g'(x)$ for $x > a$

Now if $x = 0$, $\log_a(1 + x) = \log_a 1 = 0$, and

$$D_x \log_a(1 + x) = \frac{\log_a e}{1 + x}.$$

Since $1 + x > \log_a e$, for $x > 0$, and $1 < a < e$, we have

$$1 > \frac{\log_a e}{1 + x}.$$

Thus, by Theorem 7, $x > \log_a(1 + x)$ if $x > 0$, and $1 < a < e$.

13.7 DIFFERENTIATION FORMULAS

In this final section we list the derivatives of some of the most common functions. This list together with the Chain Rule and the Differentiation Theorems of 13.3 will be sufficient to find the derivatives of any of the **elementary functions**.

An elementary function by definition is any function which can be formed by applying the operations of addition, multiplication, division, exponentiation, composition, and inverse to any of the constant, identity, trigonometric, and exponential functions. Thus the functions

$$x \to \frac{x^3 + x}{x + 1}, \quad x \to \sin^{-1} x, \quad x \to \sqrt{(\tan x)}, \quad \text{and} \quad x \to \sin(\tan^{-1} x)$$

are all elementary functions.

In the following list some of the formulas have already been derived in the text while others are left to the reader as exercises.

(i)
$$\frac{dx^n}{dx} = nx^{n-1} \text{ for } n \in R,$$

(ii)
$$\frac{da^x}{dx} = (\log_e a) a^x, \quad \text{for } a > 0,$$

(iii)
$$\frac{d \sin x}{dx} = \cos x,$$

(iv)
$$\frac{d \cos x}{dx} = -\sin x,$$

(v)
$$\frac{d \tan x}{dx} = \sec^2 x,$$

(vi)
$$\frac{d \sec x}{dx} = \sec x \tan x,$$

(vii)
$$\frac{d \cot x}{dx} = -\csc x,$$

(viii)
$$\frac{d \csc x}{dx} = -\csc x \cot x,$$

(ix)
$$\frac{d \log_a x}{dx} = \frac{\log_a e}{x} \text{ for } a > 0,$$

(x)
$$\frac{d \sin^{-1} x}{dx} = \frac{1}{\sqrt{(1 - x^2)}},$$

(xi)
$$\frac{d \cos^{-1} x}{dx} = \frac{-1}{\sqrt{(1 - x^2)}},$$

(xii)
$$\frac{d \tan^{-1} x}{dx} = \frac{1}{1 + x^2}.$$

13.8 EXERCISES

1. Show that the function f with $f(x) = [x]$ has $f'(x) = 0$ at every x where
$$[x] \neq x,$$

and that the derivative is undefined when
$$[x] = x.$$
Remember that $[x]$ stands for the greatest integer in x. See section 8.3.

2. Find the derivative function f' of f in each of the following cases.

(a)　　$f(x) = 3x^5 - 2x^4$;　　　　　(b)　　$f(x) = \sqrt[5]{2(x^2 - 1)}$;

(c)　　$f(x) = \dfrac{x^2}{x^3 - 1}$;　　　　　(d)　　$f(x) = e^{-x^2 + x}$;

(e)　　$f(x) = \tan x$;　　　　　(f)　　$f(x) = \sec x$;

(g)　　$f(x) = \cot x$;　　　　　(h)　　$f(x) = \csc x$;

(i)　　$f(x) = \sin^{-1}x$;　　　　　(j)　　$f(x) = \tan^{-1}x$;

(k)　　$f(x) = e^{ax}\sin bx$;　　　　　(l)　　$f(x) = e^{x\sin x}$.

3. Prove that the set of functions which are differentiable at every point of R forms a ring under the operations $+$ and $\cdot$ defined below:
$$(f + g)(x) = f(x) + g(x); \quad (f \cdot g)(x) = f(x) \cdot g(x).$$

4. (a) If $f(x)$ and $g(x)$ are continuous functions for $x > 0$, and if $\lim\limits_{x \to +\infty} f(x) = \lim\limits_{x \to +\infty} g(x) = \infty$, then show that
$$\lim_{x \to +\infty} f \circ g(x) = +\infty;$$

(b) Show that for any $n \in N - \{0\}$
$$\lim_{x \to +\infty} x^{1/n} = +\infty.$$

(c) Use part (b) to show that
$$\lim_{x \to +\infty} x^a = +\infty$$

for any real $a > 0$.

5. Prove Expression (4) of Theorem 1 in 13.3.

6. Show that if f is a monotone increasing function on R and if $\lim\limits_{x \to \infty} f(x) = \infty$, then
$$\lim_{x \to \infty} f^{-1}(x) = \infty$$

where f^{-1} is the inverse of the function f, so that $f^{-1} \circ f = f \circ f^{-1} = $ the identity function.

7. Prove that

(a)　　$\dfrac{d\cos^{-1}x}{dx} = -\dfrac{1}{\sqrt{(1 - x^2)}}$;

(b)　　$\dfrac{d\cot^{-1}x}{dx} = \dfrac{-1}{1 + x^2}$.

8. Find the derivative of each of the following expressions involving inverse trigonometric functions.

(a)　　$\sin^{-1}\dfrac{x^2}{2}$　　　　　(b)　　$\tan^{-1}\dfrac{x - 1}{x + 1}$

(c)　　$x(\sin^{-1}x)^2 + 2\sqrt{(1 - x^2)}\sin^{-1} x$.

(d)　　$\cos^{-1}\sqrt{x}$　　　　　(e)　　$\cot^{-1}x^2$

(f)　　$\sec^{-1} x$

9. Prove that the sequence $\{a_n\}$ where

$$a_n = 1 + \frac{1}{2} + \frac{1}{3} + \dots + \frac{1}{n} - \log n \quad \text{converges.}$$

Note: The limit of this sequence is denoted by $\gamma = 0.5772 \dots$ and is called Euler's constant. A long-standing problem (as yet unsolved) is to prove that γ is irrational.

10. By using the power series representations of $\sin x$ and $\cos x$ establish that
$$\sin (x + y) = \sin x \cos y + \cos x \sin y.$$

11. Prove that
$$\sin x < x$$
for all $x > 0$.

12. Prove that
$$\log x < x$$
for all $x > 0$.

13. Prove that
$$1 + x + \frac{x^2}{2!} < e^x$$
for $x > 0$.

14. Using the method of proof of Theorem 5 of 13.6, prove that if $f'(x) = 0$ for all $x \in (a, b)$ then for any closed subinterval $[c, d]$ of (a, b), $f(x)$ is a fixed constant value for all $x \in [c, d]$.

15. Use 14 to show that if $f'(x) = 0$ for all $x \in (a, b)$ then there is a constant k such that
$$f(x) = k$$
for all $x \in (a, b)$.

16. Show that
$$x \sin x + \cos x - x^2$$
has exactly two real zeros.

17. Prove that

(a) $\quad \log(1 + x) > x - \dfrac{x^2}{2},\qquad\qquad x > 0.$

(b) $\quad \log \dfrac{1 + x}{1 - x} > 2\left(x + \dfrac{x^3}{3}\right),\qquad 0 < x < 1.$

(c) $\quad \left(1 + \dfrac{x}{n}\right)^n < e^x < \left(1 - \dfrac{x}{n}\right)^{-n},\qquad n \in N, \quad x > 0.$

14

Integration and Antidifferentiation

14.1   INTEGRATION

In Chapter 11 we discussed antidifferences and sums of sequences, that is, functions defined on the discrete set $N = \{0, 1, 2, ...\}$ or some subset of N. In this chapter we discuss the analogs of the antidifference and the sum for functions defined on an interval $[a, b]$. The names of these analogs are **antiderivative** and **integral**, respectively.

This chapter is by no means meant to be a complete treatment of the mechanical procedures of finding antiderivatives and integrals, but only an introduction to these concepts.

To make the analogy clear, it is useful to think of a sum $\Sigma_{k=1}^{n} a_k$ for $a_k \geqslant 0$ as the area under the step function f (that is, between f and the

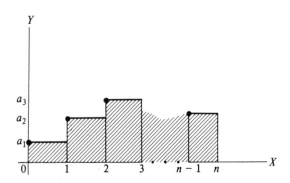

Figure 14–1 $\displaystyle\sum_{k=1}^{n} a_k$ is the area of the shaded region

x-axis) where

$$f(x) = a_k \text{ for } k - 1 \leqslant x < k \quad (k = 1, 2, \ldots, n)$$

as indicated in Figure 14–1. The analog for a continuous function f on an interval $[a, b]$ with $f(x) \geqslant 0$ is the area under the graph of f as in Figure 14–2.

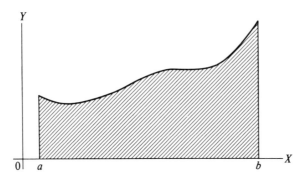

Figure 14–2

For an arbitrary continuous function f it is not possible to arrive at an exact value for this area in quite as straightforward a manner as one can in the situation depicted in Figure 14–1. Indeed, it is not immediately obvious how to define the notion of area in this case.

However, to estimate the "area" under $f(x)$ (see Figure 14–3), divide the interval $[a, b]$ into n equal subintervals by inserting points between a and b.

$$(1) \quad a < a + \frac{b-a}{n} < a + 2\frac{b-a}{n} < \cdots < a + (n-1)\frac{b-a}{n} < b.$$

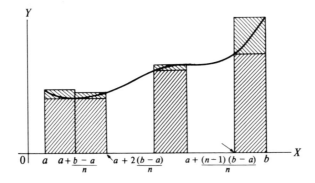

Figure 14–3

As indicated in Figure 14–4, we denote the kth closed interval by

$$I_k = \left[a + (k-1)\frac{b-a}{n}, a + k\frac{b-a}{n} \right];$$

there are two points $x_1, x_2 \in I_k$ where $f(x)$ assumes its extreme values:

(2) $\qquad m_k = f(x_1) \leqslant f(x) \leqslant f(x_2) = M_k \quad \text{for all} \quad x \in I_k.$

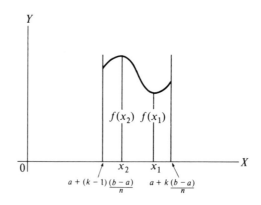

Figure 14–4

This follows from Lemma 2 of 13.6 since $f(x)$ is continuous on $[a, b]$, and consequently on each I_k. Now for each I_k (see Figure 14–3)

$$m_k \left(\frac{b-a}{n} \right) \leqslant M_k \left(\frac{b-a}{n} \right).$$

Thus if we add up the contributions for each I_k, we obtain

(3) $\qquad d_n = \sum_{k=1}^{n} m_k \left(\frac{b-a}{n} \right) \leqslant \sum_{k=1}^{n} M_k \left(\frac{b-a}{n} \right) = D_n.$

Now we can write

$$m \leqslant f(x) \leqslant M, \quad \text{for all} \quad x \in [a, b].$$

We know that m and M both exist by Lemma 2 of 13.6 (see Figure 14–5). Thus the sequences $\{d_n\}$ and $\{D_n\}$ are respectively bounded above and below by $M(b-a)$ and $m(b-a)$. Consequently l.u.b. d_n and g.l.b. D_n both exist (by OC and Theorem 1 of 8.3).

We shall prove that l.u.b. $d_n = $ g.l.b. D_n, but in order to do this we need the following two lemmas.

Lemma 1. l.u.b. $d_n \leqslant$ g.l.b. D_n.

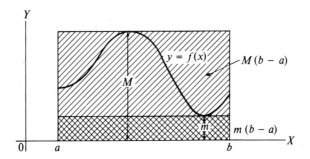

Figure 14–5

Proof. Assume that this is not the case and obtain a contradiction. If g.l.b. $D_n <$ l.u.b. d_n then

$$\exists n_0 \in N, d_{n_0} > \text{g.l.b. } D_n \text{ (definition of l.u.b.).}$$

Further

$$\exists m_0 \in N, D_{m_0} < d_{n_0} \text{ (definition of g.l.b.).}$$

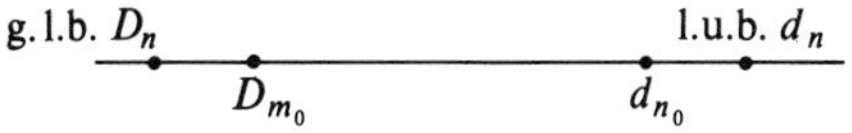

g.l.b. D_n l.u.b. d_n

D_{m_0} d_{n_0}

Figure 14–6

Since the set

$$\left\{ \frac{b-a}{m_0 n_0}, \frac{2(b-a)}{m_0 n_0}, \ldots, \frac{m_0 n_0 (b-a)}{m_0 n_0} \right\}$$

contains both the sets

$$\left\{ \frac{b-a}{n_0}, \frac{2(b-a)}{n_0}, \ldots, \frac{n_0 (b-a)}{n_0} \right\}$$

and

$$\left\{ \frac{b-a}{m_0}, \frac{2(b-a)}{m_0}, \ldots, \frac{m_0 (b-a)}{m_0} \right\},$$

it follows that

$$D_{m_0 n_0} \leqslant D_{m_0} \quad \text{and} \quad d_{n_0} \leqslant d_{m_0 n_0}.$$

Thus

$$D_{m_0 n_0} \leqslant D_{m_0} < d_{n_0} \leqslant d_{m_0 n_0}$$

and so

$$D_{m_0 n_0} < d_{m_0 n_0}.$$

Since, from (3),

$$d_{mono} \leqslant D_{mono},$$

the lemma is proved.

Lemma 2. *Let f be a continuous function on $[a, b]$. Then for every $\varepsilon > 0$ there exists an n such that*

$$M_k - m_k < \frac{\varepsilon}{b - a}$$

on each I_k, $k = 1, 2, \ldots, n$.

Proof. Again, as in Lemma 1, we assume that this is not the case and obtain a contradiction.

Suppose for some ε, say ε_1, there is no n such that $M_k - m_k < \varepsilon_1/(b - a)$ on each I_k, $k = 1, 2, \ldots, n$. Let c_1 be the mid-point of $[a, b]$. Then the lemma is false in either $[a, c_1]$ or $[c_1, b]$. (If it were not, the lemma would be true.) We denote the interval in which the theorem is false by J_1. If it is false in both $[a, c_1]$ and $[c_1, b]$ then let J_1 be the left interval.

Now denote the mid-point of J_1 by c_2. Repeat the above procedure denoting the interval in which the lemma is false (or the left one if it is false in both) by J_2. This process can be continued indefinitely since by our assumption there is no n with the required property.

Now consider the set S of left end-points of the closed intervals $J_1, J_2, J_3, \ldots$. Since S is contained in $[a, b]$ it is bounded and hence by OC has a least upper bound in $[a, b]$. Denote l.u.b. S by l. Note that the set $\{J_i\}$ is a nested set of intervals (cf. 8.5): each interval contained in its predecessor. Consequently l is always contained in each of the J_i. Also note that the length of J_i is $(b - a)/2^i$.

However, since f is continuous at $x = l$ we can always choose some symmetric interval, say $[l - \delta, l + \delta]$, such that $M_k - m_k < \varepsilon_1/(b - a)$ (if $l = a$ or b choose the appropriate half interval).

Now if we choose n so that $(b - a)/2^n < \delta$ we will have $J_n \subseteq [l - \delta, l + \delta]$. But this leads to a contradiction since in J_n the theorem is false and in $[l - \delta, l + \delta]$ it is true; hence the lemma is proved.

Remark 1. This lemma is not true if the closed interval in the statement of the lemma is replaced by an open (or half-open) interval. For, under those circumstances, l may not be in the open interval and a contradiction will not be obtained. As an example of this situation consider $f(x) = 1/x$ on $(0, 1]$. This function is continuous on $(0, 1]$, but since $l = 0$ no contradiction is obtained and the lemma is not true.

Lemma 3. *If f is continuous on $[a, b]$*

$$\lim_n (D_n - d_n) = 0.$$

Proof. From (3) we have

(4)
$$D_n - d_n = \sum_{k=1}^{n} (M_k - m_k)\frac{(b-a)}{n}.$$

(5)
$$0 \leqslant D_n - d_n < \sum_{k=1}^{n} \frac{\varepsilon}{b-a}\left(\frac{b-a}{n}\right) = \varepsilon$$

(for all sufficiently large n, by Lemma 2). Hence

$$\lim_{n} (D_n - d_n) = 0.$$

We use these lemmas to prove

Theorem 1. l.u.b. d_n = g.l.b. D_n.

Proof. From Lemma 1 and the definitions of l.u.b. and g.l.b, we have

$$d_m \leqslant \text{l.u.b. } d_n \leqslant \text{g.l.b. } D_n \leqslant D_m \text{ for all } m \in N.$$

Therefore
$$0 \leqslant \text{g.l.b. } D_n - \text{l.u.b. } d_n \leqslant D_m - d_m, \text{ for all } m \in N.$$

Hence

(6) $$\lim_{m} |\text{g.l.b. } D_n - \text{l.u.b. } d_n| \leqslant \lim_{m} (D_m - d_m), \quad \text{(Exercise 15(c) of 11.9)}.$$

Since $|\text{g.l.b. } D_n - \text{l.u.b. } d_n|$ is independent of m, we have

$$\lim_{m} |\text{g.l.b. } D_n - \text{l.u.b. } d_n| = |\text{g.l.b. } D_n - \text{l.u.b. } d_n|$$
$$= \text{g.l.b. } D_n - \text{l.u.b. } d_n.$$

Thus by (6) and Lemma 3
$$\text{g.l.b. } D_n = \text{l.u.b. } d_n.$$

Now suppose we remove the restriction that $f(x) \geqslant 0$ for $x \in [a, b]$. Then we still have (3), and so

$$d_n \leqslant \sum_{k=1}^{n} M_k \frac{b-a}{n}$$

$$\leqslant M \sum_{k=1}^{n} \frac{b-a}{n}$$

$$= M(b-a).$$

Thus $d_n \leqslant D_n \leqslant M(b-a)$. Similarly it can be shown that $D_n \geqslant m(b-a)$. Consequently l.u.b. d_n and g.l.b. D_n exist. If the reader examines the proof of Theorem 1 he will see that it does not require the assumption $f(x) \geqslant 0$, so the result of the theorem is true without restriction on the positivity of f.

We now make the following definition.

Definition 1.   *The **integral from** a **to** b of a function f defined on $[a, b]$, and denoted by $\int_a^b f(x)\, dx$, is*

(7) $$\int_a^b f(x)\, dx = \text{l.u.b. } d_n = \text{g.l.b. } D_n,$$

whenever these last two are, in fact, defined and equal.

Remark 2.   Theorem 1 asserts that $\int_a^b f(x)\, dx$ exists for functions f continuous on $[a, b]$ ($\int_a^b f(x)\, dx$ exists for a larger class of functions, but we shall not pursue the matter here). If f is non-negative and if $\int_a^b f(x)\, dx$ exists, then the above discussion makes it clear that $\int_a^b f(x)\, dx$ is the appropriate definition for the area bounded by $x = a$, $x = b$, $y = 0$, and $y = f(x)$.

Remark 3.   The variable x in $\int_a^b f(x)\, dx$ is a dummy or bound variable (cf. the remarks in 1.4, 10.6, 11.4). Consequently we can also write $\int_a^b f(t)\, dt$ or $\int_a^b f(u)\, du$ and be assured that $\int_a^b f(x)\, dx = \int_a^b f(t)\, dt = \int_a^b f(u)\, du$. The value of (7) depends only on the numbers a and b (sometimes called the **lower** and **upper limits**, respectively, of the integral) and the function f is usually called the *integrand*. The notation $\int_a^b f(x)\, dx$ (due to Leibniz, the inventor of integration, to stand for the common value of l.u.b. d_n and g.l.b. D_n) is suggestive of the procedure by which the integral was defined. The symbol dx is useful in a technique for evaluating integrals called the method of substitution.

Let us now try to evaluate some integrals.

Example 1.   Find

$$\int_0^1 x\, dx.$$

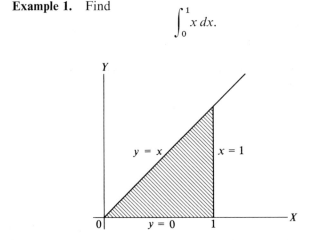

Figure 14–7

Since $x \geqslant 0$ for $x \in [0, 1]$ we can interpret $\int_0^1 x\, dx$ as the area between $x = 0$, $x = 1$, $y = 0$, $y = x$. Now this region is a triangle and we see that its area is $\frac{1}{2}$, that is, $\int_0^1 x\, dx = \frac{1}{2}$. Let us confirm this result by a formal calculation using Definition 1.

On the interval

$$I_k = \left[\frac{k-1}{n}, \frac{k}{n}\right], \quad M_k = \frac{k}{n}.$$

Hence, by (3)

$$D_n = \sum_{k=1}^{n} M_k \cdot \frac{1}{n} = \frac{1}{n^2} \sum_{k=1}^{n} k.$$

In section 2.5 it was proved that

$$\sum_{k=1}^{n} k = \frac{n(n+1)}{2}.$$

Hence

$$D_n = \frac{1}{n^2} \frac{n(n+1)}{2} = \frac{1}{2} + \frac{1}{2n}.$$

Now

$$\text{g.l.b. } D_n = \tfrac{1}{2}.$$

Similarly, on I_k, $m_k = (k-1)/n$, and

$$d_n = \sum_{k=1}^{n} \frac{k-1}{n} \cdot \frac{1}{n} = \frac{1}{n^2} \left(\sum_{k=1}^{n} k - \sum_{k=1}^{n} 1 \right)$$

$$= \frac{1}{n^2} \left(\frac{n(n+1)}{2} - n \right)$$

$$= \frac{1}{2} - \frac{1}{2n}.$$

Thus l.u.b. $d_n = \frac{1}{2}$.

Since g.l.b. $D_n = $ l.u.b. $d_n = \frac{1}{2}$,

$$\int_0^1 x\, dx = \tfrac{1}{2} \quad \text{by Definition 1.}$$

Remark 4. In the preceding example, we see that

$$\lim_{n} d_n = \text{l.u.b. } d_n$$

and

$$\lim_{n} D_n = \text{g.l.b. } D_n.$$

It can be shown that if f is integrable on $[a, b]$ then

$$\lim_n d_n = \lim_n D_n = \int_a^b f(x)\, dx$$

(see *A Modern Introduction to Calculus*, page 127, by W. Maak).

Example 2. Find $\int_0^1 x^2\, dx$.

As in Example 1 the function is continuous and increasing, so it assumes its maximum value in the interval $[(k-1)/n,\, k/n]$ at the right end-point. Thus if $f(x) = x^2$,

$$M_k = f\left(\frac{k}{n}\right) = \frac{k^2}{n^2},$$

and so

$$\int_0^1 x^2\, dx = \lim_n \sum_{k=1}^n \left(\frac{k^2}{n^2}\frac{1}{n}\right) = \left(\lim_n \frac{1}{n^3}\sum_{k=1}^n k^2\right). \qquad \text{(See Remark 4.)}$$

From Example 10 in 11.2 we obtain

$$\int_0^1 x^2\, dx = \lim_n \frac{n(n+1)(2n+1)}{6n^3} = \frac{1}{3}.$$

Example 3. The preceding examples involved the evaluation of integrals of continuous functions. We now consider an example in which the function is discontinuous and, as we shall show, not integrable.

Suppose f is the function with domain $[0, 1]$ defined by

$$f(x) = \begin{cases} 0, & x \in ([0,1] \cap Q) \\ 1, & x \in [0,1] - ([0,1] \cap Q) \end{cases}$$

Then by Corollaries 1 and 2 of 8.3, for any k, $m_k = 0$, $M_k = 1$. Thus by (3)

$$d_n = \sum_{k=1}^n m_k \cdot \frac{1}{n} = 0, \text{ for all } n \in N$$

and

$$D_n = \sum_{k=1}^n M_k \cdot \frac{1}{n} = 1, \text{ for all } n \in N.$$

Since l.u.b. $d_n = 0$, g.l.b. $D_n = 1$, by Definition 1, f is not integrable. It should not be supposed that all discontinuous functions are not integrable. For example, the step function illustrated in Figure 14–1 is integrable on any real interval.

The following theorem is helpful for evaluating integrals; its proof, which we omit, can be found in any standard text on calculus as well as in *Theory of Functions of a Real Variable*, by E. W. Hobson, Dover, Vol. I, Chapter VI.

Theorem 2.   *Let f and g be continuous functions on* $[a, b]$. *Let c be a real number such that* $a < c < b$, *and let d, e be two arbitrary real numbers. Then*

$$(8) \qquad \int_a^b \big(df(x) + e\,g(x)\big)\,dx = d \int_a^b f(x)\,dx + e \int_a^b g(x)\,dx,$$

$$(9) \qquad \int_a^b f(x)\,dx = \int_a^c f(x)\,dx + \int_c^b f(x)\,dx,$$

and, if $f(x) \leqslant g(x)$ *for all* $x \in [a, b]$,

$$(10) \qquad \int_a^b f(x)\,dx \leqslant \int_a^b g(x)\,dx.$$

Finally, if $f(x) = c$ *for all* $x \in [a, b]$, *then*

$$(11) \qquad \int_a^b f(x)\,dx = \int_a^b c\,dx = (b - a)\,c.$$

$$(12) \qquad \left| \int_a^b f(x)\,dx \right| \leqslant \int_a^b |f(x)|\,dx.$$

For the sake of completeness we adopt the convention

$$\int_a^a f(x)\,dx = 0,$$

because

$$\lim_{b \to a+} \int_a^b f(x)\,dx = 0.$$

The reader is asked to prove this in Exercise 2 of 14.6. We also adopt the convention that

$$\int_a^b f(x)\,dx = - \int_b^a f(x)\,dx$$

(See Exercise 4 of 14.6.) and so (8) to (11) hold for *any* a, b, c.

14.2   AREAS

The remark was made following Definition 1 that if $f(x) \geqslant 0$ on $[a, b]$, then we adopt the definition that $\int_a^b f(x)\,dx$ is the area enclosed by $x = a$, $x = b$, $y = 0$, and $y = f(x)$. Let us denote, in general, the area enclosed by $x = a$, $x = b$, $y = 0$, and $y = f(x)$ by $A_a^b (f)$. Then in the case $f(x) \geqslant 0$ on $[a, b]$

$$A_a^b (f) = \int_a^b f(x)\, dx.$$

Let us now examine the problem of finding area if $f(x) \leqslant 0$ on $[a, b]$. (Figure 14–8.) Then, by Theorem 2 (10) of 14.1 with $g(x) = 0$,

$$\int_a^b f(x)\, dx \leqslant 0.$$

Since we adopt the convention that the area enclosed by a curve is a positive number, it is clear that $\int_a^b f(x)\, dx$ does not represent the area in this case. We adopt the following geometric procedure.

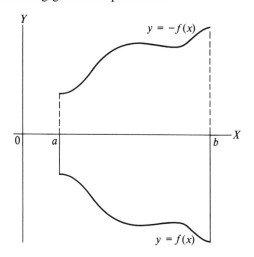

Figure 14–8   The graph of f is reflected in $y = 0$

If we reflect the graph of f in the line $y = 0$, forming the graph of $-f$, it is evident that the region enclosed by $x = a$, $x = b$, $y = 0$, $y = -f(x)$ is congruent to the region enclosed by $x = a$, $x = b$, $y = 0$, $y = f(x)$, and consequently has the same area. That is,

$$A_a^b (-f) = A_a^b (f).$$

But

$$A_a^b (-f) = \int_a^b -f(x)\, dx \quad \text{(by our previous reasoning)}$$

$$= -\int_a^b f(x)\, dx \quad \text{(by Theorem 2(8) of 14.1, with } e = 0, d = -1)$$

Thus if $f(x) \leqslant 0$ on $[a, b]$

(1)
$$A_a^b(f) = -\int_a^b f(x)\, dx.$$

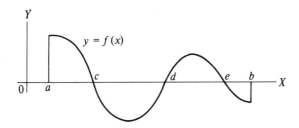

Figure 14–9

Now suppose f is such that

$$f(x) \geqslant 0 \quad x \in [a, c],$$
$$f(x) \leqslant 0 \quad x \in [c, d],$$
$$f(x) \geqslant 0 \quad x \in [d, e],$$
$$f(x) \leqslant 0 \quad x \in [e, b].$$

Then by Theorem 2(8) of 14.1

$$\int_a^b f(x)\, dx = \int_a^c f(x)\, dx + \int_c^d f(x)\, dx + \int_d^e f(x)\, dx + \int_e^b f(x)\, dx.$$

Now
$$A_a^b(f) = A_a^c(f) + A_c^d(f) + A_d^e(f) + A_e^b(f).$$

But
$$A_c^d(f) = -\int_c^d f(x)\, dx$$

$$A_e^b(f) = -\int_e^b f(x)\, dx \quad \text{(by 1)}.$$

Thus

(2) $\quad A_a^b(f) = \int_a^c f(x)\, dx - \int_c^d f(x)\, dx + \int_d^e f(x)\, dx - \int_e^b f(x)\, dx.$

Example 1. Find

$$\int_0^1 (3x^2 - 2x)\, dx.$$

From Theorem 2 of 14.1 it follows that

$$\int_0^1 (3x^2 - 2x)\, dx = 3 \int_0^1 x^2\, dx - 2 \int_0^1 x\, dx,$$

and from Examples 1 and 2 of 14.1 we obtain

$$\int_0^1 (3x^2 - 2x)\, dx = 3(\tfrac{1}{3}) - 2(\tfrac{1}{2})$$

$$= 1 - 1 = 0.$$

Let us now find the area enclosed by $y = 3x^2 - 2x$, $x = 0$, $x = 1$, $y = 0$ (note the line $x = 0$ reduces to a point in this case—see Figure 14–10). By (2)

$$A_0^1 (3x^2 - 2x) = -\int_0^{2/3} (3x^2 - 2x)\, dx + \int_{2/3}^1 (3x^2 - 2x)\, dx.$$

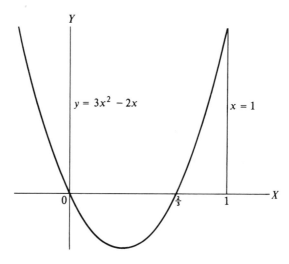

Figure 14–10

Now using Theorem 2(8) of 14.1 and the same procedures as in Examples 1 and 2 of 14.1 we find

$$-\int_0^{2/3} (3x^2 - 2x)\, dx = \int_{2/3}^1 (3x^2 - 2x)\, dx = \frac{4}{27}.$$

Thus

$$A_0^1 (3x^2 - 2x) = \frac{4}{27} + \frac{4}{27} = \frac{8}{27}.$$

14.3 ANTIDERIVATIVES AND THE FUNDAMENTAL THEOREM OF CALCULUS

As we have seen in Section 14.1, the process of integration requires the evaluation of least upper bounds and greatest lower bounds of certain sums. Even in the simple cases considered in that section this was not an easy task. In this section the whole problem is put in a different light. We prove the Fundamental Theorem of Calculus. This theorem relates integration and differentiation, thus enabling us to use already acquired differentiation techniques in the process of integration.

To prove the Fundamental Theorem we first prove the following theorem which has an obvious geometric interpretation. In geometric terms the theorem states that there is some rectangle on $[a, b]$ as base, whose area is equal to the area enclosed by $x = a$, $x = b$, $y = 0$, $y = f(x)$ (see Figure 14–11). The proof depends on the assumption that f is a continuous function.

Theorem 1. *The Mean Value Theorem for Integrals.*

If f is continuous on $[a, b]$ then there exists a $c \in [a, b]$ such that

(1)
$$\int_a^b f(x)\, dx = f(c)\, (b - a).$$

Proof. By Lemma 2 of 13.6, since f is continuous, there exist numbers m, M such that
$$m \leqslant f(x) \leqslant M \text{ for all } x \in [a, b].$$

By Theorem 2(10) of 14.1 with $g(x) = M$
$$\int_a^b f(x)\, dx \leqslant \int_a^b M\, dx$$

$$= M\, (b - a).$$

Similarly
$$m\, (b - a) \leqslant \int_a^b f(x)\, dx,$$

or

(2)
$$m\, (b - a) \leqslant \int_a^b f(x)\, dx \leqslant M\, (b - a).$$

We can exclude the case $b = a$, for then (1) is trivially true. Now divide through (2) by $b - a > 0$ and we get

$$m \leqslant \frac{1}{b - a} \int_a^b f(x)\, dx \leqslant M.$$

Now, by Theorem 2 of 12.9, since f is continuous on $[a, b]$ there is a $c \in [a, b]$ such that

$$f(c) = \frac{1}{b - a} \int_a^b f(x)\, dx,$$

or

$$\int_a^b f(x)\, dx = f(c)\,(b - a).$$

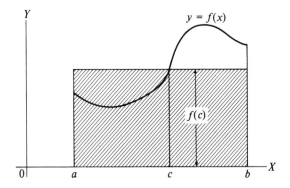

Figure 14–11   The area of the shaded rectangle is equal to the area enclosed by $x = a$, $x = b$, $y = 0$, $y = f(x)$

For continuous functions more complicated than x or x^2 and combinations of these, the techniques of Examples 1, 2, and 4 of 14.1 become extremely laborious. Fortunately the following useful theorem (which is closely related to Theorem 4, the Fundamental Theorem) provides another approach to the process of integration.

Theorem 2.   *If f is continuous on $[a, b]$ then the function F on $[a, b]$ defined by*

$$F(x) = \int_a^x f(t)\, dt$$

satisfies

$$F'(x) = f(x) \text{ for } x \in (a, b).$$

Proof.   For any $x \in (a, b)$

$$\frac{F(x + h) - F(x)}{h} = \frac{1}{h} \left[\int_a^{x+h} f(t)\, dt - \int_a^x f(t)\, dt \right]$$

Since

$$\int_a^{x+h} f(t)\, dt = \int_a^x f(t)\, dt + \int_x^{x+h} f(t)\, dt \quad \text{by (9) of 14.1}$$

$$\frac{F(x+h) - F(x)}{h} = \frac{1}{h} \int_x^{x+h} f(t)\, dt.$$

Now by Theorem 1, we have

$$\int_x^{x+h} f(t)\, dt = hf(\xi)$$

where $\xi \in [x, x+h]$.

Thus

$$\frac{F(x+h) - F(x)}{h} = f(\xi).$$

Since f is continuous

$$\lim_{h \to 0} f(\xi) = f(x).$$

Thus $\lim_{h \to 0} \dfrac{F(x+h) - F(x)}{h} = F'(x)$ exists, and $F'(x) = f(x)$.

Remark 1. Integrals of the form $\int_a^x f(t)\, dt$ are usually called **indefinite integrals,** while integrals like $\int_a^b f(t)\, dt$ are called **definite integrals.** In an indefinite integral the necessity of using a dummy variable (such as t, above) different from the upper limit, x, is apparent.

Remark 2. Theorem 2 also establishes that the indefinite integral of a continuous function is differentiable and, consequently, continuous.

Let us now see how we can use Theorem 2 to evaluate integrals. As an example we shall examine $F(x) = x^3/3$. We see that $F'(x) = f(x) = x^2$. Does this mean, by Theorem 2, that

$$(3) \qquad\qquad \frac{x^3}{3} = \int_a^x t^2\, dt\,?$$

We have no reason to suppose that (3) is true since we see that any $F(x)$ of the form $F(x) = x^3/3 + c$, where c is an arbitrary constant, also possesses the property that $F'(x) = x^2$. Thus we might just as well say that

$$(4) \qquad\qquad \frac{x^3}{3} + c = \int_a^x t^2\, dt.$$

In fact, (4) is correct for a certain choice of c, and the procedure of using Theorem 2 to evaluate integrals hinges on a method for determining the correct c. Before we investigate this we discuss an important related concept, the **antiderivative**.

Definition 1. *If $F'(x) = f(x)$ for all x in an open interval (a, b), then F is called an **antiderivative** of f in (a, b).*

As has been observed above, if F is an antiderivative of f, then so also is $F + c$ an antiderivative of f, where c is an arbitrary constant.

Theorem 3. *If F_1 and F_2 are antiderivatives of f, then*

(5) $$F_1(x) = F_2(x) + c.$$

Proof. Since by hypothesis

$$F_1'(x) = F_2'(x) = f(x),$$

it follows that

$$\frac{d(F_1 - F_2)(x)}{dx} = 0.$$

Theorem 5 of 13.6 then insures the existence of a real number c such that

$$(F_1 - F_2)(x) = F_1(x) - F_2(x) = c.$$

Hence

$$F_1(x) = F_2(x) + c.$$

Therefore if we obtain one antiderivative F of f, all others can be manufactured merely by adding constant functions to F. Note how closely these results resemble the results of 11.2 for the antidifference.

To pass from the antiderivative to the integral also resembles the process of passing from the antidifference to the sum. (See 11.2.) From Theorem 2 we know that

(6) $$F(x_0) = \int_a^{x_0} f(t)\, dt$$

where F is one of the antiderivatives of f. Suppose we have determined one such antiderivative, say F_0. Then F can be written

(7) $$F(x) = F_0(x) + c$$

for some constant c.

Now

$$F(a) = \int_a^a f(t)\, dt = 0,$$

and so from (7)

$$F(a) = F_0(a) + c = 0.$$

Thus

$$c = -F_0(a)$$

and

$$\int_a^{x_0} f(t)\, dt = F(x_0) = F_0(x_0) - F_0(a).$$

Therefore for $x_0 = b$

$$\int_a^b f(t)\, dt = F_0(b) - F_0(a).$$

The above considerations provide the proof of the following theorem (usually called the Fundamental Theorem of Calculus since it connects the two apparently unrelated processes of differentiation and integration).

Theorem 4. *The Fundamental Theorem of Calculus*

If $\dfrac{dF}{dx} = f(x)$ *on* (a, b) *where* f *is continuous on* $[a, b]$, *then*

$$\int_a^b f(t)\, dt = F(b) - F(a).$$

Corollary 1.

$$\int_a^b f'(x)\, dx = f(b) - f(a).$$

Let us now apply Theorem 4 to the evaluation of some integrals.

Example 1.

(8) $$\int_a^b t^2\, dt = \frac{b^3}{3} - \frac{a^3}{3}.$$

We have found

$$\frac{d\, x^3/3}{dx} = x^2.$$

Therefore $x^3/3$ is an antiderivative of x^2 on (a, b). Thus (8) follows from Theorem 4 if we let

$$F(x) = \frac{x^3}{3}.$$

Example 1 can be generalized to give the following result.

Theorem 5. *If $n \neq -1$, then $x^{n+1}/(n+1)$ constitutes an antiderivative of x^n and*

(9)
$$\int_a^b x^n \, dx = \frac{b^{n+1}}{n+1} - \frac{a^{n+1}}{n+1}.$$

Proof. First note that if $F(x) = x^{n+1}/(n+1)$

$$\frac{dF(x)}{dx} = \frac{(n+1)x^n}{n+1} = x^n.$$

Then applying Theorem 4 we obtain (9).

Theorem 5 can be used in Examples 1 and 2 of 14.1 to give a very succinct solution. It is clear that the case $n = -1$ must be excluded for then $x^{n+1}/(n+1)$ is not defined. However, if $0 \notin [a, b]$, then $1/x$ is continuous on $[a,b]$ and

$$\int_a^b \frac{dx}{x}$$

exists. In addition, from (4) in 13.4, it follows that $\log_e x$ is an antiderivative of $1/x$. Therefore if $0 < a < b$, then

(10)
$$\int_a^b \frac{dx}{x} = \log_e b - \log_e a.$$

On the other hand, if $a < b < 0$, then it is evident from symmetry (see Figure 14–12) that

(11)
$$\int_a^b \frac{dx}{x} = -\int_{-b}^{-a} \frac{dx}{x} = -\log_e (-a) + \log_e (-b)$$

$$= \log_e |b| - \log_e |a|.$$

(See Exercises 4, 5 of 14.6.)

Thus from (10) and (11) it follows that *for any interval $[a, b]$ not including zero*

$$\int_a^b \frac{dx}{x} = \log_e |b| - \log_e |a|.$$

Example 2.
$$\int_a^b \sin x \, dx = \cos a - \cos b.$$

Indeed,
$$\frac{d \cos x}{dx} = - \sin x.$$

and so $-\cos x$ is an antiderivative of $\sin x$. Thus by Theorem 4

$$\int_a^b \sin x\, dx = (-\cos b) - (-\cos a) = \cos a - \cos b.$$

Similarly

$$\int_a^b \cos x\, dx = \sin b - \sin a.$$

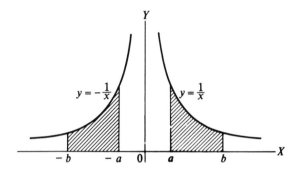

Figure 14–12

Example 3.

$$\int_0^x \sec t\, dt = \log_e (\sec x + \tan x).$$

By the Chain Rule we find

$$\frac{d}{dx} \log_e (\sec x + \tan x) = \frac{d/dx\, (\sec x + \tan x)}{\sec x + \tan x}$$

$$= \frac{\sec^2 x + \sec x \tan x}{\sec x + \tan x} \qquad \text{(from 13.7)}$$

$$= \sec x.$$

Thus by Theorem 4

$$\int_0^x \sec t\, dt = \log_e (\sec x + \tan x) - \log_e 1$$

$$= \log_e (\sec x + \tan x).$$

Example 4.

$$\int_a^b e^x\, dx = e^b - e^a.$$

By Theorem 2 and Corollary 2 of 13.2

$$\frac{d}{dx} e^x = e^x.$$

Thus by Theorem 4

$$\int_a^b e^x \, dx = e^b - e^a.$$

Example 5.

$$\int_0^x te^{t^2} \, dt = \tfrac{1}{2}(e^{x^2} - 1).$$

By the Chain Rule

$$\frac{d}{dt} (\tfrac{1}{2}e^{t^2}) = \tfrac{1}{2}e^{t^2} \cdot 2t = te^{t^2}.$$

Thus

$$\int_0^x te^{t^2} \, dt = \tfrac{1}{2}(e^{x^2} - 1).$$

It is an interesting fact that the integral $\int_0^x e^{t^2} \, dt$ cannot be evaluated in terms of elementary functions (that is, polynomials, trigonometric functions, exponentials or compositions, inverses, and algebraic combinations of these). This shows that integration is a more difficult process than differentiation.

14.4   INTEGRATION BY PARTS

We have seen in 14.3 that $\int_a^b f(x) \, dx$ can be evaluated provided we can find a function F such that $F'(x) = f(x)$. This function F is called an antiderivative of f. Although in principle this enables one to evaluate any integral, in practice it may not be easy to find a function F for a given f. For example, as we have remarked above, if $f(x) = e^{x^2}$, $F(x)$ although it is completely defined by $\int_0^x e^{t^2} \, dt$, cannot be expressed in terms of elementary functions.

In this section we consider a procedure called integration by parts, that enables one to evaluate many integrals.

Theorem 1.   *Let f', g' be continuous functions on $[a, b]$. Then*

(1) $$\int_a^b f(x) \, g'(x) \, dx = f(b) \, g(b) - f(a) \, g(a) - \int_a^b f'(x) \, g(x) \, dx.$$

Proof.   By Theorem 1 of section 13.3 we have

(2) $$(fg)' \, (x) = f'(x) \, g(x) + f(x) \, g'(x).$$

Now by Corollary 1 of 14.3

$$\int_a^b (fg)' (x) \, dx = f(b) \, g(b) - f(a) \, g(a).$$

Hence, by (2)

$$\int_a^b f(x) \, g'(x) \, dx = f(b) \, g(b) - f(a) \, g(a) - \int_a^b f'(x) \, g(x) \, dx.$$

Example 1.   Evaluate

$$\int_0^1 x \sin x \, dx.$$

We apply (1) with $f(x) = x$, $g'(x) = \sin x$.
Now we can choose $g(x) = -\cos x$ and we obtain

$$\int_0^1 x \sin x \, dx = 1 \, (-\cos 1) - 0 \, (-\cos 0) - \int_0^1 1 \, (-\cos x) \, dx$$

$$= -\cos 1 + \int_0^1 \cos x \, dx$$

$$= -\cos 1 + \sin 1 - \sin 0 \quad \text{(Example 2 of 14.3)}$$

$$= \sin 1 - \cos 1.$$

Since $\sin 1 = 0.84$, $\cos 1 = 0.54$ (approximately), we have

$$\int_0^1 x \sin x \, dx = 0.3 \text{ (approximately)}.$$

In the following example we use integration by parts twice to obtain the solution.

Example 2.   Evaluate

$$\int_1^2 x^2 \, e^x \, dx.$$

Apply (1) with $f(x) = x^2$, $g(x) = e^x$. Then

$$\int_1^2 x^2 \, e^x \, dx = 2^2 \cdot e^2 - 1^2 \cdot e^1 - \int_1^2 2xe^x dx$$

(3) $$= 4e^2 - e^1 - 2 \int_1^2 xe^x \, dx.$$

Now apply (1) with $f(x) = x$, $g(x) = e^x$ to the integral in (3). This gives

$$\int_1^2 xe^x \, dx = 2e^2 - 1 \cdot e^1 - \int_1^2 1 \cdot e^x \, dx$$

$$= 2e^2 - e^1 - (e^2 - e^1)$$

$$= e^2.$$

Thus

$$\int_1^2 x^2 e^x \, dx = 4e^2 - e^1 - 2e^2$$

$$= 2e^2 - e.$$

Using $e = 2.72$, $e^2 = 7.39$ (approximately) we find

$$\int_1^2 x^2 e^x \, dx = 12.1 \text{ (approximately)}.$$

Example 3.   Evaluate

$$\int_0^\pi \sin^2 t \, dt.$$

Apply (1) with $f(t) = \sin t$, $g(t) = -\cos t$. Then

$$\int_0^\pi \sin^2 t \, dt = \sin \pi(-\cos \pi) - \sin 0(-\cos 0) - \int_0^\pi \cos t(-\cos t) \, dt$$

(4) $$= \int_0^\pi \cos^2 t \, dt.$$

Now in (4) put $\cos^2 t = 1 - \sin^2 t$ and we have

$$\int_0^\pi \sin^2 t \, dt = \int_0^\pi (1 - \sin^2 t) \, dt$$

$$= \int_0^\pi dt - \int_0^\pi \sin^2 t \, dt.$$

Hence

$$2 \int_0^\pi \sin^2 t \, dt = \pi.$$

$$\int_0^\pi \sin^2 t \, dt = \frac{\pi}{2}.$$

14.5 IMPROPER INTEGRALS AND THE INTEGRAL TEST

We end this chapter with the development of a powerful test for the convergence (or divergence) of a series. This test is called the *integral test*. In order to discuss this we need to consider integrals like $\int_a^\infty f(x)\,dx$. An integral of this type is called an **improper** or **infinite** integral. These integrals (together with two other types like $\int_{-\infty}^b f(x)\,dx$ and $\int_{-\infty}^\infty f(x)\,dx$ which we will not need here) have many important applications in mathematics but we shall consider here only their use in developing the integral test.

Definition 1.

$$\int_a^\infty f(x)\,dx = \lim_{b \to \infty} \int_a^b f(x)\,dx$$

provided this limit exists.

This definition is reminiscent of the definition

$$\sum_{k=0}^\infty a_k = \lim_n \sum_{k=0}^n a_k$$

for infinite series. If the limit in Definition 1 exists then $\int_a^\infty f(x)\,dx$ is called a **convergent** improper integral. Otherwise it is called a **divergent** improper integral. Let us now consider some examples of improper integrals.

Example 1.

$$\int_1^\infty \frac{dx}{x} = \lim_{b \to \infty} \int_1^b \frac{dx}{x} = \lim_{b \to \infty} (\log_e |b| - \log_e |1|)$$

is a divergent improper integral because $\lim_{b \to \infty} \log_e b = \infty$, by Theorem 2 (4) of 12.8. Hence,

(1)
$$\int_1^\infty \frac{dx}{x} = \lim_{b \to \infty} \int_1^b \frac{dx}{x} = \infty\,;$$

that is,

$$\int_1^\infty \frac{dx}{x}$$

is a divergent improper integral.

Example 2.   As an example of a convergent improper integral, consider

$$\int_1^\infty \frac{dx}{x^{1+a}}$$

where $a \in R$ and $a > 0$. In this case,

$$\int_1^\infty \frac{dx}{x^{1+a}} = \lim_{b \to \infty} \int_1^b \frac{dx}{x^{1+a}} = \lim_{b \to \infty} \left[\frac{-1}{a} \left(\frac{1}{b^a} - 1 \right) \right]$$

$$= \frac{1}{a} \lim_{b \to \infty} \left(1 - \frac{1}{b^a} \right)$$

$$= \frac{1}{a} - \frac{1}{a} \lim_{b \to \infty} \frac{1}{b^a}.$$

Now by Exercise 4 of 13.8 it follows that

$$\lim_{b \to \infty} b^a = \infty.$$

Hence from (3) in 12.2,

$$\lim_{b \to \infty} \frac{1}{b^a} = 0,$$

and so

(2) $$\int_1^\infty \frac{dx}{x^{1+a}} = \frac{1}{a}.$$

Thus the left side of (2) is a convergent improper integral.

An application of Theorem 2 in 14.3 together with (1) and (2) yields the following Theorem.

Theorem 1.   *For any real number a,*

(3) $$\int_1^\infty x^a \, dx = \begin{cases} \infty & \text{if } a \geqslant -1, \\ \dfrac{-1}{a+1} & \text{if } a < -1. \end{cases}$$

Proof.   The cases where the integral is divergent follow from (1) above and (9) in 14.3 and the cases where the integral is convergent follow from (2) above and (9) in 14.3.

We now establish the integral test.

Theorem 2.   *If f is a positive, continuous, nonincreasing function defined on the positive real numbers, then the integral $\int_1^\infty f(x)\, dx$ converges if and only if the series $\Sigma_n f(n)$ converges.*

Proof.   For each $n > 1$,

(4) $$\int_1^n f(x) \, dx = \sum_{k=1}^{n-1} \int_k^{k+1} f(x) \, dx.$$

Since f is nonincreasing $f(k+1) \leqslant f(x) \leqslant f(k)$ for all $x \in [k, k+1]$. Therefore by (10) and (11) in 14.1,

(5) $$f(k) = \int_k^{k+1} f(k)\, dx \geqslant \int_k^{k+1} f(x)\, dx \geqslant \int_k^{k+1} f(k+1)\, dx = f(k+1)$$

for all $k = 1, 2, ..., n$.

In Figure 14–13, the shaded region below the curve indicates $\int_k^{k+1} f(x)\, dx$, the lightly shaded region is $\int_k^{k+1} f(k+1)\, dx$, and the complete shaded region is $\int_k^{k+1} f(k)\, dx$.

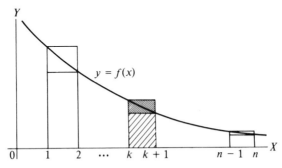

Figure 14–13

From (4) and (5) we find

(6) $$\sum_{k=1}^{n-1} f(k) \geqslant \int_1^n f(x)\, dx \geqslant \sum_{k=1}^{n-1} f(k+1).$$

The result (6) is geometrically obvious since $\sum_{k=1}^{n-1} f(k)$ is the sum of the areas of the taller rectangles in Figure 14–13; $\int_1^n f(x)\, dx$ is the area enclosed by $x = 1$, $x = n$, $y = 0$, $y = f(x)$; $\sum_{k=1}^{n-1} f(k+1)$ is the sum of the areas of the shorter rectangles. If the integral converges, then

$$\lim_n \int_1^n f(x)\, dx = \lim_{b \to \infty} \int_1^b f(x)\, dx = s$$

exists, and the increasing sequence $\{S_n\}$ where

$$S_n = \sum_{k=1}^{n-1} f(k+1)$$

has the property $S_n \leqslant s$ for all $n \geqslant 1$. Thus $\lim S_n$ exists, and $\sum_n f(n)$ converges. On the other hand, if $\sum_{n=1}^{\infty} f(n)$ exists, then since $f(x) \geqslant 0$,

$$\int_1^n f(x)\, dx < \sum_{b=1}^{\infty} f(b+1)$$

for all $n > 1$. Hence

$$\lim_{b \to \infty} \int_1^b f(x)\, dx$$

exists and is less than $\sum_{n=1}^\infty f(n+1)$.

Remark 1. Since the convergence or divergence of the series $\sum_n f(n)$ is not affected by the deletion of any initial set of terms in the series (although, of course, the sum, if the series is convergent, *is* affected) Theorem 2 can be restated as

Theorem 2′. *If f is a positive continuous nonincreasing function defined on the positive real numbers, then the integral $\int_m^\infty f(x)\, dx$ ($m \in N$) converges if and only if the series $\sum_{n=m}^\infty f(n)$ converges.*

We now prove the following Corollary which extends the results of Examples 2 and 3 in 11.6.

Corollary 1. *The series*

(7) $$\sum_n \frac{1}{n^\alpha}$$

converges if and only if $\alpha > 1$.

Proof. Although this result was partly proved in Chapter 11, we will not use the results of Chapter 11 in our present proof. By Theorem 1

$$\int_1^\infty \frac{dx}{x^\alpha}$$

converges for $\alpha > 1$ and diverges otherwise. Therefore the Corollary follows directly from Theorem 2.

The power of Theorem 2 is further indicated by the following example.

Example 3. The series

(8) $$\sum_n \frac{1}{n\,(\log_e n)^\alpha}$$

is convergent if $\alpha > 1$, divergent if $\alpha \leqslant 1$. Let us first consider the case $\alpha = 1$. We have

$$\frac{d \log_e (\log_e x)}{dx} = \frac{1}{x \log_e x}$$

and so, by the Fundamental Theorem of Calculus

$$\int_2^n \frac{dx}{x \log_e x} = \log_e (\log_e n) - \log_e (\log_e 2).$$

Since

$$\lim_{n \to \infty} \log_e n = \infty$$

and $\log_e x$ is a continuous function of x for $x > 0$,

$$\lim_{n \to \infty} \log_e (\log_e n) = \infty.$$

Therefore

(9)
$$\int_2^n \frac{dx}{x \log_e x}$$

is a divergent improper integral and by Theorem 2,

$$\sum_{n=2}^{\infty} \frac{1}{n \log_e n}$$

does *not* exist. Now if $\alpha \neq 1$, we see that

$$\frac{d}{dx} \left[\frac{(\log_e x)^{1-\alpha}}{1 - \alpha} \right] = \frac{1}{x (\log_e x)^{\alpha}}.$$

Hence

$$\int_2^n \frac{dx}{x (\log_e x)^{\alpha}} = \frac{1}{(1 - \alpha)} [(\log_e n)^{1-\alpha} - (\log_e 2)^{1-\alpha}].$$

Now if $\alpha > 1$, $\lim_n (\log_e n)^{1-\alpha} = 0$,

if $\alpha < 1$, $\lim_n (\log_e n)^{1-\alpha} = \infty$.

Thus

$$\sum_n \frac{1}{n (\log_e n)^{\alpha}}$$

is convergent if $\alpha > 1$ and divergent if $\alpha \leqslant 1$.

14.6 EXERCISES

1. Using the results of Exercise 5 in 11.9, find

$$\int_0^a x^2 \, dx.$$

2. If f is continuous on $[a, b]$ and $a < c < b$, then show that

$$\lim_{c \to a^+} \int_a^c f(x) \, dx = 0.$$

3. Find

(a) $\displaystyle\int_a^b x^{\frac{1}{2}} dx$;

(b) $\displaystyle\int_0^{\pi/2} \sin x \, dx$;

(c) $\displaystyle\int_{-\pi}^{\pi} \cos x \, dx$;

(d) $\displaystyle\int_0^1 \exp(x) \, dx$;

(e) $\displaystyle\int_0^{\pi} 2x \sin x^2 dx$;

(f) $\displaystyle\int_2^{25} \frac{2 \, dx}{x \log_e x^2}$

$$\left(\text{Hint}: \frac{2x}{x^2 \log_e x^2} = 2 \frac{1}{x \log_e x^2} \right).$$

4. Justify the convention:

$$\int_a^b f(x) \, dx = - \int_b^a f(x) \, dx.$$

5. (a) Let f be a real function with domain D and when $x \in D$ suppose that $-x \in D$. Then f is called an **even** function if $f(-x) = f(x)$ for all $x \in D$, and an **odd** function if $f(-x) = -f(x)$. Illustrate these notions graphically and give some examples.

(b) Classify the following functions as (i) even, (ii) odd, (iii) neither: 1, x, $|x|$, $\cos x$, e^x, $\tan x$, $\sin x^2$, $|\sin x|$, $2x^2 + 5x^{-2}$, $\sqrt{(1 - x^2)}$.

(c) Let f be continuous on $[-a, a]$.

   (i) Show that $\displaystyle\int_{-a}^{a} f(x) \, dx = 2 \int_0^a f(x) \, dx$ if f is even.

   (ii) Evaluate $\displaystyle\int_{-a}^{a} f(x) \, dx$ when f is odd.

(d) Let f be continuous on D and $[a, b] \cup [-b, -a] \subseteq D$.
   (i) Show that

$$\int_a^b f(x) \, dx = - \int_b^a f(x) \, dx = - \int_{-b}^{-a} f(x) \, dx$$

   if f is odd.
   (ii) State and prove a similar result for f even.

6. Let f be an increasing function on $[a, b]$. Prove

$$D_n - d_n = (b - a) \frac{(f(b) - f(a))}{n}.$$ (Cf. 14.1.)

Thence prove that a bounded monotonic function on $[a, b]$ is integrable.

7. (a) Let f, g be two functions such that $f(x) \geqslant g(x)$ on $[a, b]$. Show that the area enclosed by $x = a$, $x = b$, $y = f(x)$, $y = g(x)$ is

$$\int_a^b (f(x) - g(x)) \, dx.$$

(b) Find the area enclosed by the given curve in each of the following cases.

(i) $y = 0, \quad y = 2x - x^2$.

(ii) $y = x^4 - 2x^2, \quad y = 2x^2$.

(iii) $\sqrt{x} + \sqrt{y} = 1, \quad x = 0, \quad y = 0$.

(iv) $y = \dfrac{a}{2}(e^{x/a} + e^{-x/a}), \quad y = 0, \quad x = -a, \quad x = a$.

(v) $y = \tan x, \quad y = 0, \quad x = \dfrac{\pi}{4}$.

(vi) $y = |x - 1|, \quad y = x^2 - 2x, \quad x = 0, \quad x = 2$.

(vii) $y = x^{1/3}, \quad y = x^{\frac{1}{2}}, \quad x = 0, \quad x = 1$.

8. Evaluate the following integrals

(a) $\displaystyle\int_0^1 x^2 \sin x \, dx$

(b) $\displaystyle\int_0^\pi x^3 \cos x \, dx$

(c) $\displaystyle\int_1^2 x \log_e x \, dx$

(d) $\displaystyle\int_1^4 \sin(\log_e x) \, dx$

(e) $\displaystyle\int_0^1 x^3 e^x \, dx$

(f) $\displaystyle\int_1^3 x^3 (\log_e x)^2 \, dx$

9. Prove by the method of Examples 1 and 2 of 14.1 that

$$\int_0^1 x^3 \, dx = \frac{1}{4} \qquad \text{(see Exercise 12 of 2.7)}.$$

10. Let f be continuous on $[a, b]$. If $\int_a^b f(x) \, dx = 0$ show that there exists at least one c in (a, b) such that $f(c) = 0$.

11. Assume that f is continuous with a continuous derivative on $[a, b]$. Deduce the mean value theorem for derivatives from the mean value theorem for integrals.

12. (a) Determine whether the following improper integrals exist.

(i) $\displaystyle\int_0^\infty e^{-x} dx$

(ii) $\displaystyle\int_0^\infty e^{-x^2} dx$ (compare with (i))

(iii) $\displaystyle\int_0^\infty \sin x \, dx$

(iv) $\displaystyle\int_1^\infty \dfrac{dx}{1 + x^3}$

(v) $\displaystyle\int_0^\infty \dfrac{dx}{1 + e^x}$

(vi) $\displaystyle\int_1^\infty \dfrac{\sin x}{x^2} \, dx$

(b) Find approximate values for those improper integrals of (a) that exist.

13. Use the integral test to determine which of the following are convergent.

(a) $\displaystyle\sum_{n=1}^\infty \dfrac{1}{n^2 + 1}$

(b) $\displaystyle\sum_{n=1}^\infty \dfrac{\sin \pi/n}{n^2}$

(c) $\displaystyle\sum_{n=1}^\infty \dfrac{1}{n + 1}$

(d) $\displaystyle\sum_{n=2}^\infty \dfrac{1}{n \log n^2}$

Bibliography

T. M. Apostal: *Calculus* (Vols. I and II), Blaisdell, 1961.

G. Birkhoff and S. Mac Lane: *A Survey of Modern Algebra,* Macmillan, 1965.

B. Brainerd, D. A. Clarke, M. J. Liebovitz, R. A. Ross, and G. A. Scroggie: *Differential Calculus (Topics in Mathematics,* Vol. 2), Ryerson Division, McGraw-Hill, Toronto, 1966.

R. Courant: *Differential and Integral Calculus,* Blackie, 1937.

R. Courant and Fritz John: *Introduction to Calculus and Analysis,* Vol. 1, Interscience, New York, 1965.

R. Courant and H. Robbins: *What is Mathematics?,* Oxford, 1948.

R. L. Goodstein: *Boolean Algebra,* Macmillan, 1963.

P. R. Halmos: *Naive Set Theory,* Van Nostrand, Princeton N.J., 1960.

E. W. Hobson: *Theory of Functions of a Real Variable,* Vol. 1, Dover, 1957.

K. Knopp: *Theory and Application of Infinite Series,* Blackie, 1928.

K. Knopp: *Infinite Sequences and Series,* Dover, 1956.

W. Maak: *An Introduction to Modern Calculus,* Holt, Rinehart, and Winston, 1963.

E. Mendelson: *Introduction to Mathematical Logic,* Van Nostrand, Princeton N.J., 1964.

I. Niven: *Irrational Numbers,* J. Wiley, 1956.

W. Rudin: *Principles of Mathematical Analysis,* McGraw-Hill, 1953.

R. L. Wilder: *Introduction to the Foundations of Mathematics* (2nd ed.), J. Wiley, 1965.

ANSWERS TO SELECTED EXERCISES

SECTION 1.7

1. (i) $\{x \in I \mid 2 \nmid x\}$.

   (ii) $\{x \in N \mid x = 2 \text{ or } (x > 2 \text{ and there is no } y, 1 < y < x, \text{ such that } y \mid x)\}$.

   (iii) $\{x \in Q \mid \text{there is a } y \in Q \text{ such that } x = \sqrt{y}\}$.

   (iv) $\{x \in R \mid x \notin Q \text{ and there is a } y \in I, y < 0, \text{ such that } x = \sqrt[3]{y}\}$.

   (v), (vii) $\{x \in R \mid x \neq x\}$.

   (vi) $\{x \in R \mid x = \pi \text{ or } x = \sqrt{\pi}\}$.

   (viii) $\{x \in R \mid x > \pi \text{ or } x < \sqrt{\pi}\}$.

   (ix) $\{x \in R \mid \sqrt{\pi} < x < \pi\}$.

   (x) $\{x \in R \mid x = x\}$.

3. $\emptyset, \{1\}, \{2\}, \{3\}, \{4\}, \{1, 2\}, \{1, 3\}, \{1, 4\}, \{2, 3\}, \{2, 4\}, \{3, 4\}, \{2, 3, 4\}, \{1, 3, 4\}, \{1, 2, 4\}, \{1, 2, 3\}, \{1, 2, 3, 4\}$; 2^n.

4. $\emptyset, \{\{1\}\}, \{\{1, 2\}\}, \{N\}, \{\{1\}, \{1, 2\}\}, \{\{1\}, N\}, \{\{1, 2\}, N\}, F$; none contain 1; four contain $\{1\}$.

SECTION 2.7

1. All but **2** and **6**.

2. (a) **3, 4** of exercise 1. a^b is another.
   (b) **1** of exercise 1. a^b.

4. **5** and **7** of exercise 1 in either order.

13. (iii), (iv) are total orderings.

15. (a) $A = \{\text{even natural numbers}\}$, $B = \{\text{odd natural numbers}\}$, l.u.b. $A = 1$.
    (b) 0 is least; $N - \{0\}$ has no least, but g.l.b. is 0.

SECTION 3.8

1. (a) $\mathscr{S}$ is $7 > 2$; $\mathscr{P}$ is $5 > 2$; $\mathscr{Q}$ is $3 > 2 : \mathscr{S} \wedge \mathscr{P} \wedge \mathscr{Q}$.
   (b) $\mathscr{S}$ is $7 > 5$; $\mathscr{P}$ is $3 > 2$; $\mathscr{Q}$ is $3 > 2 : \mathscr{S} \wedge \mathscr{P} \Rightarrow \mathscr{Q}$.
   (c) $\mathscr{S}$ is $6 + 5 = 11$; $\mathscr{P}$ is $1 + 0 = 0$; $\mathscr{Q}$ is $5 = 25$; ambiguous. $\neg(\mathscr{S} \vee \mathscr{P} \Rightarrow \mathscr{Q})$ or $\neg \mathscr{S} \vee (\mathscr{P} \Rightarrow \mathscr{Q})$.
   (d) $(\mathscr{S} \vee \mathscr{P}) \Rightarrow \mathscr{Q}$.

2. (a) $\mathscr{T}$. (b) $\mathscr{T}$. (d) $\mathscr{T}$.

3. $(\mathscr{S}_1 \vee \mathscr{S}_2) \wedge \neg(\mathscr{S}_1 \wedge \mathscr{S}_2)$.

5. (a) nothing. (b) and (c) both true.

11. (a) and (c) are tautologies. (b) $\mathscr{P} = \mathscr{R} = \mathscr{F}, \mathscr{Q} = \mathscr{T}$. (d) $\mathscr{P} = \mathscr{T}, \mathscr{Q} = \mathscr{F}$.

SECTION 4.9

5. (i) (b)

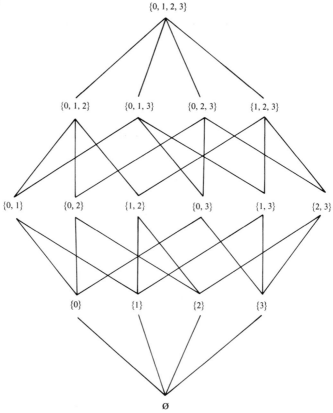

SECTION 5.6

1. (a) G3. (b) G3.

3. (b) yes.

9. yes; no.

SECTION 7.11

1. (b) $k = 0$ or $k = 1$.

2. Only (c) is onto.

3. (a) (i), (ii), (iii), and (vi).
   (b) (i) and (vi). The inverses are $f^{-1}(x) = \sqrt[3]{x}$ and $K^{-1}(x) = x$.
   (c) In (ii), f is monotone, and hence one-one, for values of x approximately between -1.8 and 1.1. In (iii), f is monotone for $x \geqslant 0$.

4.

	$f \circ g$	$g \circ f$
(a)	$x \neq 0$	$x \neq 0$
(b)	$x \neq 0$ (sic)	$x \neq 0$
(c)	$x \geqslant -1$	$\mathscr{R}$
(d)	$\mathscr{R}$	$\mathscr{R}$
(e)	$\mathscr{R}$	$\mathscr{R}$

5. (b) $[\{1\}, \{2\}, \{3\}, \{4, 5\}, \{6\}]$.

16. There are infinitely many. E.g. express any $x \in N$ in the form $x = 2^{m(x)} \cdot n(x)$ where $2 \nmid n(x)$; then $f(x) = (m(x), n(x))$ is one-one from N onto $N \times O$ ($O = \{$odd natural numbers$\}$). Let p map O one-one onto N; then the required function is $g(x) = (m(x), p(n(x)))$.

SECTION 8.7

13. (a) 41/333. (b) 17/33. (c) 115/333.

SECTION 9.11

6. (a) $f^{-1}(x) = 10^x$.
   (b) $f^{-1}(x) = \sin^{-1} x$ where $y = \sin x$ if and only if $\sin^{-1} y = x$, and $0 \leqslant x \leqslant \pi/2$.
   (c) none.
   (d) $f^{-1}(x) = f(x) = x$.

10. $A + B = 2 + i + 2j + 2k$. $AB = -1 + i + 3j + k$. $BA = -1 + i + j + 3k$ $\bar{A} = 1 - j - k$. $\bar{B} = 1 - i - j - k$. $A^{-1} = \bar{A}/3$. $B^{-1} = \bar{B}/4$. $A/B = (3 - i - j + k)/4$.

12. (a) I_3

+	0	1	2
0	0	1	2
1	1	2	0
2	2	0	1

·	0	1	2
0	0	0	0
1	0	1	2
2	0	2	1

I_4

+	0	1	2	3
0	0	1	2	3
1	1	2	3	0
2	2	3	0	1
3	3	0	1	2

·	0	1	2	3
0	0	0	0	0
1	0	1	2	3
2	0	2	0	2
3	0	3	2	1

I_5

+	0	1	2	3	4
0	0	1	2	3	4
1	1	2	3	4	0
2	2	3	4	0	1
3	3	4	0	1	2
4	4	0	1	2	3

·	0	1	2	3	4
0	0	0	0	0	0
1	0	1	2	3	4
2	0	2	4	1	3
3	0	3	1	4	2
4	0	4	3	2	1

(b) Mod 4, 2 is a zero divisor. There are none mod 5. Mod 6, 2, 3 and 4 are zero divisors.

13. (a) If $m \neq n$, the degree of $f - g$ is $\max(m, n)$. If $m = n$, it is $\leqslant m$.

SECTION 10.11

9. $\forall x\, \mathcal{P}(x) \Rightarrow \forall x\, \mathcal{Q}(x)$ only if $P \subseteq Q$, but *not* conversely.
$\forall x\, [\mathcal{P}(x) \Rightarrow \mathcal{Q}(x)]$ if and only if $P \subseteq Q$
$\{x \in A \mid \mathcal{P}(x) \Rightarrow \mathcal{Q}(x)\} = \overline{P} \cap \overline{Q}.$

SECTION 11.9

1. (a), (b), (c), and (e) are bounded; (d) and (f) are unbounded.

2. (a) $3n^2 + 3n + 1$. (b) 2^n. (c) $1/(n + 1)(n + 2)$.
(d)
$$\begin{cases} \dfrac{n!}{(n + 1 - r)!\,(r - 1)!} & \text{if } n \geqslant r \\[2mm] 1 & \text{if } n = r - 1 \\[2mm] 0 & \text{if } n < r - 1 \end{cases}$$
(e) $-1/n(n + 1)$. (f) $4n^3 + 6n^2 + 4n + 1$.

4. (a) $\sin na + 2(n + 1)\sin(a/2)\cos((2n + 1)a/2)$.
(b) $(-n^2 + n + 2)/2^{n+1}$.

5. (a) $n^3/3 + n^2/2 + n/6$. (b) $2 + (n - 1)2^{n+1}$. (c) $(\sin(n + 1)a)/2\sin(a/2)$.
(d) $1 - 1/(n + 1)$.

6. (a). (b) and (e). Converges to 1. (c) diverges. (b) converges to 1/2.

17. All of them.

SECTION 12.10

1. (a) 1. (b) ∞. (c) 0. (d) 3. (e) 0.

3. (e) 1. (g) ∞. all the rest, 0.

4. (a) 0. (b) 1. (c) 0. (d) 0. (e) undefined. (f) 0.

SECTION 13.8

2. (a) $15x^4 - 8x^3$.   (g) $-\csc^2 x$.

    (b) $2x/25(x^2 - 1)^{24/25}$.   (h) $-\csc x \cot x$.

    (c) $(-x^4 - 2x)/(x^3 - 1)^2$.   (i) $1/\sqrt{(1 - x^2)}$.

    (d) $(1 - 2x)\,e^{-x^2+x}$.   (j) $1/(1 + x^2)$.

    (e) $\sec^2 x$.   (k) $e^{ax}(a \sin bx + b \cos bx)$.

    (f) $\sec x \tan x$.   (l) $e^{x \sin x}(\sin x + x \cos x)$.

8. (a) $2x/\sqrt{(4 - x^4)}$.   (b) $1/(1 + x^2)$.   (c) $2 + (\sin^{-1}x)^2$ (d) $-1/\sqrt{(1 - x^2)}$.

    (e) $-1/(1 + x^2)$.   (f) $1/|x|\sqrt{(x^2 - 1)}$.

SECTION 14.6

1. $a^3/3$.

3. (a) $2(b^{3/2} - a^{3/2})/3$.   (b) 1.   (c) 0.   (d) $e - 1$.   (e) $1 - \cos \pi^2$.   (f) $\log_e 2$

5. (b) Even, odd, even, even, neither, odd, even, even, even, even. (c) (ii), 0.

7. (b) (i) 4/3.   (ii) 128/15.   (iii) 1/6.   (iv) $a^2(e^2 - 1)/e$.   (v) $(\log_e 2)/2$.   (vi) $(7 + 5\sqrt{5})/6$.   (vii) 1/12.

8. (a) $\cos 1 + 2 \sin 1 - 2$.   (b) $12 - 3\pi^2$.   (c) $2 \log_e 2 - 3/4$.

    (d) $1/2 + 2 \sin(\log_e 4) - 2 \cos(\log_e 4)$.   (e) $36 - 32e$.

    (f) $5/2 + 81(\log_e 3)^2/4 - 81(\log_e 3)/8$.

12. (i) 1. (ii) $\sqrt{\pi}/2$. (iii) does not exist. (iv) $0.3735507 \cdots$. (v) $\log_e 2$. (vi) $0.50407 \cdots$.

13. (a) and (b) are convergent.

LIST OF SYMBOLS

Note. With some exceptions, italic letters are used to denote numbers, functions, and sets; generally, capitals denote sets, and lowercase denotes numbers or functions (again, with some exceptions). Script letters always denote relations or sentences.

Symbol	Name	Page
$\{a_n\}$	sequence	92
$f \circ g$	composite function	94
i_A	identity function	95
f^{-1}	inverse function	97
$[x]$	greatest integer in x	116
$|x|$	absolute value	117
$\lim_n a_n$	limit of a sequence	120
$A^n\,(R^n)$	set of ordered n-tuples (of reals)	143
$x \to f(x)$	mapping notation for functions	145
$\vDash$	logical consequence	167 to 169
$\forall, \exists$	quantifiers	180
Δ	difference operator	202
Δ^{-1}	antidifference operator	204
$\lim\sup, \lim\inf$	superior limit, inferior limit	212
$\sum_{i=p}^{q} a_i, \sum_n a_n, \sum_{n=1}^{\infty} a_n$	summation notation, series	219, 224
$(a, b), [a, b]$	open and closed intervals	251
$(a, b], [a, b)$	semi-open intervals	251
$N_x, N_{x,\varepsilon}, N_a^*$	neighborhoods	251, 252
$\lim_{x \to a} f(x)$	limit of a function	252
$\lim_{x \to a+} f(x), \lim_{x \to a-} f(x)$	one sided limits	255
$\lim_{x \to \infty} f(x), \lim_{x \to -\infty} f(x)$	infinite limits	263
$\dfrac{df}{dx}, \dfrac{dy}{dx}, f', Df$	derivative	290
$\displaystyle\int_a^b f(x)\,dx$	integral	324
$A_a^b(f)$	area	328
$\displaystyle\int_a^\infty f(x)\,dx$	improper integral	341

Axioms

A1 to A3, M1 to M3, D, C1, C2	9
O1 to O4	12
O'1 to O'4	14
OA, OM	15
OW	16
A4	50
OM'	55
M4	63
OD	66
OC	114

INDEX